Climate Justice and Public Health

Climate Justice and Public Health

Realities, Responses, and Reimaginings for a Better Future

EDITED BY

Rajini Srikanth and Linda Thompson

University of Massachusetts Press

Amherst and Boston

This book will be made open access within three years of publication thanks to Path to Open, a program developed in partnership between JSTOR, the American Council of Learned Societies (ACLS), University of Michigan Press, and The University of North Carolina Press to bring about equitable access and impact for the entire scholarly community, including authors, researchers, libraries, and university presses around the world. Learn more at https://about.jstor.org/path-to-open/.

ISBN 978-1-62534-803-6 (paper); 804-3 (hardcover)

Designed by Sally Nichols
Set in Minion Pro
Printed and bound by Books International, Inc.

Cover design by adam b. bohannon
Cover photos: (top) unknown photographer, *USAID Coastal City Adaptation Project, in Quelimane city, Mozambique*, 2014. Public Domain USGov; (bottom) photo by Jessie F. Delos Reyes. *USAID workers support a mangrove planting activity as part of a climate change adaptation strategy in the Philippines*, 2011. Public Domain, USGov.

Library of Congress Cataloging-in-Publication Data

Names: Srikanth, Rajini, editor. | Thompson, Linda, 1955– editor.
Title: Climate justice and public health : realities, responses, and
   reimaginings for a better future / edited by Rajini Srikanth and Linda
   Thompson.
Description: Amherst : University of Massachusetts Press, [2024] | Includes
   bibliographical references and index. |
Identifiers: LCCN 2023046512 (print) | LCCN 2023046513 (ebook) | ISBN
   9781625348036 (paperback) | ISBN 9781625348043 (hardcover) | ISBN
   9781685750756 (ebook)
Subjects: LCSH: Medical climatology. | Climatic changes—Health aspects. |
   Climate justice. | Environmental health.
Classification: LCC RA793 .C585 2024 (print) | LCC RA793 (ebook) | DDC
   613/.11—dc23/eng/20231229
LC record available at https://lccn.loc.gov/2023046512
LC ebook record available at https://lccn.loc.gov/2023046513

British Library Cataloguing-in-Publication Data
A catalog record for this book is available from the British Library.

To our planet and the wellness of
current and future generations

CONTENTS

**PART 4: ACQUIRING UNDERSTANDING: TWO CASE STUDIES
OF LONG-TERM IMPACTS**

ILLUSTRATIONS

FIGURES

TABLES

Climate Justice and Public Health

Climate Justice and Public Health

Realities, Responses, and Reimaginings for a Better Future

RAJINI SRIKANTH AND LINDA THOMPSON

This volume examines the intersection of climate justice and public health, a nexus that has become impossible to ignore in the twenty-first century.[1] Among the most visibly graphic representations of this intersection in the recent past are the images of "abandoned" African American residents of New Orleans after the city was devastated by Hurricane Katrina in 2005. The hurricane assaulted African American communities with especial force, while communities with higher socioeconomic status (white communities in particular) had the means to evacuate to safer and drier ground.[2] The injustice of this disparate opportunity for safety foregrounded for the nation in emphatic fashion that protection and recovery from climate disasters is not uniformly prioritized across all communities. The debacle at Flint, Michigan, and the resistance and the protests at Standing Rock against the Dakota Access Pipeline reminded the country and the world of the fragility of climate/environmental justice and health for African American communities and indigenous peoples.[3] Since March 2020, the COVID pandemic has revealed irrefutably the dire inequities endured for generations by communities of color in the sphere of public health.

Epidemics and climate disasters (such as hurricanes, floods, heatwaves, and tsunamis) make evident to societies the values and assumptions that govern their social, political, and economic structures. How a society reacts when confronted by these emergencies reveals the strengths and weaknesses in its institutions and policies. There is growing recognition among public health researchers and professionals that climate injustice and health disparities are inextricably linked.[4] Extreme heat, drought, and floods causing elevated risks of disease, food shortages, compromised freshwater resources, and poor air quality are all impacts of climate change that disproportionately affect vulnerable populations and those in under-resourced and underserved communities. Race and income determine the resilience of population groups and their ability to survive and overcome the combined risks of climate injustice and health inequities. Different communities and demographic groups have varied exposure to these risks as well as differential resources to cope. Research suggests that the drivers behind climate injustice and health disparities are fundamentally the same—social inequities and the institutional power structures that favor specific groups and therefore neglect or jeopardize others.

There is a crucial need for broad systemic changes in our health systems, transportation infrastructure, and the production and distribution of energy. Since March 2020, when the United States was forced to confront the seriousness of the coronavirus, radio shows have hosted discussions and interviews, and newspapers have run stories, about the need to reimagine our society and consider how we might reassess our priorities as a people. The COVID-19 pandemic exacerbated the gross inequities and stark injustices that have been a constant reality in the lives of the vast majority of the population both in the United States and the world. The pandemic jolted us into examining how our lives and our societies are structured, and it has forced us to consider the decisions we make and actions we take both as individuals and as societies by examining and interrogating the language we use, the laws we construct, the policies we implement, the incentives we offer or withhold, and the opportunities we provide (or do not) in every sphere of life. With increasing urgency, we are asking what it will take to envision and bring into reality a planet that flourishes, to connect human wellness (in all the multiple senses of the word) with the wellness of the natural world and all the myriad living beings within it. As the indigenous scholar and Anishinaabe feminist environmentalist Deborah McGregor (a contributor to this volume) asserts, we are experiencing multiple crises, and the climate crisis exacerbates all other crises. In this moment of reckoning, how do we reimagine and re-create our world, keeping climate justice and racial justice at the center of our consciousness, so that *all* life can flourish?

This volume responds to that challenge, and it does so, first, by recognizing the inadequacy in and harmful consequences of current frameworks of knowing (in which diverse areas of human endeavor and knowledge—for example, scientific, environmental, artistic, legal, philosophical, historical, and cultural—are largely viewed as separate from one and other); and secondly, by engaging in an act of radical reimagining to connect the seemingly disparate spheres of human activity and, thereby, to cultivate a desire for holistic interconnections and relationships with other living beings and natural landscapes.

NECESSARY REIMAGININGS

In 2015, the Rockefeller Foundation–Lancet Commission focused on planetary health, a global concern amplified since then by systems thinkers in the public health and sustainability worlds. The commission's report is titled "Safeguarding Human Health in the Anthropocene Epoch," and its principal articulations emphasize new frameworks of thought, knowledge, and analysis. It cites three specific challenges to be overcome:

> Firstly, conceptual and empathy failures (imagination challenges), such as an over-reliance on gross domestic product as a measure of human progress, the failure to account for future health and environmental harms over present day gains, and the disproportionate effect of those harms on the poor and those in developing nations.
>
> Secondly, knowledge failures (research and information challenges), such as failure to address social and environmental drivers of ill health, a historical scarcity of transdisciplinary research and funding, together with an unwillingness or inability to deal with uncertainty within decision making frameworks.
>
> Thirdly, implementation failures (governance challenges), such as how governments and institutions delay recognition and responses to threats, especially when faced with uncertainties, pooled common resources, and time lags between action and effect.[5]

The contributors to this volume show how and why these challenges manifest themselves in myriad areas of human endeavor, and they offer new paradigms that seek to address the shortcomings of established ways of comprehending and acting. A focus on wellness requires that we think about the ecosystem in which we are situated. Our health and wellness (not just our physical and physiological health but our emotional and spiritual health) are intertwined with the quality of the air we breathe and the water we drink, the nutritional value of the food we eat, and the stresses we experience from deprivation, hostilities, diminishment of our self-worth, and the opportunities we have (or that are withheld from us) for a fulfilled life. At this moment, there are multiple pandemics: the COVID-19 pandemic, the pandemic of racism, and the pandemic of inequity.[6] Those working in the areas of climate justice, environmental justice, antiracism, public health, and wellness recognize that these seemingly separate areas of focus are, in reality, inseparable, that to address one and not the others simply replicates incomplete interventions, and perpetuates current crises of inequity.

Climate Interactive (a systems-thinking organization that "create[s] and share[s] tools that help people see connections and drive effective and equitable climate action") has coined the term "multisolving" to underscore that climate protection and health improvement have to be approached together. But while the urgency for connected action is obvious, team members of Climate Interactive remind us of the many obstacles to multisolving: "disciplinary silos, budgetary silos, jurisdictional silos, weak community engagement skills, challenge of funding prevention vs. care, and long-term benefits within systems oriented toward short-term decision-making." These silos reveal the failure of our imagination and a refusal to reassess intellectual territoriality and recognize the dangers of fragmented thinking.[7]

For us to truly tackle climate change and respond to climate injustice, we need a transformational shift in our thinking, in our privileging humans over all other forms of life. We need to bring into the realm of science and policy concepts such as gratitude and humility. Indigenous communities have long been cautioning us against the "conquest mindset" that sees the earth and all its resources as available for extraction and commodification. We would do well to turn to Robin Wall Kimmerer, the Potawotami botanist and scientist who exhorts us to reflect on the value of gratitude: "Gratitude is founded on the deep knowing that our very existence relies on the gifts of beings [plants and trees] who can in fact photosynthesize. Gratitude propels the recognition of the personhood of all beings and challenges the fallacy of human exceptionalism—the idea that we are somehow better, more deserving of the wealth and services of the Earth than other species."[8] In a similar vein, Lorén Spears of the Narragansett Nation and the executive director of the Tomaquag Museum reminds us that indigenous peoples and indigenous knowledge are deeply rooted in place—"language, foodways, daily life, agriculture, fishing, and medicines" are all intertwined with place. She points out that there was a fundamental clash of ideologies when the settlers came, the clash between "a conquest economy and a worldview that holds land and people in balance." The history of settler colonialism is the history of the "dehumanization and commodification of all the gifts of the creator."[9] Several scholars and researchers working at the intersection of climate change and public health point to the exploitative origins of our current planetary condition. Phenomena such as the slave trade and colonialism commodified relationships and set the foundation for treating both people and natural landscapes as property to be "worked" and mined for what they would yield. Such an orientation to living beings and to our physical landscapes runs counter to the respect and humility that Robin Wall Kimmerer urges.

Rhys Jones, writing from within a Maori framework, comments on the harmful consequences of climate change on the health of indigenous peoples: "Direct impacts include death, illness, and injury due to heat waves and extreme weather events. Powerful indirect impacts on health are mediated by a complex interaction of social, environmental, and economic factors. These include shifting patterns of infectious disease, air pollution, freshwater contamination, impacts on the built environment from sea level rise, forced migration, economic collapse, conflict over scarce resources, and increasing food insecurity." Jones underscores that "indigenous peoples are subject to conditions that result in disproportionate vulnerability to climate change. These include a tendency to live in geographical locations that are particularly prone to the impacts of climate change, dependence on lands and environmental resources for basic needs and economic security (e.g., food, shelter, medicines, fuel), and experiencing economic deprivation as well as social and political marginalization."[10]

In an article in September 2020, Renee Cho makes a persuasive case for "why climate change is an environmental justice issue." Pointing to heat islands, extreme weather events, air pollution, and disrupted sanitation, among other consequences of climate change, Cho provides evidence for how these conditions affect communities of color most severely. She also turns to the climate justice essayist Mary Annaise Hegler to echo that "climate change is actually the product of racism." Hegler's argument is hard-hitting: "It started with conquest, genocides, slavery, and colonialism. . . . That is the moment when white men's relationship with living things became extractive and disharmonious. Everything was for the taking; everything was for sale. The fossil fuel industry was literally built on the backs and over the graves of Indigenous people around the globe, as they were forced off their land and either slaughtered or subjugated—from the Arab world to Africa, from Asia to the Americas."[11]

POLICY GROUNDED IN RESPECT AND HUMILITY

Sentiments like respect and humility may seem misplaced in the pragmatic world of policy, but this volume and its contributors caution that we are at a point in the history of mankind that calls for radical changes in how we engage our planet and ourselves. To persist in the misguided belief that policy cannot draw from philosophy, or science cannot attend to justice without compromising its rigor and "objectivity," is to continue on the path of species destruction through exacerbating climate injustice and public health disparities.

The NPR show Science Friday discussed the question in an episode titled "Can Genetic Modification Help Plants Survive Climate Change?" that aired on July 22, 2022. The question foregrounds the triumph of science as a human endeavor that has helped us through multiple crises. However, while recognizing the tremendous value of science in helping us tackle the problems of public health, we should also interrogate the triumphalist orientation of mankind and ask ourselves about the inherent dangers of a mindset that sees some humans as having the power to control other humans, living creatures, and natural resources. Ira Flatow, the host of Science Friday, presented the scientific problem at hand for the show that day in July: plants that are important to our food supply are negatively affected by increasing temperatures and lose the immunity to ward off diseases when the weather gets very warm. Gene editing can help with restoring their immune systems (kick-starting them). Flatow turned to Sheng Yang He, a researcher at Duke University, to provide the explanation: "A gene we call CB60g . . . functions like immune's master switch. At one temperature, it turns out this gene is not turned on for some reason." Sheng Yang and his team determined how to modify a part of this gene to be "temperature insensitive"; as a result, "the plant is able to switch on this master immune gene and then make salicylic acid, and other defense systems allow plants to actually resist pathogen[s], even at warm temperature."[12]

While this procedure represents a triumph of humanity's scientific development and also provides much-needed protection against damage to our food supply and health, procedures such as gene editing reinforce and exacerbate the anthropocentric mindset that has contributed to the current climate crisis we face with all its attendant damage to the natural resources of the planet. The question to ask is, How do we continue to draw on the benefits of science to human life while at the same time maintaining our humility in the face of our interdependence on a complex ecosystem of living things and natural resources? This is not an idle question but one that is fundamental to our being able to recover from the ravages of the extractive and exploitative approaches to our planet that have characterized human participation for millennia. Our long-term survival as a species is predicated on our being able to acknowledge our past predatory actions and to commit to reexamining our relationship to one and other and to the natural resources of our planet.

RECOGNIZING THE LONG SHADOWS OF COLONIALISM AND SLAVERY

Increasingly, those scholars, researchers, activists, and policymakers working in the areas of climate justice and public health are realizing that all the conventional approaches to mitigating climate change or all the typical climate action responses are merely temporary and short reprieves from the inexorability of the climate crisis. This crisis strikes the most disadvantaged, the most vulnerable, the most unconnected from the depredations wrought on the ecosystem. The injustice of climate change is that those who had the least part to play in causing it are the ones who are the most disproportionately affected by it. The racism that made possible conquest and genocide and that set in place colonial systems has continued into the present day, with indigenous peoples and communities of color experiencing the most deleterious impact of climate change floods, neighborhoods of heat islands, hurricanes, poor air quality, poor soil quality for the harvesting of fresh food, contaminated water, and lack of access to health care.

Robert Bullard, considered the father of environmental justice in the United States, said in an interview with PBS reporter Walter Isaacson in March 2020,

If you look at environmental justice, climate change basically overlays this whole issue of who has contributed most and who is going to be impacted the greatest. And if you look at the footprint of climate change—and climate change is more than parts per million and greenhouse gases. It also includes who is most vulnerable. Who is going to have the burden of living in these areas that are going to be hit hard, whether it's droughts or there's flooding or whether other kinds of issues. And, again, climate change, sea level rise[,] will exacerbate the inequities that already exist. If people are poor, they're in low-lying areas . . . prone to flooding, you're going to get more flooding. You're going to get more droughts and more disasters. You're going to get more of the issues of the widening gap between haves and have-nots. And you're going to get this whole piling on effect of not having, you know, a level of resilience to bounce back because of a flood.[13]

In 2018, the American Association of Geographers presented Bullard with its award for non geographers who have had a profound impact on the field of geography. Bullard's plenary talk at that year's conference (held in New Orleans on April 12) of the AAG was titled "The Quest for Environmental and Climate Justice: Why Race and Place Still Matter." The conference's theme was "Black Geographies, Public Engagement, and the Spatiality of Hazards." Bullard's plenary address drew attention, as he always does, to the intersections among environmental justice, human rights, civil rights, economic justice, self-determination, and the full spectrum of circumstances that affect a person's life, including housing, transportation, water, food, air quality, and health.[14]

The fight for environmental and climate justice is ongoing. In April 2021, a permit was revoked by the Massachusetts Department of Environmental Protection for a proposed biomass energy plant in Springfield, Massachusetts. First proposed in 2008, the plant would have been the state's only largescale biomass plant and would have burned about twelve hundred tons of waste wood per day. This was proposed in a city that was given the distinction of being ranked the "asthma capital" of the United States by the Asthma and Allergy Foundation. It is also a city where 25.5 percent of its residents are living in poverty and nearly one in five children have asthma. After the permit was eventually revoked after a long campaign by environmentalists and climate justice advocates, Caitlin Peale Sloan, interim director of the Massachusetts Chapter of the Conservation Law Foundation, said, "The last thing the asthma capital of the U.S. needs is a plant spewing air pollution and further imperiling public health." Springfield city councilor and opponent of the facility, Jesse Lederman, praised the decision: "The days of polluters being rubber stamped in communities like ours are over. For too long communities like ours have been targeted by out-of-town developers seeking to get rich at the expense of public health and [the] environment of our children, seniors, and all residents, leading to generations of concentrated pollution and health and environmental inequities."[15]

Nadia Kim makes a powerful case for the compelling activism of Filipinx and Latinx women in Los Angeles, fighting the "emotional apathy" of the bureaucrats of the oil companies and government agencies to the air pollution caused by oil refineries and resulting in high levels of asthma in their communities.[16] Her study is particularly noteworthy for its spotlighting of the emotional force that the activists draw on; Kim brings the politics of affect into the center of the struggle, arguing that the rage and sorrow of the women—the public "performance" of emotion—is necessary to shake the "stonefaced" (3) men of the corporations and municipal agencies. The women's emotional testimonials can evoke "shame" (86), says Kim, and the women activists used this understanding in "strategizing their own emotive responses and displays" (75), bringing visibility to their anguish while also showing the contrasting indifference of the bureaucrats and corporate capitalists and so shaming them.

The episode of the podcast Killer Combination: Climate, Health, and Poverty that aired on February 12, 2021 opens with the question, What happens when climate, public health, and poverty converge? To illuminate the dangerous nexus, the episode begins in rural Alabama, where households don't have public waste systems and need to rely on septic systems, the costs of which have to be borne by residents who are often too poor to pay for this fundamental right. Moreover, they are criminalized for this inability and arrested for not being able to install their own septic systems.[17]

That type of injustice—where those who have suffered systemic racism are deemed delinquent for their lack of resources—is replicated on a global scale. The prime minister of Barbados, Mia Mottley, excoriated the leaders of Western nations and the global monetary lending and policymaking bodies at the climate summit in Glasgow in 2021: you save your profits rather than save our lives. These profits, as historians and political economists who study the Caribbean nations show, were made through extracting the natural resources of the Caribbean and employing slave and indentured labor.[18]

Thus, what Mottley and one of her key financial advisers, Avinash Persaud, argue is that what the colonial powers and today's global economic powers are obligated to do is set right the moral failings of the past. At a World Trade Organization speech, her inaugural presidential lecture, she made on March 23, 2022, Mottley reminded the attendees of the double standards of the current global order. "The global order is not working," she begins (time signature 13:19–13:20). "Too many people in this world live in conditions of hunger, of poverty, of indignity, and of inequality [13:37–13:45]. . . . Greed continues regrettably to motivate too many; that we are more concerned with generating profits than with saving people is perhaps the greatest condemnation that can be made of our generation globally. We continue to have a world that is segregated regrettably between those who came first and in whose image the global order is now set. . . . It is an embalming of the old colonial order that existed at the time of the establishment of these institutions" (14:37–15:20). She goes on to say that that after World War II, the "advanced countries of the world agreed that Germany did not have to service debt in excess of 5 percent of its exports, but in the island nations that debt is in excess of 30 percent. I ask simply, 'Where is the justice that we speak of?'" (19:20–19:44).[19] Today, Ukraine's debt is forgiven, given the hostilities it is facing. But the same understanding is not made available to small island developing states such as hers that continue to face the double crises of the pandemic and climate change.

Power. Who has it? Why do they have it? How can they be made to recognize the destructions caused by their power and, therefore, to be held morally accountable for taking corrective action? The answers to these questions about power require honesty and courage, particularly among those who currently hold the power. Climate justice and equitable public health are deeply intertwined with histories and systems of power. Confronting these histories and bringing science and technology "into conversation" with these histories is an essential step toward addressing the gross inequities in the material realities of the lives of the majority of humankind. No aspect of human endeavor is separate from others and, therefore, conquest and greed cannot be isolated from the incredible contributions made by our species.

SCOPE OF THIS BOOK

This volume provides an interdisciplinary examination of the intersection of climate change, public health, and inequality. It includes historical perspectives on race, class, and public health; analyses of likely climate change physical impacts and policies across different groups; and perspectives from various communities about the risks of climate change and environmental injustice on communities. The contributions illuminate the following sets of issues, both in the United States and around the globe:

- Efforts to foreground fact-based assessments of climate change's impacts on health within the volatile political climate that has snarled science and policy discussions.
- Grassroots movements to advance climate justice and health equity, an intersectional effort that is propelled by a set of principles that demand sustainable production, the protection of Mother Earth, and the involvement of communities most affected in crafting policy solutions.
- Innovative policy work at the level of city, county, or region that embraces climate change mitigation and adaptation as an integral part of health equity policies and efforts to change the narrative and infuse equitable solutions into climate-driven health policies and programs.

Our aspiration is to reach the current and the next generation of public health researchers and public health officers, health care professionals, social workers, community organizers, climate justice and health equity policy experts, governmental leaders, scholars of climate and environmental justice, community organizers and activists for environmental justice and wellness, and undergraduate and graduate students. We, along with Bullard, place our faith and hope in today's change agents and in their capacity for envisioning holistically how humankind can live with respect and humility on this planet, ensuring the longevity and wellness of all living beings and the natural landscapes. This book seeks to enlarge our collective consciousness so that the intersections among all facets of human endeavor are foregrounded.

Our volume offers the rich texture of multiple interrelated perspectives that complicate and amplify one another and simultaneously show

how much valuable insight can be gained from cultivating an approach that centers vulnerable and marginalized populations in charting climate action and public health responses. Climate justice and public health are themselves interdisciplinary fields. Each has become the concern of researchers, scholars, and policymakers in the last fifteen years. The chapters included in this collection span anthropology, indigenous studies, history, geography, nature conservancy, gerontology, infant care, nursing, geographical information systems, psychiatry, urban planning, and community development.

SUSTAINABLE SOLUTIONS REQUIRE RADICAL COLLABORATION

The impetus for this book emerged from the confluence of several forces: the COVID pandemic, the realities of asymmetrical health care access in the urban setting of Boston, initiatives to study how the city of Boston can prepare itself for the inevitable sea-level rise that will occur in the near future, the physical location of University of Massachusetts Boston (UMass Boston), the dire need in Boston during the pandemic for nurses and those trained in health care delivery, and the urban, public, community-engaged mission of UMass Boston. UMass Boston is the only public research university in the city of Boston, which has an impressive array of elite private institutions. Students from UMass Boston come from the neighborhoods and communities of the greater Boston Metropolitan Area, and the majority of them remain in this area even after graduation. In fact, UMass Boston is considered to provide a large proportion of the workforce of the city.

Our students are primarily commuters who work part- or full-time, so the experience of living and working in the city's neighborhoods is what they carry into the classrooms, and that knowledge intersects with the academic material of their courses. Learning, for them, is not a theoretical exercise; for many of our students, the knowledge they acquire in their courses helps them understand the conditions of their communities and what needs to be done to address shortfalls and deficiencies. For our students, poor access to health care, challenges of excessive heat given the lack of green cover in their neighborhoods, and increased likelihood of flooding from sea-level rise are among the immediate and urgent realities of their daily lives. They experience climate crisis and climate injustice every day, and they feel the impact of an inequitable public health system on their bodies and minds and on those of their family members.

As a result, several colleges and research institutes at the university came together to set up the Sustainable Solutions Lab to underscore that climate justice and public health were intertwined not only with each other but also with multiple other areas of knowledge and action. The Sustainable Solutions Lab is an interdisciplinary partnership among six colleges/schools and four institutes within University of Massachusetts Boston: the College of Liberal Arts, the College of Management, the Manning College of Nursing and Health Sciences, the McCormack School of Policy and Global Studies, the School for the Environment, the College of Education and Human Development, the Institute for Asian American Studies, the Institute for New England Native American Studies, the Mauricio Gastón Institute for Latino Community Development and Public Policy, and the William Monroe Trotter Institute for the Study of Black History and Culture.

The Sustainable Solutions Lab's mission is to understand the disproportionate impacts of climate change on marginalized populations and to work with them to develop sustainable and equitable solutions. The collaborative and cross-disciplinary connections among the researchers and scholars of the lab underscore that to meaningfully pursue "wellness" in a society requires the insights of multiple and seemingly disparate fields of knowledge. Many of the contributors to this volume are connected to the lab and attend its regular convenings. Further, the university's new urban public health degree is particularly attuned to issues of health disparity and the structural circumstances that contribute to inequity of wellness.

We highlight the Sustainable Solutions Lab as a model for intracampus collaboration that maintains a thriving community of interest across scholars and practitioners in diverse fields. The paradigm it offers is grounded in the values that we have discussed above centering justice and racial equity, interrogating and rejecting extractive and exploitative mindsets, and embracing humility and gratitude. We are also deeply grateful to the contributors to this volume who are external to UMass Boston. They are engaged in their own robust communities of practice and networks of collaboration and are working toward meaningful disruptions of unjust structures and policies in their locations. We hope that these clusters of transformative thought and action can multiply to create an unstoppable force for climate justice and equity in public health.

ORGANIZATION AND OVERVIEW OF CHAPTERS

The fourteen chapters in the volume are organized into four sections, followed by a conclusion and a coda that offers one type of pathway into the future.

Part 1: Changing Paradigms for Wellness

We begin with Deborah McGregor and colleagues, who ask the question, What does it mean to live well with the earth in the face of climate/ecological crisis? Their chapter, "Indigenous Peoples and Well-Being in the Context of Climate Change," asserts that the revival of indigenous languages is crucial to protecting against the deleterious impacts of climate change. The assault on indigenous communities, resulting in the destruction of their cultures (including languages) and ways of living with natural resources, has stripped them of the knowledge of restorative traditional practices. This chapter, with a focus on Canada, makes the case that language retention and revitalization initiatives are in fact critical to adaptation and resilience of indigenous communities.

Surili Sutaria Patel and Adrienne Hollis's chapter, "Connecting the Dots: Climate Change, Social Justice, and Public Health," makes the obvious but often neglected observation that decades of inequitable social, political, and economic practices have severely disadvantaged particular communities in the United States, especially those that are Black and Brown. They state forcefully, "To stand on the right side of history, we must name and address the many isms of the climate dilemma, including structural racism, sexism, and classism. While there are many other isms to be cited and dismantled, these three, if addressed with dignity, can help the millions of people who have been left behind to reclaim their narrative."

Laura Peters and colleagues, in their chapter, "Intersecting Participatory Action Research and Remote Sensing for Climate Change, Health, and Justice," begin with "positive health" as a foundation for justice and show how "two distinct leading empirical methodologies with emancipatory potentials—participatory action research and satellite remote sensing—are connected within a multidimensional justice framing encompassing space, time, and relationships." They emphasize partnerships that combine grassroots and "top-down" approaches "to build healthier futures as the climate changes." They draw on efforts across diverse locations around the globe, including the Amazonian Brazil, Malawi, Vietnam, and Australia.

Clair Cooper's chapter, "Nature-Based Solutions, Social and Health Inequalities, and Climate Justice in European Cities," trains a sharp eye on the "complex and intertwined relationship between different features that influence and characterize the operationalization of NBS, social and economic determinants of health, and consequences for climate justice." While nature-based solutions might seem to present a viable and desirable approach, Cooper uses geometric data techniques to understand where nature-based solutions incorporate ground-level community-centered realities and where they do not. The multilayered relationships among health inequity, urban conditions, and nature-based solutions are not well understood, argues Cooper, and she cautions that many nature-based solutions only marginally engage the residents of the communities they are meant to benefit. She calls for a focus by government officials and others engaged in nature-based solutions to consider how to genuinely collaborate with the most marginalized groups of urban residents.

Part 2: Implementing and Assessing Interventions

In the second section, the chapters address the adaptive practices various groups are using to address conditions of climate injustice. Barbara Sattler's chapter, "Climate Change and Evolving Roles for Nurses," highlights the initiative taken by nurses to draw attention to the medical field's complicity in creating dangerous and toxic climate and environmental conditions. Her chapter features nurse researchers and nurse activists who recognized the urgent need to reduce plastics, protect agricultural workers from the impact of excessive heat, disseminate guidance on nutritional food sources and encourage the cultivation of community gardens, and change purchasing practices within hospitals to reduce their carbon footprint. These nurses are inspirational in the systemic thinking they brought to change the conditions in their hospitals and communities.

José Martínez-Reyes and Camille Martinez's chapter, "Maya Agriculture, Green Land Grabs, and Climate Justice in Quintana Roo, Mexico," discusses how the Maya of central Quintana Roo in Mexico are dealing with the profound changes being wrought on their traditional agricultural practices, with the consequent deleterious impact on their food security and health. Interviews with Maya farmers reveal how they respond to climate injustice by reinforcing their traditional ecological knowledge and the adaptive strategies they practice to maintain a productive agriculture and a "moral ecology" with the forest.

Sajani Kandel and Antonio Raciti's chapter, "Structural Inequalities and Extreme Heat in the Boston Region: Sharing Adaptation Lessons from Dorchester," examines the impact of extreme heat on Boston's Dorchester neighborhood, home to many low-income Black and Brown residents. They observe that they are particularly interested in heat planning initiatives undertaken in historically disenfranchised communities in order to understand how they "cope and respond to ongoing heat planning initiatives." Their interviews with Dorchester residents convey the residents' perspectives on Boston's heat-planning initiatives and reflect on the prognosis for the future of these interventions.

Most urban metropolitan areas in the United States (and globally) include individuals who experience housing instability, and for this subset of the population variations in temperature can lead to significant erosion of already precarious health conditions. Authors Kim Flike, Shoshana Aronowitz, and Teri Aronowitz, in their chapter, "Climate Change and Housing Instability," illuminate how climate events such as floods, fires, and extreme heat exacerbate conditions of vulnerability among those living in poverty. People experiencing homelessness are affected by cumulative years of neglect of basic infrastructure—such as access to potable water—in cities, and their lack of access to "power and resources" makes it impossible for them to advocate for their rights. This chapter presents three extreme weather events—wildfires, floods, and extreme temperatures (heat and cold)—and their impact on people experiencing homelessness. The authors discuss the strengths and weaknesses of the responses within the United States to address the vulnerabilities of people experiencing homelessness due to extreme weather events, and they recommend specific interventions that draw on holistic frameworks to address the inadequacies.

Part 3: Valuing Ecosystem Thinking

This third part offers spotlights on particular segments of the population that have been adversely affected by climate injustice and the policy initiatives that have been advanced to address them, with varying degrees of success. A recurring theme in this part that echoes what the earlier sections reveal is that truly effective interventions and paradigms must be conceived and implemented with a full-ecosystem mindset. In other words, fragmented approaches with disparate and disconnected facets cannot yield meaningful and positive change in people's life conditions.

In their chapter "Advancing Climate Justice to Achieve Birth Equity: An Intersectionality-Based Policy Analytic Approach," Lisa Heelan-Fancher and Laurie Nsiah-Jefferson focus on pregnant women and their developing fetuses to show the undeniable connections between climate and those conditions that "negatively affect infant health outcomes that include increased risk of preterm birth (PTB), low birth weight, stillbirths, and infant mortality." They note that "[m]aternal health is also at risk with heat exposure, with studies identifying increased incidence of maternal hypertensive disease, placental abruption, and pregnancy-related complications such as gestational diabetes and poor maternal mental health." Their particular focus is on extreme heat and its negative impact on pregnant mothers and their fetuses. Drawing on critical race theory and feminist frameworks, Heelan-Fancher and Nsiah-Jefferson insist on infusing separate spheres of inquiry with an intersectional understanding that considers race and gender simultaneously, particularly as they relate to questions of power and privilege. The intersectionality-based policy approach they advocate includes as crucial considerations "power, social justice and equity, intersecting categories, multilevel analysis, reflexivity, diverse knowledge, and alternative time and space of marginalized groups."

Older adults constitute another significant vulnerable group, and Caitlin Connelly and colleagues' chapter, "At Risk but Overlooked and Underserved: Older Adults and Vulnerabilities to Climate Change Impacts," focuses on this population and the impact of climate change on their well-being. Their chapter offers a systems view of the many gaps in policies and interventions designed to address the needs of these adults. They spotlight key vulnerability factors, including chronic conditions and care dependency, social isolation, socioeconomic disadvantage, and substandard housing. Older adults are usually not considered in public health planning, and it is not until climate disaster strikes that local and state public health departments, first responders, health care providers, and society in general consider their needs. The authors want to educate health and long-term services and supports advisers on the particular effects of climate change on their clients, and they call for building capacity in the public health sector to address the climate justice needs of older adults.

Climate disasters displace people from their homes and create vulnerable migrants. Two chapters offer analyses of the mental health conditions of migrants who are forced to leave their homes and familiar surroundings as a result of extreme weather events (hurricanes and resulting floods). Rosalyn Negrón and colleagues' chapter, "Climate Displacement and Migration after Hurricane Maria: Implications for Puerto Ricans' Mental Health," considers the mental health impact of Hurricane Maria on both those who had to or made the decision to evacuate (approximately two hundred thousand) and those who chose to stay. The authors examine the failures of the state—at both the local and the federal levels—to provide support for residents so that they could make fully informed and thoughtful decisions. The seventy-two Puerto Ricans they interviewed over the period of a year consisted of a majority who chose to stay back on the island and twenty-five who migrated to the United States. The authors write that climate change and migration are closely intertwined, in the case of Puerto Rico, with colonialism and racialized hierarchies, such as manifest in the case of post-Maria mobilities. The long shadows of colonialism and racialized hierarchies are typically elided over in policymaking or in immediate interventions following an extreme climate event. Negrón and colleagues' findings point to the urgent need for coordinated responses of multiple sectors of society that take into consideration the intricate entanglements of history, politics, economics, government agencies at all levels, and civil society organizations. For those who left Puerto Rico for the United States, the receiving locations will need to consider and address the reality that for many climate migrants the loss of community and social networks exacerbates disaster-related traumas.

In a similar vein and pointing to the crucial need for multiple-agency cooperation and collaboration, Janice Hutchinson offers an overview of various studies on the mental health of New Orleanians (primarily Black New Orleanians) in the immediate period as well as years after Hurricane Katrina in 2005. Her chapter, "Hurricane Katrina, PTSD, Preparedness,

Resilience, and Recovery," looks beyond what the Diagnostic and Statistical Manual of Mental Disorders tells us about post-traumatic stress disorder to consider studies on resilience and recovery and how these outcomes might be facilitated. Her chapter also underscores the need for systems-level responses and considers whether the Arkansas model is an appropriate prototype for desired responses to extreme climate events. Priorities of the Arkansas model include ensuring "physical safety, confidentiality, nonjudgmental listening, and empathy; they also helped develop coping skills, connected services with families, gave information regarding assistance, and integrated survivors as quickly as possible into the Arkansas community. There was also flexibility in providing services." The Arkansas model is considered the "intervention of the future," and Hutchinson invites us to think about its strengths and to consider whether it is possible to replicate it.

Part 4: Acquiring Understanding: Two Case Studies of Long-Term Impacts

Part 4 examines two locations—Vieques (in Puerto Rico) and Bangladesh—to study the long-term and downstream impacts of climate change. Both locations underscore that the deleterious consequences of ill-considered actions persist through time; addressing the conditions resulting from these actions requires first a willingness to take responsibility for the original actions and then to engage in crafting the corrective and necessary interventions.

Lorena Estrada-Martínez and colleagues focus on the long-term effects of colonial/imperial practices on Vieques in their chapter, "Consequences of and Responses to Compounded Vulnerabilities Rooted in Colonialism: The Case of Vieques, Puerto Rico." They trace the impact of the US Navy's use of the island as a training ground for military activities (including the deployment of bombs) for over sixty years on the environment, the island's natural resources, and public health. Compared to the rest of Puerto Rico, Vieques has escalated levels of many illnesses, including cardiovascular conditions and cancer, and this chapter examines the connection between contamination of the land and air in Vieques and health. Protracted grassroots activism led to the cessation of military training on Vieques, and this type of activism is also currently responding to address new contaminant pathways from extreme climate events such as hurricanes and flooding. The chapter analyzes Vieques' political, military, and socioeconomic history, including social and health impacts of land expro-priation and occupation.

Reazul Ahsan, Prokriti Nokrek, and Afrida Asad's chapter, "Climate Change and Long-Term Impact on Women's Reproductive Health: An Unfolding and Untold Social Crisis in the South Coast Region of Ban-gladesh," brings us to a country that lies in a floodplain and whose resi-dents are not unfamiliar with extreme weather events. However, in recent decades, the impact of global warming has significantly increased the level of risk of countries in the Global South, most of which are themselves not responsible for the excessive burning of fossil fuels that contribute to global warming. One of the long-term impacts of climate change in Ban-gladesh is the increase of salinity levels in the soil and in the groundwater table. Shrimp farming has become a particularly dangerous occupation for Bangladeshi women whose reproductive health is adversely affected by the increased salinity of the waters in which they are required to stand for several hours in order to farm for shrimp. Their reproductive organs' exposure to highly concentrated saltwater leads to urinary and vaginal tract infections. This study uses a qualitative focus-group approach to assess the health and social impact on the marginalized community of female workers in the south coast region of Bangladesh as a result of cli-mate change. The authors emphasize the need for multidisciplinary and multisector collaboration among climate scientists, health care providers, local government officials, development organizations, affected commu-nities, and national and international governance organizations.

Following the conclusion, we offer a coda, "Creating Hope and Systems Thinking in the Next Generation of Climate Stewards and Public Health Workers." The coda provides hope for the future. It describes an initiative at University of Massachusetts Boston that was launched by a National Endowment for the Humanities Connections Grant in 2018. The project, "Living with the Urban Ocean: Inquiring, Imagining, Embracing," is a place-based curriculum that is grounded in the campus' location on Bos-ton Harbor and its archipelago of thirty islands. Students from all majors are shown the importance of bringing the humanities into conversation with the sciences to understand how history, culture, and values have pro-found and long-term impacts on the ways in which we engage with our natural world, including Boston Harbor and its thirty islands, and with one another. Eighty-one percent of graduates of UMass Boston remain in

Massachusetts, so the university plays a crucial role in cultivating the next generation of systems-level thinkers and actors in the city of Boston and the region.

We close this introduction with a gesture toward the future and drawing on an utterance of hope from the Oneida poet Roberta Hill Whiteman:[20]

> I keep faith that the spirits will keep you aware
> that we are
> related to everything here.
> —Roberta Hill Whiteman, from "Wherever in Winter"

NOTES

1 The Environmental Protection Agency's September 2021 Request for Applications reads, "Cumulative Health Impacts at the Intersection of Climate Change, Environmental Justice, and Vulnerable Populations/Lifestages: Community-Based Research for Solutions. It is anticipated that a total of approximately $8.1 million will be awarded under this announcement, depending on the availability of funds, quality of applications received and other applicable considerations. The EPA anticipates funding approximately six awards under this RFA. Requests for amounts in excess of a total of $1,350,000, including direct and indirect costs, will not be considered. The total project period requested in an application submitted for this RFA may not exceed 3 years." https://www.epa.gov/system/files/documents/2021-09/shc-climate-change-and-ej-rfa_-final.pdf.

From the Yale Center on Climate Change and Health: "Clinic in Climate Justice, Law, and Public Health." See the text: "The clinic is an innovative collaboration between Yale School of Public Health and Vermont Law School and includes faculty and students from both Yale and Vermont Law School. It operates year-round, with enrollment in both fall and spring semesters. Interdisciplinary student teams carry out applied projects at the intersection of climate justice, law and public policy, and public health. Each team works with a partner organization to study, design, and implement a project, typically through community-based participatory research practices. The course affords the opportunity to have a real-world impact by applying concepts and competencies learned in the classroom."

Past projects include 'Climate Justice for Persons Vulnerable to Heat Stress in Northern Manhattan' (partner organization: WE ACT for Environmental

Justice), 'Health and Environmental Justice in Alabama: Pollution Impacts in Two Low-Income African American Communities' (partner organizations: Ashurst Bar/Smith Community Organization, Tuskegee University, Black Belt Citizens Fighting for Health and Justice, and University of Maryland–College Park, and EJ Clinic at Vermont Law School), and 'Managing Non-Communicable Diseases after Natural Disasters in the Caribbean' (partner organization: Eastern Caribbean Health Outcomes Research Organization." https://ysph.yale.edu /yale-center-on-climate-change-and-health/policy-and-public-health-prac-tice/clinic-in-climate-justice-law-and-public-health/.

2   The Pew Research Center's report by Carroll Doherty, "Remembering Katrina: Wide Racial Divide over Government's Response," August 27, 2015, https://www .pewresearch.org/fact-tank/2015/08/27/remembering-katrina-wide-racial -divide-over-governments-response/.

  Jean Ait Belkhir and Christiane Charlemaine, "Race, Gender and Class Lessons from Hurricane Katrina," *Race, Gender & Class* 14, no. 1/2 (2007): 120–52, JSTOR, accessed October 10, 2022, http://www.jstor.org/stable/41675200.

  Spike Lee, When the Levees Broke: A Requiem in Four Acts, documentary film about the destruction wrought on New Orleans by Hurricane Katrina, focusing on the challenges endured by New Orleanians following the disaster of the hurricane, first broadcast on HBO (2006).

3   Susan J. Masten, Simon H. Davies, and Shawn P. Mcelmurry, "Flint Water Crisis: What Happened and Why?," *Journal-American Water Works Association* 108, no. 12 (2016): 22–34; Lindsey J. Butler, Madeleine K. Scammell, and Eugene B. Benson, "The Flint, Michigan, Water Crisis: A Case Study in Regulatory Failure and Environmental Injustice," *Environmental Justice* 9, no. 4 (2016): 93–97; Benjamin J. Pauli, *Flint Fights Back: Environmental Justice and Democracy in the Flint Water Crisis* (Boston, MA: MIT Press, 2019); Nick Estes, *Our History Is the Future: Standing Rock versus the Dakota Access Pipeline, and the Long Tradition of Indigenous Resistance* (New York: Verso Books, 2019). Josué Rivas, "Solidarity in Standing Rock," *World Policy Journal* 34, no. 4 (2017): 62–75; Guillaume Proulx and Nicholas Jon Crane, "'To See Things in an Objective Light': The Dakota Access Pipeline and the Ongoing Construction of Settler Colonial Landscapes," *Journal of Cultural Geography* 37, no. 1 (2020): 46–66.

4   Susan L. Prescott and Alan C. Logan, "Planetary Health: From the Wellspring of Holistic Medicine to Personal and Public Health Imperative," *Explore* 15, no. 2 (2019): 98–106.

5   See the Rockefeller Foundation's site, https://www.rockefellerfoundation.org/ report/safeguarding-human-health-anthropocene-epoch/. The Lancet Commission's superb and comprehensive report can be found at https://www.thel-ancet.com/action/showPdf?pii=S0140-6736%2815%2960901-1.

6 I. U. Iruka, S. M. Curenton, J. Sims, K. A. Escayg, and N. Ibekwe-Okafor, RAPID-EC, "Black Parent Voices: Resilience in the Face of the Two Pandemics— COVID-19 and Racism: Researchers Investigating Sociocultural Equity and Race (RISER) Network," University of North Carolina, Frank Porter Graham Child Development Institute, 2021.

7 Climate Interactive can be found at https://www.climateinteractive.org/about/. For an explanation of the obstacles, see especially page 11 of the study "Multi-solving at the Intersection of Health and Climate," https://www.climateinter active.org/wp-content/uploads/2018/01/Multisolving-at-the-Intersection-of -Health-and-Climate.pdf.

8 Robin Wall Kimmerer, "Returning the Gift," Center for Humans & Nature, October 1, 2013, https://humansandnature.org/earth-ethic-robin-kimmerer/.

9 Lorén Spears expressed these sentiments in her presentation at a virtual sym-posium titled "New England Traditional Ecological Knowledge and Climate Justice" on July 28, 2022, hosted by the Sustainable Solutions Lab at University of Massachusetts Boston.

10 Rhys Jones, "Climate Change and Indigenous Health Promotion," *Global Health Promotion* 26, no. 3, suppl. (April 2019): 73–81, https://journals.sagepub.com /doi/epub/10.1177/1757975919829713.

11 Renee Cho, "Why Climate Change Is an Environmental Justice Issue," from Columbia Climate School's *State of the Planet*, September 22, 2020, https://news .climate.columbia.edu/2020/09/22/climate-change-environmental-justice/.

12 "Can Genetic Modification Help Plants Survive Climate Change?," interview with professor Shang Yang He, Science Friday, https://www.sciencefriday.com /segments/genes-plants-heat-climate-change/. Transcript available at site.

13 Amanpour and Company, "Pollution Is Segregated," a conversation with Robert Bullard, https://www.youtube.com/watch?v=gU-D3YkOe-w, time signature: 3:48–4:48.

14 Robert Bullard's keynote address, "The Quest for Environmental and Climate Justice: Why Race and Place Still Matter," https://www.youtube.com/watch?v =9KtKzldxLAM, time signature: 10:15–1:01:13.

15 Miriam Wasser, "Mass. Revokes Air Permit for Controversial Biomass Facility in Springfield," wbur, updated April 2, 2021, https://www.wbur.org/news/2021/04 /02/springfield-biomass-permit-revoked.

16 Nadia Y. Kim, *Refusing Death: Immigrant Women and the Fight for Environmen-tal Justice in LA* (Stanford, CA: Stanford University Press, 2021), 3; see especially "Introduction" (1–35) and "Emotions as Power" (73–103).

17 Climate One, Killer Combination: Climate, Health, and Poverty, February 12, 2021, https://www.climateone.org/audio/killer-combination-climate-health-and -poverty.

18 See Mark Harvey, "Slavery, Indenture and the Development of British Indus-
    trial Capitalism," *History Workshop Journal* 88 (Autumn 2019): 66–88, https://
    doi.org/10.1093/hwj/dbz027; Tayyab Mahmud, "Cheaper Than a Slave: Inden-
    tured Labor, Colonialism and Capitalism," *WHITTIER Law Review* 32, no. 2
    (2013): 215.

19 Mia Mottley, "Inaugural Presidential Lecture," delivered at the World Trade Orga-
    nization, March 23, 2022, https://www.youtube.com/watch?v=2IDy3KlRZFc&t
    =1215s.

20 Roberta Hill Whiteman, "Wherever in Winter," in *Returning the Gift: Poetry
    and Prose from the First North American Native Writers' Festival*, ed. Joseph
    Bruchac (Tucson: University of Arizona Press, 1994), 320–21.

Part 1
Changing Paradigms for Wellness

CHAPTER 1

Indigenous Peoples and Well-Being in the Context of Climate Change

DEBORAH MCGREGOR, CASEY BAS, MAHISHA SRITHARAN, KESHBIR BRAR, NATASHA VERHOEFF, HILLARY MCGREGOR, AND STEVEN WHITAKER

The land is there, outside our windows, under our feet, all around us, thinking, feeling, conversing and offering its teachings. When we start to really listen, to the land and to the people whose identities and traditions are fundamentally shaped through long dialogue with the land, transformation follows.
—Sean Blenkinsop and Mark Fettes, "Land, Language, and Listening"

Indigenous peoples are among the first to face the direct consequences of climate change, due to their dependence upon, and close relationship with, the environment and its resources.
—UNDESA, Climate Change

The climate crisis threatens Indigenous peoples—their ways of life, cultures, traditions, lands, and languages—more than any other group on the planet. We know this to be true, "even though indigenous peoples contribute the least to greenhouse emissions."[1]

The ongoing injustice that perpetuates this tragic irony clearly must be addressed, for the benefit of not only Indigenous peoples themselves but of the entire planet. For the loss of Indigenous cultures, unconscionable as that is, also means the loss of incredible knowledge systems that are critical to the development of a sustainable global society.[2] In short, the health of the planet is inextricably tied to the well-being of Indigenous peoples.[3]

The United Nations reports that while Indigenous peoples "make up less than 5 percent of the world's population . . . they speak an overwhelming majority of the world's estimated 7,000 languages."[4] Moreover, while they "own, occupy or use resources on some 22% of the global land area, [this land] in turn harbours 80% of the world's biological diversity."[5] The potential loss of Indigenous cultures and languages therefore threatens the globe with not only cultural loss but biodiversity loss as well. Planetary health reports, including from the Intergovernmental Science-Policy Platform on Biodiversity and Ecosystem Services, have stated in no uncertain terms that global society *must* achieve "transformative" change across virtually *all* sectors if we are to succeed in maintaining a planetary ecosystem capable of sustaining human and many other forms of life.[6] But what would a fully "transformed" society look like? On what fundamental principles would it be based? As Sean Blenkinsop and Mark Fettes write, "The kinds of social and cultural changes needed to respond to the climate and ecological crises require new/different/incredibly flexible stories to be told in many ways and many places throughout society; but these cannot simply be invented out of thin air. They need to be grounded in the *land*."[7] Scholars have thus begun to highlight the linkages between language, biodiversity and planetary health, climate change and land, asserting that "connections between health, language and land have been slowly emerging over the last decade and a half—it appears that fostering sustainable relations between land and language also translates to greater sustainability in the human segment of the population."[8]

Understanding the connections that exist among language, land, and human health is essential for building resilience within Indigenous communities as they continue to adapt to environmental and climate crises. Recognizing this need further underlines the urgency around ensuring the revitalization of Indigenous languages in Canada and across the world, as virtually all such languages are in at least some danger (and many in immediate danger) of extinction.[9] To grasp the true significance of Indigenous languages, the knowledge they hold, and their relevance to addressing global environmental and climate challenges, it is vital to understand the links among language, the land, and the health of its peoples.

INDIGENOUS PERSPECTIVES ON HEALTH, THE IMPORTANCE OF LAND, AND THE IMPACTS OF COLONIZATION

Many Indigenous conceptions of health emphasize the importance of holism, of balance among the physical, emotional, intellectual, and spiritual elements of a person.[10] Achieving optimal health requires a deep understanding that the land is "intertwined and interconnected" with these various components of health.[11] The community recognizes that "the individual is in constant transaction with the physical environment" and "gives a central role to connections among individuals and to place in the health and well-being of the person."[12] Planetary health, defined as "the health of human civilization and the natural systems upon which it depends," is thus also entirely necessary for Indigenous well-being, as it is for all humanity.[13]

The medicine wheel represents health as an equal balance among the physical, mental, emotional, and spiritual components.[14] From the perspective of many Indigenous cultures and ways of living, health involves not only the individual but also the environment and broader ecosystems. Having a connection to the land and "place" directly affects physical, social, emotional, and spiritual well-being.[15] Studies have shown that even Indigenous youth living in urban areas who cultivate their relationships with the land can experience greater resilience to climate change impacts and related stressors.[16]

Monitoring the state of the land is a way of life for many Indigenous communities; it is a way to give back to the land, an act of reciprocity, and an adaptation strategy.[17] In turn, when this relationship with the land is severed through land loss and destruction or degradation, negative health outcomes frequently arise, and these in turn can exacerbate historical trauma and what Jacqueline Middleton and colleagues refer to as "ecological grief," a profound emotional response that may occur due to negative changes in a person's environment.[18] In one study, community members from Rigolet, Nunatsiavut, provided a clear example of this when they described nature as another person, a community member who when harmed or lost led to feelings of loss for someone in one's own family.[19]

Relating to nonhuman beings as relatives is another important aspect of the connection between the land and Indigenous health and well-being. For example, the Blackfoot Nations of the western prairies in Canada have spiritual and cultural connections with bison, a sacred animal in ceremonies and a staple food source.[20] Indigenous health is also tied to food sovereignty and access to traditional foods, medicines, and education.[21] For example, the Kwakwaka'wakw, Haida, and Nuu-chah-nulth nations in British Columbia consume many types of wild berries as food but also use them medicinally for their antibacterial properties.[22]

These teachings and other land-based skills have been systematically disrupted due to colonial systems and policies, such as the residential school system, the Sixties Scoop, and the current child welfare system.[23] As well, ongoing ecological degradation means that the environment, or rather the state of the environment, can act as both a health benefit and a health risk for Indigenous communities and their traditional food systems.[24] Traditional foods are a source of cultural, economic, and nutritional benefit for Inuit peoples, for example, but at the same time can expose communities to harmful contaminants such as lead, PCBs, and mercury. Pollutants accumulating in food chains through bioaccumulation and biomagnification have been particularly well-documented, as Indigenous peoples tend eat those animals or fish parts in which contaminates bioaccumulate.[25] A devastating example concerns the Grassy Narrows First Nation in northwestern Ontario, whose river system was (and remains) contaminated with mercury (Hg) from pulp and paper mill effluent released between 1962 and 1970. A longitudinal study published in the *Lancet* revealed "a significant association between longitudinal Hg exposure from freshwater fish consumption and premature mortality. In the First Nation community of Grassy Narrows, individuals who died prematurely before reaching 60 years old had significantly higher Hg exposure between 1970 and 1997 than those who lived longer."[26] Indigenous food sources are also facing scarcity and extinction largely due to the long-term impacts of industrial infrastructure, extraction, habitat loss, and human settlement patterns,[27] all of which are now being exacerbated by the additional pressures of the climate crisis. To date, government-sanctioned interventions have been frequently ineffective in protecting such traditional food sources. For example, the fisheries of the Mik'maq were mismanaged by the Department of Fisheries and thus depleted; today, Mik'maq are bolstering *netukulimk*, the name given to their laws of procurement, in order to sustainably manage their own traditional foods.[28]

INDIGENOUS HEALTH AND CLIMATE CHANGE

Indigenous communities are at a higher risk of exposure to the consequences of climate change and environmental dispossession, leading to poorer health outcomes than those observed in the general population.[29] In part, this is due the close ties Indigenous peoples have with the land on which they depend. Such disproportionate consequences are made even worse, however, by preexisting health inequities, as described by the

National Collaborating Centre for Indigenous Health: "Indigenous Peoples, in Canada and globally, are recognized as uniquely sensitive to the impacts of climate change because they often live in geographic regions already experiencing rapid change and because of their close relationships with and dependence on land, waters, animals, plants, and natural resources for their sustenance, livelihoods, cultures, identities, health, and well-being.... Non-climate determinants of health exacerbate these sensitivities, including a greater existing burden of health inequities compared to non-Indigenous populations and the historic and ongoing effects of colonization and socio-economic and political marginalization."[30] As one example, the climate crisis has widely disrupted ancient ecological patterns, such as animal migration, consequently hindering the ability of Indigenous peoples living in northern Canada to hunt traditional game.[31] The use of plants for food, medicine, and other needs is also vitally important. When access to such environmental resources is reduced, such as through aquaculture development or timber cutting, communities' economic, social, and cultural resources are restricted and Indigenous communities are unable to engage in relationship building, leading to negative impacts on their health and well-being.[32]

As Indigenous peoples experience increasing environmental stressors stemming from climate change, upholding and reinforcing their connections to the land will be key to their health and resilience. For Indigenous peoples and communities to heal from historical and contemporary traumas arising from colonization, land loss, displacement, and climate change, it is imperative to reconnect to the natural environment through reinvigorating cultural practices, honoring the land, and engaging in Indigenous-led cultural activism.[33]

INDIGENOUS HEALTH AND LANGUAGE

According to Canada's 2016 census, there are over seventy Indigenous languages spoken in the country.[34] To help Indigenous communities become more resilient to the impacts of climate change, it is crucial to strengthen and revitalize Indigenous languages.

Indigenous health is highly linked to connections to the land and environment.[35] Health representatives from First Nation and Inuit communities in Canada identified environmental/cultural connections as one of six primary determinants of health.[36] In Indigenous cultures, this strong connection to land is maintained through language. Ongoing colonial policies in Canada and elsewhere, including both those aimed directly at Indigenous peoples and those that actively support the degradation of Indigenous territories, are increasingly threatening the existence of those languages.

Across Canada and around the world, there exist inequitable and unsustainable power imbalances between majority and minority populations, often highlighted and perpetuated in the way the respective languages are either used and supported or denigrated within a society. As Nicholas Reo and colleagues point out, "As a minority culture encounters a politically more powerful culture, there are social and economic pressures to assimilate into that culture at the expense of Indigenous lifeways and ways of speaking. The assimilatory pressures are typically institutionalized in formal educational systems where priority is given to teaching in national languages and about national histories."[37] There are many reasons why Indigenous languages are under threat of disappearance. Larry Glorenflo and colleagues state that intergenerational transmission, the number of people who speak the language, and other factors can affect how much risk there is of a language becoming lost.[38] For many Indigenous communities in Canada, intergenerational transmission of Indigenous languages has suffered due to the external factors of colonialism, residential schools, and other racist policies and practices.[39] Canada's genocidal policies and its history of colonial practices that remain embedded in institutions and legislation have indelibly affected Indigenous peoples and their traditions, cultures, knowledge systems, and languages. Residential schools, for example, were institutions created with the primary purpose of "killing the Indian" by preventing Indigenous children from learning their language and culture and forcing them to learn and take up those of the colonizer.[40] As a result of such trauma, Indigenous peoples now have to contend "with cultural confusion, shame in not being able to voice one's mother language, and poorer health outcomes."[41]

Language is not simply a means of communication. It permeates every aspect of a culture and contains a wealth of cultural knowledge and experience. As Nicholas Reo and colleagues point out, "Indigenous languages are integral to Indigenous communities and cultures. The history of a people, as well as their cultural values and interactions with their environment, slowly accrete in language over time. For this reason, the language becomes particularly adept in supporting culturally specific activities, verbal art, and group identity."[42] Language shapes Indigenous peoples' understanding of the world and how they interact with their environments. Jenanne Ferguson and Marissa Weaselboy underscore the importance of this connection, stating that "when examining language ideologies—the beliefs, attitudes, and ideas speakers have about language—we find that many Indigenous cultures do not conceive of language as separate from culture. . . . They are nested within each other, fundamentally intertwined. To lose one's Indigenous language thus means the loss of certain cultural elements, and an ensuing connection [with] Land."[43] Similarly, Rosalie Schultz asserts that "loss of languages leads to loss of the knowledge transmitted through those languages. Indigenous languages hold knowledge of ecosystems and their sustainable management and care accumulated over hundreds of generations. Language loss is associated with biodiversity loss."[44]

Land, language, and the health and well-being of Indigenous peoples are all interlinked. As Margo Greenwood and Nicole Marie Lindsay state, "All of the myriad determinants of health for Indigenous peoples point back to the critical relationships we have with our traditional lands and territories: our systems of self-government, our languages, our cultures, our healing traditions, our relationships to the animals and plants that nourish our bodies and spirits, our ceremonies and protocols for maintaining these relationships and ensuring our collective survival."[45] These statements highlight the dire need for retaining and revitalizing Indigenous languages, many of which are on the brink of being lost. The issue goes beyond protecting languages for their own sake, as worthwhile as that certainly is. The knowledge and experience embedded within these languages, the vital lessons and understandings of how to live sustainably with the earth, are even more valuable, and may indeed hold keys for the survival of humanity itself.

LANGUAGE AND CLIMATE CHANGE

Language is not often included in analyses of climate change impacts, and even less often in discussions of strategies for minimizing its effects. Climate change issues and solutions are most often discussed using the dominant Western languages, particularly English. Thus, the most promulgated frameworks and concepts for addressing the climate crisis are derived from Western ways of thinking. It is these Western worldviews, however, that have caused so many of the disastrous circumstances we see today, including biodiversity loss, the climate crisis, and genocidal policies aimed at Indigenous peoples. Though they are being utilized in attempts to address climate change, Western perspectives on globalization and capitalism, and seeing nature, animals, and even people as resources to exploit for economic growth, are what produced the problem in the first place.[46]

Describing how language lays the foundation for the way in which we view things, Claudia Gafner-Rojas states that "Indigenous languages express the characteristic community-centered way of life of their speakers."[47] The way nature is depicted through language greatly dictates how it is valued and treated. The Western concept of ecosystem services, for example, values nature solely for the processes and products it provides to humans. As Matthias Schröter and colleagues see it, this notion "contradict[s] holistic perspectives of indigenous people."[48]

Western anthropocentric understandings of nature run counter to Indigenous worldviews. As Jenanne Ferguson and Marissa Weaselboy explain, "Indigenous teachings tell us we come from the land, and therefore the lands own us, rather than the other way around. This relationship of coming from the land reminds us how all our interactions also continue to shape us as stewards of Land. Language takes on a specific role in this stewardship, as it is one key medium or conduit by which a (human) being may also connect with Land."[49] Many Indigenous knowledge systems are founded on a holistic understanding of the world. They center not on humans but on Creation, the land, the environment, nature, and the connections between nature and all its living things. Societal recognition of the importance of such holistic understandings is what many scholars, activists, and Indigenous peoples have been advocating for, for some time. As Nylah Burton confirms, "Indigenous language revitalization requires a seismic shift in the world. It is much easier to maintain the status quo when discussing the climate crisis through the lens and the language of colonization."[50]

LANGUAGE, PEOPLE, AND PLANETARY HEALTH

There are close to seven thousand languages spoken in the world, of which "more than 4,800 occur in regions containing high biodiversity."[51] Glorenflo and colleagues state that "biodiversity is equal to if not higher in areas with more indigenous presence than areas with less."[52] In their analysis of language and biodiversity, they observe that "the co-occurrence of linguistic (and, in many ways, cultural) and biological diversity identified in this study is fortuitous in that it provides the basis for bringing together organizations and researchers focusing on biodiversity conservation and those concerned with linguistic and cultural conservation in particular regions."[53]

Nylah Burton and Emilie Ens and colleagues discuss the connections and importance of languages, particularly Indigenous languages, to climate change, biodiversity, and the land as embedded in biocultural knowledge. Indigenous languages hold an abundance of biocultural knowledge and provide Indigenous peoples with the knowledge and tools needed to address environmental and climate challenges. Thus, the loss of these languages can greatly diminish communities' resilience and their ability to address the climate crisis.[54]

LANGUAGE AND LAND

The land is an integral part of many Indigenous peoples' culture and way of life. Land is also central to Indigenous teachings, knowledge, and customs. Indigenous peoples' languages often are very much embedded in their land, as knowledge and understanding of their environments comes from the land. Therefore, Indigenous languages need to be thought of and seen as belonging to that land. As Ferguson and Weaselboy confirm, "Language reverberates within and through Land, and thus is intrinsically connected to all beings upon that land; it is not a variable easily separated, as it allows and acknowledges both communion with—and understanding of—land."[55] The intertwined relationship between Indigenous peoples and the land is seen through their knowledge systems, practices, and language: "Language is an essential part of being in sustainable relationships with Land."[56]

The Mi'kmaw, for example, have many words and phrases that speak to their strong connection to the land, including *weji-sqalia'timk*, *netuku-limk*, and *ko'kmanaq*. *Weji-sqalia'timk* translates as "where we sprouted or emerged from," which is a reference to the land as the origin of people and implies that if the land is sick, the people are sick.[57] *Netukulimk* is a value system that governs the relationship between the Mi'kmaw and nature, and states that the land must be used sustainably and that it holds people accountable for wasteful use of land. *Ko'kmanaq* translates as "our relations" and creates a relationship of respect, kinship, and reciprocity with both living things and inanimate objects. These values, which include concepts of taking only what you need and giving back and offering thanks, are conveyed through the ways in which the land is used.

In contrast, Western knowledge systems and languages engage with and talk about the land in a vastly different manner. As Sean Blenkinsop and Mark Fettes state,

> In English, "land" is a noun; it presents itself as something concrete, something separate from the people and animals that move across it, something that can be mapped and built upon, and subdivided and possessed. The phrase "traditional territory" in the land acknowledgements is perhaps a tacit recognition of the inadequacy of this way of thinking. A traditional territory does not refer to a history of ownership, but a history of reciprocal relationship, people and land shaping and adjusting to each other over a long stretch of time. Yet "territory", too, has some of the same problems as "land"—indeed, it reinforces the sense of land as the passive partner, since it is felt to be the beings that live on the land (human, bear, coyote, raven) that define, dwell on, and defend their territories.[58]

Language is a powerful tool, not only as a means of communication but also as a framework for understanding the world around us. English has become the world's most spoken language and is "the default language of international business."[59] Such business includes global debates about climate change and the environment. By communicating predominantly within the constraints of this Western perspective, decision makers perpetuate the view of land and nature as inferior to humans. According to Ferguson and Weaselboy, "Phenomenologically, Land must be experienced through Indigenous language in order to fully appreciate those layers of meaning and appreciate the nuances of what sustainable relations are within that Indigenous culture."[60] Many Indigenous languages and ways of knowing center around a holistic and relational way of seeing and understanding the world. All living and "nonliving" things are seen in connection with each other, and valuing these relationships is key to ensuring a sustainable way of living.

Environmental dispossession caused by climate change significantly challenges communities' ability to be on the land and share such knowledge. Indigenous youth, elders, and community health representatives have emphasized the importance of reconnecting to land to share traditional knowledge that promotes health and community connectedness.[61]

LANGUAGE AND INDIGENOUS CALLS TO ACTION

Safeguarding and ensuring the continued survival of Indigenous languages is vital to global sustainability and is a leading issue for many Indigenous peoples and communities. This is reflected in the numerous Indigenous declarations and reports that have emerged in recent years. Language, for example, was a key topic of discussion among attendees at the Reconnecting with Mother Earth gathering held in 2017 that brought together First Nations youth and elders to talk about climate change. The report that arose from this gathering stresses the importance of Indigenous peoples being able to speak their own language to reconnect and maintain a relationship with Mother Earth as a strategy for meeting climate challenges.[62]

The Onjisay Aki International Climate Summit, held in Manitoba, also brought together Indigenous knowledge keepers and climate leaders in 2017. From this summit emerged twelve Onjisay Aki calls to action, the third of which directly relates to language, its significance to ancestral knowledge, and "living in balance with the Earth," and reads as follows: "To support revitalization of Indigenous languages, which are foundational to stewardship. Indigenous languages are connected to our land-based ancestral knowledge. Critical concepts and teachings are embedded in Indigenous language, stories and songs."[63] While the remaining calls to action do not explicitly mention it, language is an integral part of achieving them as well. For example, the first four calls all fall under the theme of ancestral knowledge and include the need to ensure that Indigenous knowledge is passed down from elders to youth. Indigenous languages hold an abundance of knowledge that is vital for Indigenous communities and can help in the fight against climate change and can repair the fraught relationship with Mother Earth brought on by Western knowledge and ways of thinking about nature. As stated in the twelfth and final call to action, the purpose of all the calls is to "initiate a planetary transformation that creates a healthier future for our children, grandchildren and seven generations to come."[64]

LAND-BASED LEARNING, LANGUAGE, AND WELL-BEING

Many scholars and Indigenous peoples have spoken about the benefits of reconnecting Indigenous peoples and youth to the land. Katie Big-Canoe and Chantelle Richmond talk about the importance of environmental repossession and Indigenous peoples reconnecting to and "reclaiming their traditional lands and ways of life."[65]

Jennifer Redvers looks to land-based healing for addressing mental health and well-being, stating that

> in order to rediscover and re-establish a fundamental relationship with the land, one must first experience it directly through practical, culturally-rooted activities, languages, and interactions that return us to the land physically, emotionally, mentally, and spiritually. By bringing land connection to the forefront in the dialogue within Indigenous mental health in Canada, we can draw on the resilience of the land itself to more effectively overcome ongoing impacts of colonization and land-degradation. A healthy land relationship not only ensures the land's health, it also facilitates healthy relationships within ourselves, between one another, and within the larger world, leading to greater balance and mental wellbeing for future generations.[66]

Chantal Viscogliosi and colleagues highlight that "it is essential to work with communities to support the implementation of intergenerational actions involving elders in order to overcome existing health and social challenges based on a holistic approach favouring cultural security."[67] Intergenerational learning and knowledge exchange is crucial to language revitalization, as elders can pass on language to young people in their community. Language revitalization is also rooted in land education and land-based learning. Eve Tuck and colleagues state that "land education puts Indigenous epistemological and ontological accounts of land at the center, including Indigenous understandings of land, Indigenous language in relation to land, and Indigenous critiques of settler colonialism. It attends to constructions and storying of land and repatriation by Indigenous peoples, documenting and advancing Indigenous agency and land rights."[68] Drawing on Sandra Styres' article on "land as first teacher," Blenkinsop and Fettes explain that "land teaches not simply by offering a place to live, or the food and material resources necessary for survival, but more fundamentally by showing us and letting us experience, continually and in myriad ways, what living relationships look like and feel like and how they weave together to make greater, more complex, self-sustaining and adaptive wholes."[69] To create frameworks adept at addressing the challenges of the climate crisis, "they need to be grounded in the land." Learning from and off the land is part of many Indigenous peoples' knowledge systems. Blenkinsop and Fettes highlight that "Indigenous cultures across North America emphasize the importance of respectful listening and silent, attentive observation; these are values instilled from the earliest age. Intriguingly, there is also a long tradition in environmental education, natural history and even field biology that has placed a premium on quiet as a way to more deeply connect with, carefully observe and better understand the more-than-human world."[70]

LANGUAGE REVITALIZATION EFFORTS

Canada has recently made efforts to recognize and provide support for Indigenous language revitalization. In 2015, the Truth and Reconciliation Commission added five new calls to action that pertain to language and culture. These calls to action (numbers 13 to 17) are:

1. To acknowledge that Aboriginal rights include Aboriginal language rights.
2. To enact an Aboriginal Languages Act.
3. To appoint, in consultation with Aboriginal groups, an Aboriginal Languages Commissioner. The commissioner should help promote Aboriginal languages and report on the adequacy of federal funding of Aboriginal-languages initiatives.
4. For postsecondary institutions to create university and college degree and diploma programs in Aboriginal languages.
5. For all levels of government to enable residential school Survivors and their families to reclaim names changed by the residential school system by waiving administrative costs for a period of five years for the name-change process and the revision of official identity documents, such as birth certificates, passports, driver's licenses, health cards, status cards, and social insurance numbers.[71]

In Canada, the Indigenous Languages Act was passed in 2019, the purpose of which is "to support the reclamation, revitalization, maintaining and strengthening of Indigenous languages in Canada."[72] Masud Khawaja discusses how across the country there have been efforts made by post-secondary institutions to aid in creating Indigenous language programs. Khawaja states that "it is hoped that the barriers to Indigenous language education would be dismantled, especially at the institutional level. By doing so, Indigenous students may access language education through thoughtfully designed programs, strengthening collective cultural identity as a result."[73]

Many Indigenous communities are making efforts with the support of governments to revitalize and preserve their languages. For example, the Michif Language Project is a joint effort between Heritage Canada and the BC Métis Federation, which aims "to pursue the language preservation efforts of the French Michif Dialect by developing a full curriculum and resource library for beginner-level learners."[74] It is crucial for Indigenous communities to have access to resources and supports to create programs or work toward finding ways to revitalize Indigenous languages. Many communities have also found ways to ensure that their languages are not lost. A project called the Biigtigong Language Project created by the Biigtigong Nishnaabeg, an Ojibway First Nation, has used technology and created a website to document their language. To connect youth to Indigenous languages there are many efforts focused on teaching language and culture through land-based activities as a way of passing on knowledge through environmental repossession.[75] Land-based activities for teaching youth Indigenous knowledge, language, and culture draw on Indigenous traditional practices of sharing knowledge and stories through experiences on the land.

Although the government of Canada has taken steps recently to ensure the revitalization of Indigenous languages and to provide greater support for preserving and safeguarding them, there is much more the government needs to do to try to undo the harms of colonialism that have forced the erasure of Indigenous peoples' languages and cultures. Many scholars have highlighted how interconnected Indigenous peoples' health and well-being are to their culture, language, knowledge, and land.[76] Language and Indigenous knowledge are interconnected, and both play an important role in addressing the climate crisis. As Susan Chiblow states, "Indigenous worldviews and ways of being have been recognized internationally as vital to addressing humanity's environmental crisis. With the Indigenous worldview embedded in the languages, the revitalization of the Indigenous languages in Canada will benefit all life's beings in addressing the crisis."[77] Although the interconnected nature of language, knowledge, health, and climate change has been discussed by scholars, it has not been effectively translated into policy or government initiatives. Language revitalization and government funding are focused on reconciliation and cultural efforts and do not address climate justice and health as reasons for Indigenous language revitalization. It is important to not only recognize these connections but also understand them in order to provide the right levels of support for the language revitalization of Indigenous communities.

CONCLUSION

Indigenous ontologies, knowledge, and languages offer different ways of knowing from Western approaches and are crucial to addressing climate change.[78] We have seen as well that the survival of such knowledge, and the languages that both contain and transmit it, is integrally connected with the health of Indigenous peoples. Indigenous peoples' health is in turn entirely dependent on the heath of the land.

Indigenous languages thus cannot feasibly be separated from the people, their culture, and their land; all of these elements are inextricably interconnected. These vital linkages need to be further explored to help address the environmental and climate crises. In practical terms, this also becomes a discussion around Indigenous rights. As Siham Drissi states, "When the rights of indigenous peoples are protected—and particularly their rights to land, territories and resources—their culture thrives and nature thrives."[79] The currently predominant Western frameworks and concepts are not adequate to truly address the real causes of climate change and ensure the well-being of Indigenous communities. While Western knowledge will of necessity be involved in addressing such crises, the failure of this knowledge to make substantial progress thus far means that new solutions are required, and many of these must come from Indigenous communities and be embedded in their beliefs, knowledge, and languages.[80]

NOTES

1 UNDESA, *Climate Change.*

2 Mihi Ratima, Debbie Martin, and Treena Delormier, "Indigenous Voices and Knowledge Systems—Promoting Planetary Health, Health Equity, and Sustainable Development Now and for Future Generations," *Global Health Promotion* 26, no. 3 (2019), https://journals.sagepub.com/doi/full/10.1177/175797 5919838487; Sumudu A. Attapatu, Carmen G. Gonzalez, and Sara L. Seck, eds., *The Cambridge Handbook of Environmental Justice and Sustainable Development* (Cambridge: Cambridge University Press, 2021); Sione Tu'itahi, Huti Watson, Richard Egan, Margot W. Parkes, and Trevor Hancock, "Waiora: The Importance of Indigenous Worldviews and Spirituality to Inspire and Inform Planetary Health Promotion in the Anthropocene," *Global Health Promotion* 28, no. 4 (2021): 73–82.

3 Jennifer Redvers, "'The Land Is a Healer': Perspectives on Land-Based Healing from Indigenous Practitioners in Northern Canada," *International Journal of Indigenous Health* 15, no. 1 (2020): 90–107, https://doi.org/10.32799/ijih. v15i1.34046.

4 United Nations (UN), n.d., *International Day of the World's Indigenous Peoples, 9 August,* https://www.un.org/en/observances/indigenous-day/background.

5 Douglas J. Nakashima, Kirsty Galloway McLean, Hans Thulstrup, et al., *Weathering Uncertainty: Traditional Knowledge for Climate Change Assessment and Adaptation* (Paris: UNESCO / Darwin, Australia: UNU, 2012), 120, https://www. researchgate.net/publication/259609164_Weathering_Uncertainty_Tradition al_Knowledge_for_Climate_Change_Assessment_and_Adaptation.

6 IPBES, Eduardo S. Brondizio, Josef Settele, Sandra Diaz, and Hien T. Ngo, eds., IPBES Secretariat, *Global Assessment Report of the Intergovernmental Science-Policy Platform on Biodiversity and Ecosystem Services* (Bonn, Germany: IPBES Secretariat), 1144, https://www.ipbes.net/sites/default/files/inline/files/ipbes_ global_assessment_report_summary_for_policymakers.pdf.

7 Blenkinsop and Fettes, "Land, Language and Listening," 1040 (emphasis added), https://doi.org/10.1111/1467-9752.12470.

8 Eriel Tchekwie Deranger, Rebecca Sinclair, Beze Gray, et al., "Decolonizing Climate Research and Policy: Making Space to Tell Our Own Stories, in Our Own Ways," *Community Development Journal* 57, no. 1 (2022): 52–73; Blenkinsop and Fettes, "Land, Language and Listening," 1040; Larry J. Gorenflo, Suzanne Romaine, Russell A. Mittermeier, and Kristen Walker-Painemilla, "Co-occurrence of Linguistic and Biological Diversity in Biodiversity Hotspots and High Biodiversity Wilderness Areas," *Proceedings of the National Academy of Sciences* 109, no. 21 (2012): 8032–37, https://doi.org/10.1073/pnas .1117511109; Margo Greenwood and Nicole Marie Lindsay, "A Commentary

on Land, Health, and Indigenous Knowledge(s)," *Global Health Promotion* 26, no. 3, suppl. (2019): 82–86, https://doi.org/10.1177/1757975919831262; Nicholas J. Reo, Sigvanna Meghan Topkok, Nicole Kanayurak, et al., "Environmental Change and Sustainability of Indigenous Languages in Northern Alaska," *Arctic* 72, no. 3 (2019): 215–228, http://dx.doi.org.ezproxy.library.yorku.ca/10.14430/arctic68655; Jenanne Ferguson and Marissa Weaselboy, "Indigenous Sustainable Relations: Considering Land in Language and Language in Land," *Current Opinion in Environmental Sustainability* 43 (2020): 1–7, https://doi.org/10.1016/j.cosust.2019.11.006.

9 Benjamin T. Wilder, Carolyn O'meara, Laurie Monti, and Gary Paul Nabhan, "The Importance of Indigenous Knowledge in Curbing the Loss of Language and Biodiversity," *BioScience* 66, no. 6 (2016): 499–509, https://doi.org/10.1093/biosci/biw026.

10 Joshua Ostapchuk, Sherilee Harper, Ashlee Cunsolo Willox, and Victoria L. Edge, "Exploring Elders' and Seniors' Perceptions of How Climate Change Is Impacting Health and Well-being in Rigolet, Nunatsiavut/ᖅᑭᒥᖅᕈᓂᖅ ᐃᓐᓇᐃᑦ ᐊᒻᒪᓗ ᐃᓄᑐᖅᑲᐃᑦ ᐃᓱᒪᔾᔪᓯᖏᓐᓂᒃ ᓇᒍᓚᑦ, ᓄᓇᑦᓯᐊᕗᒻᒥ ᓯᓚᐅᑉ ᐊᓯᔾᔨᐸᓪᓕᐊᓂᖓᓂᒃ ᐊᑦᑐᐃᓂᖅᑲᖅᑎᓪᓗᒍ ᐃᓗᓯᖅᒥᒃ ᐊᒻᒪᓗ ᖅᑲᓄᐃᙱᓂᖏᓐᓂᒃ," *International Journal of Indigenous Health* 9, no. 2 (2015): 6–24.

11 Sherilee L. Harper, Victoria L. Edge, James Ford, et al., "Climate-Sensitive Health Priorities in Nunatsiavut, Canada," *BMC Public Health* 15 (2015): 605, https://doi.org/10.1186/s12889-015-1874-3.

12 Harper et al., "Climate-Sensitive Health Priorities," 1–18.

13 Whitmee, Sarah, Andy Haines, Chris Beyrer, Frederick Boltz, Anthony G. Capon, Braulio Ferreira De Souza Dias, Alex Ezeh, et al., "Safeguarding Human Health in the Anthropocene Epoch: Report of The Rockefeller Foundation–Lancet Commission on Planetary Health," *Lancet* 386, no. 10007 (2015): 1973–2028, https://doi.org/10.1016/S0140-6736(15)60901-1. Horton, Richard, Robert Beaglehole, Ruth Bonita, John Raeburn, Martin McKee, and Stig Wall, "From Public to Planetary Health: A Manifesto." *Lancet* (British edition) 383, no. 9920 (2014): 847–847.

14 Sean A. Hillier, Abdul Taleb, Elias Chaccour, and Cécile Aenishaenslin, "Examining the Concept of One Health for Indigenous Communities: A Systematic Review," *One Health* 12 (2021): 100248.

15 Álvaro Fernández-Llamazares, María Garteizgogeascoa, Niladri Basu, et al., "A State-of-the-Art Review of Indigenous Peoples and Environmental Pollution," *Integrated Environmental Assessment and Management* 16, no. 3 (2020): 324–41.

16 Andrew R. Hatala, Chinyere Njeze, Darrien Morton, et al., "Land and Nature as Sources of Health and Resilience among Indigenous Youth in an Urban Canadian Context: A Photovoice Exploration," *BMC Public Health* 20 (2020): 1–14.

17 Alexandra Sawatzky, Ashlee Cunsolo, Andria Jones-Bitton, et al., "'The Best Scientists Are the People That's Out There': Inuit-Led Integrated Environment and Health Monitoring to Respond to Climate Change in the Circumpolar North," *Climatic Change* 160 (2020): 45–66.

18 Jacqueline Middleton, Ashlee Cunsolo, Andria Jones-Bitton, et al., "Indigenous Mental Health in a Changing Climate: A Systematic Scoping Review of the Global Literature," *Environmental Research Letters* 15, no. 5 (2020): 053001; Fernández-Llamazares et al., "State-of-the-Art Review."

19 Harper et al., "Climate-Sensitive Health Priorities," 605.

20 Montesanti, Stephanie Rose, and Wilfreda E. Thurston, "Mapping the Role of Structural and Interpersonal Violence in the Lives of Women: Implications for Public Health Interventions and Policy," BMC Women's Health 15, no 1 (2015): 100, https://doi.org/10.1186/s12905-015-0256-4.

21 King, Hayden, Shiri Pasternak, and Riley Yesno, "Land Back: A Yellowhead Institute Red Paper," Yellowhead Institute, 2019, https://redpaper.yellowheadinstitute.org/wp-content/uploads/2019/10/red-paper-report-final.pdf.

22 King et al., "Land Back."

23 Agata Durkalec, Chris Furgal, Mark W. Skinner, and Tom Sheldon, "Climate Change Influences on Environment as a Determinant of Indigenous Health: Relationships to Place, Sea Ice, and Health in an Inuit Community," *Social Science & Medicine* 136 (2015): 17–26; King et al., "Land Back."

24 Durkalec et al., "Climate Change Influences."

25 Fernández-Llamazares et al., "State-of-the-Art Review," 324–341; Government of Canada, *Contaminants in Canada's North: State of Knowledge and Regional Highlights,* Canadian Arctic Contaminants Assessments Report, 2017, https://science.gc.ca/eic/site/063.nsf/eng/h_97661.html.

26 Aline Philibert, Myriam Fillion, and Donna Mergler, "Mercury Exposure and Premature Mortality in the Grassy Narrows First Nation Community: A Retrospective Longitudinal Study," *Lancet Planetary Health* 4, no. 4 (2020): e141–e148, https://www.thelancet.com/journals/lanplh/article/PIIS2542-5196(20)30057-7/fulltext.

27 King et al., "Land Back."

28 King et al., "Land Back."

29 Katie Big-Canoe and Chantelle A. M. Richmond, "Anishinabe Youth Perceptions about Community Health: Toward Environmental Repossession," *Health & Place* 26 (2014): 127–35, https://doi.org/10.1016/j.healthplace.2013.12.013.

30 National Collaborating Centre for Indigenous Health (NCCIH), "Climate Change and Indigenous Peoples' Health in Canada," 113; Peter Berry and Rebekka Schnitter, "Health of Canadians in a Changing Climate: Advancing Our Knowledge for Action," *Ottawa, ON: Government of Canada* 10 (2022): 329522, https://changingclimate.ca/health-in-a-changing-climate/chapter/2-0.

31 Hillier et al., "Examining the Concept."

32 Diana Lewis, Lewis Williams, and Rhys Jones, "A Radical Revision of the Public Health Response to Environmental Crisis in a Warming World: Contributions of Indigenous Knowledges and Indigenous Feminist Perspectives," *Canadian Journal of Public Health* 111 (2020): 897–900.

33 Freeman, Bonnie M., "Promoting Global Health and Well-Being of Indigenous Youth through the Connection of Land and Culture-Based Activism," supplement, *Global Health Promotion* 26, no. 3 (2019): 17–25, https://journals.sagepub.com/doi/full/10.1177/1757975919831253

34 Statistics Canada, *Census in Brief: The Aboriginal languages of First Nations People, Métis and Inuit,* Statistics Canada, October 25, 2017, https://www12.statcan.gc.ca/census-recensement/2016/as-sa/98-200-x/2016022/98-200-x2016022-eng.cfm.

35 Greenwood and Lindsay, "Commentary," 82–86.

36 C. A. Richmond and N. A. Ross, "The Determinants of First Nation and Inuit Health: A Critical Population Health Approach," *Health & Place* 15, no. 2 (2009): 403–11; Tobias, Joshua K., and Chantelle A. M. Richmond, "'That Land Means Everything to Us as Anishinaabe . . .': Environmental Dispossession and Resilience on the North Shore of Lake Superior," *Health & Place* 29 (2014): 26–33.

37 Reo et al., "Environmental Change," 216.

38 Glorenflo at al., "Co-occurrence of Linguistic and Biological Diversity," 8032–37.

39 National Collaborating Centre for Aboriginal Health, "Culture and Language as Social Determinants of First Nations, Inuit, and Métis Health," National Collaborating Centre for Aboriginal Health, 2016, 12, https://www.ccnsa-nccah.ca/docs/determinants/FS-CultureLanguage-SDOH-FNMI-EN.pdf.

40 National Collaborating Centre for Aboriginal Health, "Culture and Language as Social Determinants," 12.

41 National Collaborating Centre for Aboriginal Health, "Culture and Language as Social Determinants," 11.

42 Reo et al., "Environmental Change," 218.

43 Ferguson and Weaselboy, "Indigenous Sustainable Relations," 3.

44 Rosalie Schultz, "Closing the Gap and the Sustainable Development Goals: Listening to Aboriginal and Torres Strait Islander People," *Australian and New Zealand Journal of Public Health* 44, no. 1 (2020): 12, https://doi.org/10.1111/1753-6405.12958.

45 Greenwood and Lindsay, "Commentary," 85.

46 James Fairhead, Melissa Leach, and Ian Scoones, "Green Grabbing: A New Appropriation of Nature?," *Journal of Peasant Studies* 39, no. 2 (2012): 237–61, https://doi.org/10.1080/03066150.2012.671770.

47 Claudia Gafner-Rojas, "Indigenous Languages as Contributors to the Preservation of Biodiversity and Their Presence in International Environmental Law," *Journal of International Wildlife Law & Policy* 23, no. 1 (2020): 49, https://doi.org/10.1080/13880292.2020.1768693.

48 Matthias Schröter, Emma H. Van der Zanden, Alexander P. E. van Ouden-hoven, et al., "Ecosystem Services as a Contested Concept: A Synthesis of Critique and Counter-Arguments," *Conservation Letters* 7, no. 6 (2014): 521, https://doi.org/10.1111/conl.12091.

49 Ferguson and Weaselboy, "Indigenous Sustainable Relations," 1–7.

50 Nylah Burton, "Learning a New Language Can Help Us Escape Climate Catastrophe," *Vice News*, April 15, 2021, https://www.vice.com/en/article/v7e4wx/learning-indigenous-languages-climate-crisis.

51 Glorenflo et al., "Co-occurrence of Linguistic and Biological Diversity," 8035.

52 Glorenflo et al., "Co-occurrence of Linguistic and Biological Diversity," 8036.

53 Glorenflo et al., "Co-occurrence of Linguistic and Biological Diversity," 8036.

54 Burton, "Learning a New Language"; Emilie J. Ens, Petina Pert, Philip A. Clarke, et al., "Indigenous Biocultural Knowledge in Ecosystem Science and Management: Review and Insight from Australia," *Biological Conservation* 181 (2015): 133–49.

55 Ferguson and Weaselboy, "Indigenous Sustainable Relations," 2.

56 Ferguson and Weaselboy, "Indigenous Sustainable Relations," 3.

57 Lewis, Williams, and Rhys Jones, "Radical Revision."

58 Blenkinsop and Fettes, "Land, Language, and Listening," 1036.

59 Berlitz Corporation, "The Most Spoken Languages in the World," February 5, 2023, https://www.berlitz.com/en-uy/blog/most-spoken-languages-world.

60 Ferguson and Weaselboy, "Indigenous Sustainable Relations," 3.

61 Karsten Hueffer, Mary Ehrlander, Kathy Etz, and Arleigh Reynolds, "One Health in the Circumpolar North," *International Journal of Circumpolar Health* 78, no. 1 (2019): 1607502.

62 Deborah McGregor, *Taking Care of Each Other: Taking Care of Mother Earth* (Toronto, Canada: Chiefs of Ontario, 2018), 91.

63 Onjisay Aki, *Onjisay Aki International Climate Calls to Action*, Onjisay Aki International Climate Summit, 2017, http://onjisay-aki.org/onjisay-aki-international-climate-calls-action.

64 Aki, *International Climate Calls to Action*.

65 Big-Canoe and Richmond, "Anishinabe Youth Perceptions," 133.

66 Redvers, "The Land Is a Healer," 102.

67 Chantal Viscogliosi, Hugo Asselin, Suzy Basile, et al., "Importance of Indigenous Elders' Contributions to Individual and Community Wellness: Results from a Scoping Review on Social Participation and Intergenerational Solidarity," *Canadian Journal of Public Health* 111 (2020): 678, https://doi.org/10.17269/s41997-019-00292-3.

68 Eve Tuck, Marcia McKenzie, and Kate McCoy, "Land Education: Indigenous, Post-colonial, and Decolonizing Perspectives on Place and Environmental Education Research," *Environmental Education Research* 20, no. 1 (2014): 13, https://doi.org/10.1080/13504622.2013.877708.

69 Sandra D. Styres, "Land as First Teacher: A Philosophical Journeying," *Reflective Practice* 12, no. 6 (2011): 717–31, https://doi.org/10.1080/14623943.2011.6010 83; Blenkinsop and Fettes, "Land, Language, and Listening," 1037.

70 Blenkinsop and Fettes, "Land, Language, and Listening," 1040.

71 Government of Canada, "Language and Culture," January 2023, https://www.rcaanc-cirnac.gc.ca/eng/1524495846286/1557513199083.

72 Government of Canada, "Indigenous languages legislation," modified June 28, 2019, https://www.canada.ca/en/canadian-heritage/campaigns/celebrate-indigenous-languages/legislation.html.

73 Masud Khawaja, "Consequences and Remedies of Indigenous Language Loss in Canada," *Societies* 11, no. 3 (2021): 89, https://doi.org/10.3390/soc11030089.

74 BC Métis Federation, "Michif Language Project," 2023, https://bcmetis.com/programs-services/michif-language-project/.

75 Big-Canoe and Richmond, "Anishinabe Youth Perceptions," 127–35; Kathleen Mikraszewicz and Chantelle Richmond, "Paddling the Biigtig: Mino Biimadisiwin Practiced through Canoeing," *Social Science & Medicine* 240 (2019): 112548, https://doi.org/10.1016/j.socscimed.2019.112548.

76 Kyle Whyte, "Indigenous Climate Change Studies: Indigenizing Futures, Decolonizing the Anthropocene," *English Language Notes* 55, no. 1 (2017): 153–62, https://doi.org/10.1215/00138282-55.1-2.153; Big-Canoe and Richmond, "Anishinabe Youth Perceptions," 127–35; Susan Chiblow and Paul J. Meighan, "Language Is Land, Land Is Language: The Importance of Indigenous Languages," *Human Geography* 15, no. 2 (2022): 206–10, https://doi.org/10.1177/19427786211022899; Redvers, "The Land Is a Healer," 90–107.

77 Chiblow and Meighan, "Language Is Land," 4.

78 Beau J. Austin, Cathy J. Robinson, Dean Mathews, et al., "An Indigenous-Led Approach for Regional Knowledge Partnerships in the Kimberley Region of Australia," *Human Ecology* 47 (2019): 577–88, https://doi.org/10.1007/s10745-019-00085-9.

79 United Nations Environment Program (UNEP) and Siham Drissi, "Indigenous Peoples and the Nature They Protect," UNEP, June 2020, http://www.unep.org/news-and-stories/story/indigenous-peoples-and-nature-they-protect.

80 Eriel Tchekwie Deranger, Rebecca Sinclair, Beze Gray, et al., "Decolonizing Climate Research and Policy: Making Space to Tell Our Own Stories, in Our Own Ways," *Community Development Journal* 57 no. 1 (2022): 52–73, https://doi.org/10.1093/cdj/bsab050. Eriel Tchekwie Deranger, "The Climate Emergency and the Colonial Response," 2021, Yellowhead Institute, https://yellowheadinstitute.org/2021/07/02/climate-emergency-colonial-response/#:text=Colonization%20caused%20climate%20change.,putting%20all%20life%20at%20risk.

CHAPTER 2

Connecting the Dots

Climate Change, Social Justice, and Public Health

SURILI SUTARIA PATEL AND DR. ADRIENNE L. HOLLIS

limate justice sits at the intersection of social justice and climate change. It is achieved when climate solutions are created
to advance social justice. Climate justice is about recognizing and correcting the disproportional effects of climate change on poor communities
and communities of color.

FIGURE 2.1. The relationship between social justice, climate justice,
and climate change. Image credit: University of Omaha Nebraska.

According to the Center for Economic and Social Justice, social justice
is "the virtue which guides us in creating those organized human interactions we call institutions." In turn, social institutions, when justly organized,
provide us with access to what is good for the person, both individually and
in our associations with others. Social justice also imposes on each of us a
personal responsibility to collaborate with others, at whatever level of the

'Common Good' in which we participate, to design and continually perfect our institutions as tools for personal and social development."[1]

In the same way people have designed the institutions in which we live, so too have we designed the solutions to help us achieve climate justice. The goal of working toward justice is to seek fair outcomes for all. The path toward justice is rooted in prioritizing people and relationships above power and profits.

The former president of Ireland Mary Robinson says that "climate justice insists on a shift from a discourse on greenhouse gases and melting ice caps into a civil rights movement with the people and communities most vulnerable to climate impacts at its heart."[2]

In this chapter, we will illustrate how natural and man-made environmental conditions unevenly jeopardize the health of some low-wealth communities of color, and how unjust structures upheld by our local politics, societies, and global community continue to harm the same communities.

An Overview of Discriminatory Practices

In the United States, conditions have been created in which some people are forced to live within both real and imaginary boundaries. In addition to racism, both individual and structural, classism, sexism, and other *isms* have contributed to the serious conditions disproportionately faced by some communities. We use the *isms* framework to demonstrate the many upstream struggles in society that privilege some populations over others, including the public health impacts of climate change.

Racism

Redlining was a common practice that began in the 1930s. Redlining occurred when federal banks refused to insure mortgages in and near African American communities. Neighborhoods in almost 240 cities across the country were mapped and color-coded, based on desirability for residential use, racial and ethnic demographics, and home prices, as well as existing amenities. They were coded green for "best," blue for "still desirable," yellow for "definitely declining," and red for "hazardous."[3]

Segregation practices continue with expulsive zoning, in which facilities that pollute the environment are intentionally sited in or near areas inhabited by people of color, guided by the NIMBY (not in my back yard) stance held by those living in more desirable areas in suburbia, where the educational system is better, as is access to healthy foods and greater economic wealth. According to the National Community Reinvestment Coalition, the effects of redlining practices initiated almost ninety years ago are still being felt by communities.[4] For example, long-term exposure to particulate matter is associated with racial segregation, as more highly segregated areas suffer higher levels of exposure.[5]

Classism
Classism exists when people are treated differently based on their social class or perceived social class.[6] Important indicators of classism include income and wealth inequality (socioeconomic status), educational level, income, and perceived power. As a result of these (and other) factors, people are treated unfairly and "ranked" by social class, usually categorized as upper-, middle-, working-, and lower-class.

Classism results in conditions in which groups of people face a myriad of barriers to well-being, including hunger, lack of access to health care or running water, homelessness, unemployment, and so on. Further layers of discrimination and unfavorable treatment are added by prejudicial attitudes based on class, in which a person's "value" is based on their economic status, job status, educational level, family history, or genetics. Those in society who are valued less than others will inherently and disproportionately face the health effects of climate change, such as exposure to heat events leading to cardiovascular or respiratory illnesses in the populations experiencing houselessness.

Sexism
One of the conditions related to sexism is gender injustice. Gender justice exists when every woman and girl lives in dignity and in freedom, without any fear. It includes sharing power and responsibility between women and men at home, in the workplace, and in the wider national and international communities.[7] According to Oxfam America, during a natural disaster women and children are more likely to experience harm. Yet women, who are often excluded from decision making, are overlooked in times of aid and relief.[8]

The multiple roles women play in the household, from provider to caregiver to primary consumer, is worth studying, as the choices women make have significant implications for climate change. Women's capacity to earn a livable wage, their deep understanding of the connections between environmental conditions and the conditions of their family, and their consumer power are all affected by or contribute to the changing climate. Improving women's agency and their ability to succeed in these roles benefits their family units as well as society at large. Women of color often face further disadvantages due to the systems and policies surrounding them, such as economic and electoral injustices.

Policies and systems have existed that were (and are) designed to protect white privilege at the expense of communities of color, including instances in which local, state, and federal policy mandated segregation in the past. In addition, discrimination can occur on the basis of sex or gender identification or perceived class.

Environmental Conditions Overburden Populations of Color

One aspect of dismantling climate injustices involves naming the conditions and structures that uphold the various types of discrimination. These include environmental conditions, such as contaminants, food insecurity, air pollution, and water quality, that harm the health of communities. Harmful environmental conditions are disproportionately felt by communities of color. The following are a few examples.

Environmental Conditions
A number of chemicals contaminate the water, land, and air, and negatively affect both the environment and public health. The majority of contaminants enter the environment via industrial and commercial facilities, oil and chemical spills, and nonpoint sources (i.e., roads, parking lots, storm drains, wastewater treatment plants, and sewage systems).[9]

Controlling for all other factors, race is the single most important determinant of who bears the burden of society's pollution. More than half of people who live within two miles of a toxic waste facility are people of color.[10] People of color are nearly twice as likely to live in a fenceline community near industrial facilities. In addition, people of color are exposed to 38 percent more outdoor nitrogen dioxide (produced by vehicle exhaust and power plants) than Whites.[11] In large urban areas, low-income Whites are exposed to less pollution than even the highest-income Blacks, Asians, and Hispanics.[12]

Research has shown that proximity to a hazardous waste site correlates with race. African American families are more likely to live close to these facilities. As of December 9, 2021, there were 1,322 Superfund sites on the National Priorities List and fifty-one sites proposed to the list, for a total of 1,373 sites in the United States.[13] These sites include previous manufacturing facilities, processing plants, mining sites, and landfills. In addition to Superfund sites, there are also eighty-two sites earmarked for action through the Environmental Protection Agency's Superfund alternative approach agreements.[14] This number does not include Brownfields, Resource Conservation and Recovery Act, or other sites that have yet to be addressed. Although the expectation is that these sites will not be placed on the National Priorities List, they are not excluded. Many hazardous waste sites and industrial facilities have been contaminated for decades and continue to affect people and the environment.

These sites contribute to environmental contamination and affect public health. The populations around these sites have a greater potential for experiencing adverse health conditions such as asthma, cardiovascular disease, chronic obstructive pulmonary disease, and other illnesses.[15] Furthermore, Superfund sites could reduce the life expectancy of people living in nearby communities by as much as 1.2 years.[16]

Air Pollution

Air pollution is the leading environmental risk factor for adverse health. Disease-causing air pollution remains high in pockets of America—particularly where many low-income and African Americans and other communities of color live and are disproportionately affected by indoor and outdoor air pollution. Recent studies have shown that air pollution level and income level are linked. Poorer communities suffer from bad air more than wealthy communities. Air pollution is an environmental justice issue, as minority communities often bear the burden of "hosting" pollution.

African Americans, Hispanics, Asians, and other people of color are disproportionately exposed to a regulated air pollutant called fine particulate matter (PM2.5).[17] This particular type of deadly pollution comes from a number of sources, including mining operations, agriculture, wood-burning stoves and fireplaces, vehicle emissions, wildfires, and the fossil fuel industry. Particulate matter destroys the respiratory systems of those exposed. Particulate matter can be deposited deep in lung tissues and can cause damage.[18] In addition, PM2.5 has been linked to COVID-19, a virus that continues to disproportionately affect these same communities. A number of scientists postulate that particulate matter, particularly PM2.5, allows the coronavirus to travel further—both in the air and in the lungs.[19] Communities located according to racist practices are at greater risk of other opportunistic infections, as well as of health effects due to extreme weather caused by climate change.

Water Pollution
All life needs water to function. When water is unavailable due to contamination or poor water quality, inadequate or nonexistent plumbing, or lack of access, living things die. In the United States, over 489,836 households do not have complete plumbing, over 1,160 community water systems are in violation of the Safe Drinking Water Act, and over twenty-one thousand Clean Water Act permittees are in significant noncompliance.[20]

Sand Branch, an unincorporated town in Dallas County, Texas, and a freedmen's community established in the 1800s by former slaves, has not had potable water in thirty years. The residents have never had running water and have instead relied on wells—that is, until the 1980s, when the wells became contaminated with *E. coli*. All 150 residents of Sand Branch live below the poverty line, in one of the richest cities in the United States. Yet the county has not provided water to this community. Instead, Sand Branch residents, who are 87 percent Black, rely on donated bottled water for cooking, bathing, and drinking. Sadly, there are communities like Sand Branch all around the country that do not have access to clean water, a basic necessity of life. As a result of historic disenfranchisement and systemic racism, as well as economic apartheid, these communities have seen that the government has not and will not provide needed assistance. As such, these communities are developing processes, identifying tools, and engaging with people who will assist them. The New Alpha Community Development Corporation in Florence, South Carolina, is on the forefront of addressing water issues in its local community (see box on next page).

COMMUNITY SPOTLIGHT

One community organization, the New Alpha Community Development Corporation (CDC) in Florence, South Carolina, is leading in ways to address water shortages and water quality issues in its community, where citizens have at times had to boil their water to make it potable. There is a lack of available clean water in Florence, a rural community in a town that is almost 50 percent Black. Rev. Leo Woodberry, pastor of Kingdom Living Temple and New Alpha CDC, is working to address this issue by using solar hydropanels, the sun, and the air to extract water from the environment.[21] Reverend Woodberry learned about the hydropanels during a protest by residents from Denmark, South Carolina, who were protesting because they were not able or allowed to drink their water because the city had chlorinated the water at high concentrations. As it turned out, a representative from SOURCE® (the company that provides hydropanels) showed up at the protest. Reverend Woodberry took it upon himself to learn more about SOURCE® through other community partners, then reached out to the manufacturer about setting up a demonstration project.

Community leaders such as Reverend Woodberry have realized that the government has not provided much-needed assistance. In Florence, the mayor and the city council have not been involved in the effort to provide water and have not engaged in the hydropanel demonstration project. When the panels were installed in Florence, and the mayor and city council came to the press event, the general feeling from the community was that this effort could be replicated in other places. After speaking with Reverend Woodberry, however, we discovered that he was stunned to realize that government officials were not interested in engaging with the community, nor were they interested in creating policies to support other communities in need of clean drinking water. Instead, they seemed to want to ignore the hydropanel opportunity and pretend that the technology did not exist. Rev. Woodberry reasoned that the government depended on the money they collected from the water bills, and so had no reason to provide citizens with access to free water. This is a type of economic apartheid. As Reverend Woodberry stated, "Imagine if everyone was making their own water. What would that do to the city budget? There are no financial or economic incentives to do this." The burden of access to clean drinking water falls on communities and their leaders to develop processes, identify tools and resources, and engage with people who will aid them. At the end of the day, clean water is a basic necessity of life.

Unjust Structures Worsen Climate Impacts on Health

In addition to these environmental conditions are the structural determinants of health such as economic, energy, and electoral injustices that can block a community's access to a healthy future. The structures, created at the United States' inception, are rooted in the principles of racial, class, and gender hierarchy, and "they shape the distribution of power and resources across the population, engendering health inequities along racial, class, and gender lines and intersections."[22] Below are examples of the determinants that overburden communities and how climate change exacerbates them.

Economic Injustice

Access to resources plays a major role in communities' acquiring and maintaining adaptive capacities in the face of a climate event such as a hurricane or extreme heat. Financial capital can make the difference between evacuating before a hurricane makes landfall or rebuilding after the destruction. Financial independence can afford the time off during extreme heat events to visit the doctor or refill an asthma inhaler prescription. These are examples at the individual level. At the community level, low-wealth neighborhoods "often live on the most fragile land, and they are often politically, socially, and economically marginalized, making them especially vulnerable to the impacts of climate change," Christina Chan, director of the World Resource Institute's Climate Resilience Practice, told Global Citizen.[23]

The economic impacts of climate change will challenge our global markets, and have devastating financial, social, and health impacts locally. Globally, climate change is projected to push 132 million people into extreme poverty in less than a decade, according to the World Bank.[24] At the local level, people experiencing drought may also undergo undue hardships. For example, droughts can affect the agricultural industry in a way that impedes feeding families and employing workers. This leads to a rise in the cost of basic food products and a decrease in fresh water supplies. The extreme temperatures make it unsafe for farmworkers to remain in the fields for prolonged periods. And in extreme cases, communities may need to relocate if food and work are out of reach (see figure 2.2).

History has elucidated the impact that other challenges have on communities. For example, because of COVID-19 and the resulting lockdown, over

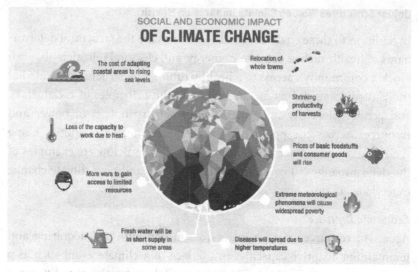

FIGURE 2.2. The social and economic impact of climate change. By Iberdrola.[25]

one hundred million people lost their jobs in 2020, which led to more than three trillion dollars in lost labor income.[26] Most of the economic impacts were felt in poor communities and communities of color. One result of COVID-19 is a worsening of the economic situation for minorities and poor people. In addition, the employment rate remains below prepandemic levels, people are unable to pay their rent or mortgages, and millions of people are suffering from food scarcity.[27] The impacts of the pandemic and the economic fallout have been widespread, but remain particularly prevalent among Black adults, Latino adults, and other people of color. These disproportionate impacts reflect harsh, long-standing inequities—often stemming from structural racism—in education, employment, housing, and health care that the current crisis has exacerbated.[28]

While improving household income equity seems like a case-by-case approach, there are great societal benefits to elevating community wealth. When a community thrives financially, it will have the resources to improve educational institutions, afford quality housing and buildings, and create green spaces—all of which are necessary and effective co-benefits of climate and health strategies. Improving neighborhood conditions is not only essential for living healthy lives but can also help overcome flooding, urban heat island effects, air pollution, and more.

Energy Injustice

Energy justice is the "principle that all people should have a reliable, safe and affordable source of energy; protection from a disproportionate share of costs or negative impacts or externalities associated with building, operating, and maintaining electric power generation, transmission, and distribution systems; and equitable distribution of and access to benefits from such systems. A global energy system that (1) fairly disseminates the costs and benefits of energy services and (2) contributes to more representative and impartial energy decision making."[29]

According to the Bureau of Labor and Statistics, Americans spend more than 60 percent of their time in their homes, and even more during the coronavirus pandemic. The dependency on energy resources to heat or cool a home is essential to health and comfort. Poor-quality housing in neighborhoods with limited greening (i.e., tree canopies, green or reflective roofing, or open green spaces) can quickly turn into urban heat islands during the summer months, increasing the cost of energy usage during extreme weather events. Renters, who have less of an incentive to invest in an air conditioning system, are more at risk of heat-related illnesses.[30] With the exception of a few municipalities (such as Dallas, Texas; Montgomery County, Maryland; and Tempe, Arizona), there is no law requiring landlords to provide air conditioning the way they must provide heat.[31] According to the American Council for an Energy-Efficient Economy, "Systemic exclusions, underinvestments, discriminatory lending practices, and limited housing choices have limited Black, Indigenous, and People of Color communities' access to efficient and healthy housing."[32]

Communities of color often experience the increased energy burdens.[33] As a result of historical bank lending practices, redlining, segregation, employment discrimination, and other racially discriminatory policies, Black and African American communities face disproportionately high energy burdens.[34] The Department of Energy defines an "energy burden" as "the percentage of gross household income spent on energy costs."[35] The median energy burden of Black households is 43 percent higher than that of white households.[36] The American Council for an Energy Efficient Economy reports that households experiencing high energy burdens are more likely to "stay caught in cycles of poverty,"[37] and that "families suffering from high energy burdens also tend to experience stress from living in constant fear of losing necessary electricity and gas service due to inability to pay their bills."[38]

Household energy is directly linked to health. In the face of climate change, the extremes of temperatures are of the greatest concern. Coping with cold stress, such as properly heating a home, can reduce cases of hypertension and stroke, cardiovascular events, and possibly death. Overcoming extreme heat, in contrast, can be done by adequately cooling the home through air conditioning when available. Opening windows for ventilation, when safe from air pollution, can also support household cooling.[39] Yet these options—heating and cooling—come at a price not everyone can afford. The demands of extreme temperatures are accelerated by climate change and will worsen in the coming years, especially for low-wealth communities of color.

Electoral Injustice

According to the International Institute for Democracy and Electoral Assistance, an electoral justice system is a "key instrument of the rule of law and the ultimate guarantee of compliance with the democratic principle of holding free, fair and genuine elections."[40] Civic participation, including exercising the right to vote, can determine the health of a community.[41]

Climate advocates are aware that in order to protect communities against the threats of environmental contaminants, air pollution, and poor air quality, they need to protect their right to vote, because the communities worst hit by environmental justice challenges are the same ones first and worst affected by climate events and burdens. While Jim Crow–era tactics are of the past, contemporary voter suppression maneuvers are happening across the country. State legislatures, boards of election, and other governmental entities are creating new voter identification laws, reducing the number of polling places and purging voter rolls. Then there are gerrymandering, intimidation, and other anecdotal tactics that play a concrete role in impeding voter turnout.[42]

Climate change worsens access to voting. In the aftermath of Hurricane Katrina, we learned that Black families that were displaced were less likely to return.[43] They required mail-in ballots or early voting options. Flooding from severe storms can also restrict access to the polls, as do the health conditions caused by extreme heat. Simply put, getting voters of color to

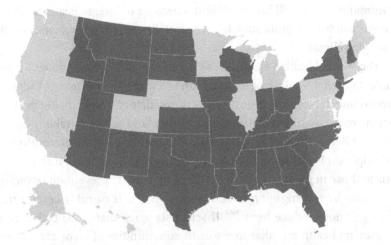

FIGURE 2.3. Map of twenty-nine states that have enacted restrictive voting laws since Shelby County v. Holder. Source: Brennan Center analysis of publicly available data from 2013–2023.

the polls could mean supporting clean energy and climate policies that promote wellness and opportunities for all.

If every vote does not count, then why are so many states creating new voter restriction laws? For the first time, in 2016, fourteen states enacted new voting restrictions in a presidential election. In addition, twenty-five states have created new voting restrictions since 2010 (see figure 2.3).

Discouraging and disenfranchising voters by way of inaccessible voter registration processes or limited ballot access, can have grave consequences on the communities that want to speak up for themselves and for democracy as a whole.

Climate Change Narratives Exacerbates Existing Environmental Threats to Public Health

Communities of color, particularly Black and African American communities, bear a disproportionate burden of the negative health and environmental impacts from a warming planet, including more deaths from extreme heat and property loss from flooding in the wake of sea level rise.

American Indians and Alaska Natives are 48 percent more likely than other groups to live in areas that will be inundated by flooding from sea level rise under that scenario; Latinos are 43 percent more likely to live in communities that will lose work hours because of intense heat; and Black people will suffer significantly higher mortality rates as a result of a rapidly changing climate.[44]

The World Health Organization's definition of the precautionary principle states that "in the face of uncertain but suggestive evidence of adverse environmental or human health effects, regulatory action should prevent harm from environmental hazards, particularly for vulnerable populations."[45] Bernard Goldstein explains, "No matter how the precautionary principle evolves, the value of acting in a precautionary manner is obvious to those in public health. It is a form of primary prevention, avoiding problems by not engaging in activities until it is reasonably certain that they will not produce harm."[46] If scientists agree that climate change is worsening health and that low-wealth communities of color are disproportionately affected,[47] why are we not correcting course?

The current US narrative around climate justice does not reflect the collective experiences of those who are most affected by climate change. False narratives can perpetuate foundations of inequity and exclusion based on a perceived racial, class, and gender social order. The danger in upholding dominant but false narratives is that some communities are excluded and entire populations dehumanized. A report by the DeBeaumont Foundation put it best: "Narrative change is a key part of building a shared reality of interconnectedness and equality."[48]

Climate narratives can serve as motivators for action, and are most appropriately delivered by those who have experienced a climate event firsthand. Research shows that the narratives that are the most activating are those that offer opportunities for social and economic development.[49]

Conclusion

For far too long, both local champions and world leaders have admonished that climate change and social justice are inextricably linked. So too are the solutions. While it is important to define climate justice, it is equally imperative to celebrate climate narratives that create healthy environmental conditions, uphold social justice in our systems, and improve the public's health, especially for those communities most at risk.

We notice that the impact of our influence is most evident at the structural level. As we address the structures and conditions that shape our lives, we begin advancing opportunities for access to health—leveling the playing field for health. The first structure we focus on is racism, then we ensure opportunities for disregarded communities and families to build generational wealth. We are beginning to see the impact of our work in policy and with community groups. What gives us hope is when we begin to see partners in government agencies and academic institutions and even leaders in the field talk about and acknowledge that structures, like racism, have created barriers to achieving health potential. We have seen our recommendations reflected in talking points, research designs, and policy shifts. We aspire to continue to shift the narrative from "We should address structural racism" to "We can and must address structural racism in all climate adaptation and mitigation interventions—together."

The resistance comes from people in power and with influence who do not understand the intersectionality of this work and the "sum of us" concept presented by Heather McGhee: when we help others thrive, we all thrive.[50] Or, as Martin Luther King Jr said, "Injustice anywhere is a threat to justice everywhere."[51] The biggest challenges in this realm are political access and education, along with funding and policy limitations. In addition, we are guilty of ignoring the voices of those directly affected by climate change instead of listening to them and incorporating their traditional knowledge. The first step to overcoming barriers is to work with all affected or potentially affected groups as equal partners, especially affected communities, and to provide opportunities for their voices to be heard.

While the shift is incredibly slow, we recognize that some philanthropic organizations are beginning to invest in work around health equity and justice, particularly for disenfranchised and underfunded community-based entities. In addition, some federal agencies are exploring ways to incorporate equity into their work, and a few environmental nongovernmental organizations are beginning to identify and address barriers to equity and inclusion.

The late archbishop Desmond Tutu once said, "Reducing our carbon footprint is not just a technical scientific necessity; it has also emerged as the human rights challenge of our time."[52] His call to action, like many others long before and those still to come, has begged for a future that values human health over profits. According to the Guardian, Tutu also said, "There is a word we use in South Africa that describes human relationships: Ubuntu. It says: I am because you are. My successes and my failures are bound up in yours. We are made for each other, for interdependence. Together, we can change the world for the better."[53]

Our actions, or inaction, around climate change will define this decade. To stand on the right side of history, we must name and address the many isms of the climate dilemma, including structural racism, sexism, and classism. While there are many other isms to be cited and dismantled, these three, if addressed with dignity, can help the millions of people who have been left behind to reclaim their narrative. And we can collectively prioritize people and right relationships for a climate just future for generations to come.

NOTES

1 "Defining Economic and Social Justice," Center for Economic and Social Justice, https://www.cesj.org/learn/definitions/defining-economic-justice-and-social-justice.

2 "Climate Justice," Sustainable Development Goals, United Nations, https://www.un.org/sustainabledevelopment/blog/2019/05/climate-justice.

3 Shuo Jim Huang and Neil Jay Sehgal, "Association of Historic Redlining and Present-Day Health in Baltimore," *PLoS One* 17, no. 1 (January 19, 2022): e0261028, https://doi.org/10.1371/journal.pone.0261028.

4 Bruce Mitchell and Juan Franco, *HOLC 'Redlining' Maps: The Persistent Structure of Segregation and Economic Inequality*, National Community Reinvestment Coalition technical report, March 2018, https://ncrc.org/holc/.

5 Yan Wang, Itai Kloog, Brent A. Coul, et al., "Estimating Causal Effects of Long-Term PM2.5 Exposure on Mortality in New Jersey," *Environ Health Perspect* 124 (2016): 1182–88.

6 "But What Is Class," About Class, Class Action, http://classism.org/about-class.

7 "Woman's Rights and Gender Justice," Oxfam, https://www.oxfamamerica.org/explore/issues/womens-rights-and-gender-justice/gender-justice.

8 "Climate Justice and Gender Justice: An Essential Pairing to Get Resilience Right," National Democratic Institute, https://www.ndi.org/our-stories/climate-justice-and-gender-justice-essential-pairing-get-resilience-right.

9 "Contaminants in the Environment," National Ocean Service, https://oceanservice.noaa.gov/observations/contam/.

10 Robert Bullard, Paul Mohai, Robin Saha, and Beverly Wright, *Toxic Wastes and Race at Twenty, 1987–2007: A Report Prepared for the United Church of Christ Justice and Witness Ministries* (Cleveland, OH: United Church of Christ, 2007).

11 University of Minnesota, "People of Color Live in Neighborhoods with More Air Pollution than Whites, Groundbreaking U.S. Study Shows," *Science Daily*, April 15, 2014, https://www.sciencedaily.com/releases/2014/04/140415181327.htm.

12 University of Minnesota, "People of Color."

13 "Superfund National Priorities List (NPL)," US Environmental Protection Agency, https://www.epa.gov/superfund/superfund-national-priorities-list-npl.

14 "Superfund Alternative Approach," US Environmental Protection Agency, https://www.epa.gov/enforcement/superfund-alternative-approach#sites.

15 David E. Kramar, Aaron Anderson, Hayley Hilfer, et al., "A Spatially Informed Analysis of Environmental Justice: Analyzing the Effects of Gerrymandering and the Proximity of Minority Populations to U.S. Superfund Sites," *Environmental Justice* 11 (2018): 29–39.

16 Amin Kiaghadi, Hanadi S. Rifai, and Clint N. Dawson. "The Presence of Superfund Sites as a Determinant of Life Expectancy in the United States," *Nature Communications* 12 (2021): https://doi.org/10.1038/s41467-021-22249-2.

17 Christopher W. Tessum, David A. Paolella, Sarah E. Chambliss, et al., "PM2.5 Polluters Disproportionately and Systemically Affect People of Color in the United States," *Science Advances* 7 (2021): https://10.1126/sciadv.abf4491.

18 "Particle Pollution," American Lung Association, accessed December 2021, https://www.lung.org/clean-air/outdoors/what-makes-air-unhealthy/particle-pollution.

19 Leonardo Setti, Fabrizio Passarini, Gianluigi De Gennaro, et al., "SARS-Cov-2RNA Found on Particulate Matter of Bergamo in Northern Italy: First Evidence," *Environ Research* 188 (September 2020): doi: 10.1016/j.envres.2020 .109754; Antonio Frontera, Claire Martin, Kostantinos Vlachos, and Giovanni Sgubin, "Regional Air Pollution Persistence Links to COVID-19 Infection Zoning," *Journal of Infection* 81, no. 2 (April 2020): 318–56.

20 J. Tom Mueller and Stephen Gasteyer, "The Widespread and Unjust Drinking Water and Clean Water Crisis in the United States," *Nature Communications* 12 (2021): https://doi.org/10.1038/s41467-021-23898-z.

21 Curtis Bunn, "'Water from the Heavens': South Carolina Pastor's Crusade for Clean Drinking Water," *NBC News*, March 23, 2021, https://www.nbcnews.com/news/nbcblk/water-heavens-south-carolina-pastors-crusade-clean-drinking-water-rcna479.

22 Jola Crear-Perry, Rosaly Correa-de-Araujo, Tamara Lewis Johnson, et al., "Social and Structural Determinants of Health Inequities in Maternal Health," *Journal of Women's Health* 30, no. 2 (2021): 230–35, https://doi.org/10.1089/jwh.2020 .8882.

23 Joe McCarthy, "Why Climate Change and Poverty Are Inextricably Linked," *Global Citizen*, February 19, 2020, https://www.globalcitizen.org/en/content/climate-change-is-connected-to-poverty/.

24 Stephane Hallegatte and Brian Walsh, "COVID, Climate Change and Poverty: Avoiding the Worst Impacts," *World Bank blog*, October 7, 2020, https://blogs.worldbank.org/climatechange/covid-climate-change-and-poverty-avoiding-worst-impacts.

25 "How Is Climate Change Affecting the Economy and Society?," Economic Impacts of Climate Change, Iberdrola, https://www.iberdrola.com/sustainability/impacts-of-climate-change.

26 Felix Richter, "COVID-19 Has Caused a Huge Amount of Lost Working Hours," *World Economic Forum*, February 4, 2021, https://www.weforum.org/agenda/2021/02/covid-employment-global-job-loss/.

27 Center on Budget and Policy Priorities, "Tracking the COVID-19 Economy's Effects on Food, Housing and Employment Hardships," COVID Hardship Watch, https://www.cbpp.org/research/poverty-and-inequality/tracking-the-covid-19-economys-effects-on-food-housing-and.

28 Center on Budget and Policy Priorities, "Tracking the COVID-19 Economy's Effects."

29 "Energy Justice and Climate Change: Key Concepts for Public Health," American Public Health Association, https://apha.org/-/media/files/pdf/topics/climate/energy_justice_key_concepts.ashx.

30 Rachel McMonagle, Ivana Castellanos, and Surili Sutaria Patel, "Advancing Energy Justice as a Climate and Public Health Solution," *Journal of World Medical & Health Policy* 13, no. 1 (2021): https://doi.org/10.1002/wmh3.420.

31 Suzy Khimm and Joshua Eaton, "As Deadly Heat Waves Spread, Access to Air Conditioning Becomes a Lifesaving Question," *NBC News*, August 20, 2021, https://www.nbcnews.com/news/us-news/deadly-heat-waves-spread-access-air-conditioning-becomes-lifesaving-question-n1277213.

32 Ariel Drehobl, Lauren Ross, and Roxana Ayala, *How High Are America's Residential Energy Burdens?*, September 2020, American Council for an Energy-Efficient Economy, https://www.aceee.org/sites/default/files/pdfs/u2006.pdf.

33 Constantine E. Kontokosta, Vincent J. Reina, and Bartosz Bonczak, "Energy Cost Burdens for Low-Income and Minority Households," *Journal of the American Planning Association* 86, no. 1 (2020): 89–105, https://doi.org/10.1080/0194 4363.2019.1647446.

34 Drehobl, Ross, and Ayala, *Residential Energy Burdens.*

35 "Low-Income Community Energy Solutions," State and Local Solution Center, US Department of Energy, last accessed December 2021, https://www.energy.gov/eere/slsc/low-income-community-energy-solutions.

36 Drehobl, Ross, and Ayala, *Residential Energy Burdens.*

37 Drehobl, Ross, and Ayala, *Residential Energy Burdens.*

38 "Lifting the High Energy Burden in America's Largest Cities: How Energy Efficiency Can Improve Low-Income and Underserved Communities," American Council for an Energy Efficient Economy, https://www.aceee.org/sites/default/files/publications/researchreports/u1602.pdf.

39 Sonal Jessel, Samantha Sawyer, and Diana Hernández, "Energy, Poverty, and Health in Climate Change: A Comprehensive Review of an Emerging Literature," *Frontiers in Public Health* 7 (2019): 357.

40 International Institute for Democracy and Electoral Assistance, Electoral Justice: An Overview of the International IDEA Handbook, 2010, https://www.idea.int/sites/default/files/publications/chapters/electoral-justice-handbook/electoral-justice-handbook-overview.pdf.

41 "Civic Participation," US Health and Human Services, Office of Disease Prevention and Health Promotion. https://www.healthypeople.gov/2020/topics-objectives/topic/social-determinants-health/interventions-resources/civic-participation.

42 Tamara Toles O'Laughlin and Peggy Shepard, "Voter Suppression Is a Climate Justice Issue—and 2020 Is the Tipping Point," Grist Fix Solutions Lab, October 30, 2020, https://grist.org/fix/voter-suppression-climate-justice-issue/.

43 Toles O'Laughlin and Shepard, "Voter Suppression."

44 US Environmental Protection Agency, *Climate Change and Social Vulnerability in the United States: A Focus on Six Impacts*, Social Vulnerability Report, Climate Change Impacts and Risk Analysis, http://www.epa.gov/cira/social-vulnerability-report.

45 Marco Martuzzi, Joel A. Tickner, World Health Organization Regional Office for Europe, "The Precautionary Principle: Protecting Public Health, the Environment and the Future of Our Children," 2004, https://apps.who.int/iris/handle/10665/346211.

46 Bernard D. Goldstein, "The Precautionary Principle Also Applies to Public Health Actions," *American Journal of Public Health* 91, no. 9 (September 2001): 1358–61, https://doi.org/10.2105/AJPH.91.9.1358.

47 "Policy Brief for the United States of America: 2021," Lancet Countdown on Health and Climate Change, https://www.lancetcountdownus.org/wp-content/uploads/2021/10/USA-2021-English-Lancet-Countdown-Policy-Brief.pdf.

48 "Healing through Policy: Creating Pathways to Racial Justice," De Beaumont Foundation, https://debeaumont.org/wp-content/uploads/2021/10/PolicyPapers_Combined-Document_FINAL.pdf.

49 Jochen Hinkel, Diana Mangalagiu, Alexander Bisaro, and J. David Tàbara, "Transformative Narratives for Climate Action," *Climatic Change* 160 (2020): 495–506, https://doi.org/10.1007/s10584-020-02761-y.

50 Heather McGee, *The Sum of Us: What Racism Costs Everyone and How We Can Prosper Together* (New York: Penguin Random House, 2021).

51 Martin Luther King, "Letter from a Birmingham Jail," accessed on Nov 12, 2023, https://www.csuchico.edu/iege/_assets/documents/susi-letter-from-birmingham-jail.pdf.

52 Hinkel et al., "Transformational Narratives."

53 "Desmond Tutu: We Fought Apartheid: Now Climate Change Is Our Global Enemy," *Guardian*, September 20, 2014, https://www.theguardian.com/commentisfree/2014/sep/21/desmond-tutu-climate-change-is-the-global-enemy.

CHAPTER 3

Intersecting Participatory Action Research and Remote Sensing for Climate Change, Health, and Justice

LAURA E. R. PETERS, ILAN KELMAN, JAMON VAN DEN HOEK,

EIJA MERILÄINEN, AND GEORDAN SHANNON

We have one habitable planet, and our health depends on it. The planet is currently undergoing rapid and substantial change, but across the globe, people and their health are not affected evenly by the impacts of climate change. Poverty and marginalization mediate the relationship between climate change and health,[1] demonstrating prevalent injustices. As such, adverse health outcomes are not an inevitable consequence of climate change but result from a culmination of political decisions that are routinely made far away and without input from the people most affected, and without much understanding of or interest in their perceptions and lived experiences, or the varied factors that affect their health.

Climate change and health justice gaps are thus ever widening. Attention tends to focus on attribution and accountability, emphasizing *injustices*, but more work needs to be done to explicitly conceptualize and pursue actions explicitly supporting justice. To contribute to addressing this challenge, this chapter examines methodologies for justice-based pathways for health within a changing climate. Following a brief outline of community positive health as the basis for just futures, two distinct leading empirical methodologies with emancipatory potentials—participatory action research and satellite remote sensing—are connected within a multidimensional justice framing encompassing space, time, and relationships. Based on this discussion, conclusions indicate the importance of combining bottom-up and top-down partnerships to build healthier futures as the climate changes.

POSITIVE HEALTH AS A BASIS FOR JUSTICE

The World Health Organization defines "health" as "a state of complete physical, mental and social well-being and not merely the absence of disease or infirmity."[2] This conceptualization of health as the presence of multifaceted well-being, or "positive health," is reflective of traditions and practices from diverse cultures around the world. For example, Aboriginal and Torres Strait Islander communities in Australia, First Nations communities in Canada, and communities across the Andes hold understandings of health that emphasize the interconnectedness of people and the interdependencies among people and their environments.[3]

Health is not explained solely by risks. Positive health expands beyond the absence of what is undesirable to encompass the presence of conditions and practices that support and sustain health and well-being (i.e., salutogenic factors), including collective actions that communities take to pursue their health goals. Positive health at the community level is not defined objectively or universally but rather normatively in line with local values.[4] As such, the most powerful and meaningful measures of health may not align neatly with global health indicators used to monitor health phenomena and their determinants across contexts.

This framing opens perspectives on how climate change may affect the multiple determinants of positive health. For example, as climate change affects agriculture,[5] decreased access to nutrition and degraded land-based livelihoods undermine local capabilities to experience well-being and pursue health goals. Changing weather disrupts the sense of place and identity of peoples founded on traditional knowledges, such as the Inuit, with consequences to their mental health.[6] However, communities also have long histories of adapting to changes in their environments. Communities can and do take charge of their own health through diverse forms of knowledge and action, provided they have the resources and opportunities to do so.

Solutions for climate change must not only avoid negative health outcomes but also support positive health processes. Top-down health interventions, especially when formulated as illness care rather than positive health care, often further exclude affected communities along with their values, priorities, and goals, and can even deepen dependencies and worsen health outcomes over the long term, thereby perpetuating injustices. For instance, expecting local and Indigenous communities in Amazonian

Brazil to comply with conservation goals imposed on the basis of climate or other environmental change can impede their ability to meet their health and well-being subsistence needs and make decisions about sustainable resource use in areas designated as protected by outside agencies.[7]

Combining techniques and data can better reflect local needs and capabilities for health, better meld top-down and bottom-up approaches to ensure that localized health impacts of global environmental change are understood and addressed, and thus better achieve the justice required for health in a changing climate. To illustrate, this chapter details the potentials and limitations of two distinct methods, participatory action research and satellite remote sensing, that are commonly viewed as providing emancipatory opportunities, and discusses these within a multidimensional justice framework of space, time, and relationships.

METHODS TO PURSUE HEALTH JUSTICE IN A CHANGING CLIMATE

Innovations in participatory action research (PAR) and satellite remote sensing (RS) have been separately advanced by different research communities that are optimistic about the ability of each to produce robust and actionable knowledge on climate change and health. However, an overenthusiasm for a particular method's potential has led to insufficient attention paid to their weaknesses. While PAR and RS may each produce valuable insights, these methods tend to conform with purely bottom-up and top-down approaches, respectively. This chapter brings together diverse interdisciplinary expertise developing and using these methods to consider and critique their strengths and weaknesses in tandem, with the aim of identifying where they may be used to complement each other to champion change needed for justice-based futures.

Participatory Action Research

"Participatory action research" is a catchall phrase covering research that collaborates with people to document, analyze, and take actions based on their grounded perceptions, experiences, and priorities.[8] Individual and collective experiences can provide rich insights into place-based relationships at both personal and ancestral timescales. However, climate change impacts extend beyond the embodied experience of any given person or group, and we should keep in mind that past environments have never been reliable predictors for the future. Furthermore, local marginalization and limited perspectives on histories and future potentialities as well as distal relationships between phenomena may cloud the ability to identify or adopt needed solutions. Participatory action research methodologies may not reflect the interests and values of the "local," much less the diverse voices that comprise it, and may fall short of advancing its emancipatory ideals in forging futures for positive health.

Participatory action research is a social science method and perspective that not only strives to produce knowledge (research) but also attempts to address societal issues (action).[9] This approach involves the people who are affected by a phenomenon,[10] reenvisions them as experts, and moves toward dissolving the line between the researcher and the researched.[11] Participatory action research has traditionally focused on issues connected with social justice, particularly striving to echo the voices and catalyze the actions of marginalized groups.[12] Because of its ability to facilitate critical consciousness about an issue of concern through collective knowledge making, PAR has the potential to offer insights into positive health, including salutogenic processes and capabilities.

The PAR approach has been adopted for climate change and health research to draw attention to the perspectives of those whose lives are most affected by adverse developments, beginning the reflexive research process with immediate local concerns. For instance, PAR has been used to link food security and climate change adaptation through foregrounding local concerns in Cameroonian communities dependent on forest ecosystems and providing them with resources to start adapting their agricultural practices to climate change.[13] Given that health injustices often have deep historical roots, PAR can bring an understanding of the past into present research and action on climate change affecting current and future generations. While PAR typically begins with local concerns, action is not limited to this scale, as the determinants of health are influenced by policies and actions at larger institutional scales or being conducted by outside entities.[14] For PAR to be effective at achieving climate change and health justice, it needs to translate "community" or "local" concerns into action and politics at multiple scales, sometimes also connecting across communities with shared experiences and concerns.

Ideally, PAR would change bodies of knowledge to be better attuned to different perspectives, and also would make society at large better account for patterns of marginalization. For instance, dominant knowledge production and dissemination about climate change and climate change adaptation in the context of agriculture can be inappropriate for smallholder farmers and their communities,[15] and imposing dominant models of agriculture may be ineffective or counterproductive for positive health. Participatory action research can help smallholder farmers produce and share knowledge relevant to their agricultural activities, increasing their confidence in articulating and pursuing locally beneficial climate change adaptation strategies such as agroecology in places like Malawi.[16] However, PAR may not be able to challenge all dominant narratives about climate change and adaptation: Rachel Bezner Kerr and colleagues observe that local communities might even end up blaming themselves for driving climate change, in line with the dominant narrative that connects local deforestation with climate change, while omitting other major emission sources, such as state forest sales and traffic.[17]

Participatory action research faces considerable shortcomings, and the chasm between the researcher and the researched remains.[18] For instance, the research agenda at the nexus of climate change and health may be framed by researchers and funders with local knowledge sprinkled in as a cultural supplement rather than respected as legitimate expertise.[19] Likewise, external resources brought to the PAR process may be fleeting and/or focused on the periphery of solutions identified by affected communities. As climate change is often disconnected from the primary concerns of the marginalized, PAR applied to the intersection of climate change and health can fall short of truly being a grassroots approach.[20] Adding to this, PAR relies on local investments (including time and effort) in often resource-deprived contexts to effect change, which can come with opportunity costs where the potential to overcome power arrangements is limited, and which can bring immediate risk to life and health where radical change is sought. However, PAR can also fall prey to nonprogressive and narrow local values, interests, and goals that demonstrably undermine long-term health and well-being or deepen local inequities, and people involved in the research process may not self-critically identify their contributions to problems.

Satellite Remote Sensing

Satellite remote sensing produces data that are increasingly assumed as defining climate change and its impacts, which are then connected to implications for health.[21] Quantitative RS data on environmental and climate change can provide a long-term perspective from local to global levels and are often presented as being transparent and objective. Recent decades have seen a turn toward "socializing the pixel" by interdisciplinary scholars who seek to leverage satellite imagery to address questions related to climate change impacts on health.[22] Such work benefits from the diverse sensitivities of RS data toward agricultural conditions, heat stress, or air pollution, which link climate change, health, and justice. The presumed impartiality of RS data would ideally inform a disinterested approach to justice, but the application of RS data remains highly subjective and interpretive and requires appropriate political and cultural contextualization to make them relevant for justice and positive health.[23]

Remote sensing involves the acquisition of imagery from an observational platform outside of Earth's atmosphere, free from atmospheric turbulence and other disruptions. The launch of the US Geological Survey/National Aeronautics and Space Administration Landsat-1 satellite in 1972 initiated the longest-running continual satellite monitoring system of our planet, which continues to this day following the launch of Landsat-9 in September 2021. Landsat and similar satellites have been designed to gather data primarily on terrestrial resources such as forests and crops, while other platforms are better equipped for collecting data on surface temperatures, air pollution, and many other factors relevant to human health and climate change. In addition to civil RS satellites launched and maintained by government agencies, the last decade has seen a rapid proliferation of new commercial RS platforms, which can offer very high-resolution (0.5–4 meters) daily coverage for a price per square kilometer. Commercial sensors have thus far been used more for detecting discrete features such as buildings or roadways but are increasingly used for environmental and climate change impact monitoring.

Satellites are designed to remain operational for years, and many satellites operate for a decade or more, collecting a new image of any location on Earth ranging from every day to every two weeks. The continual collection of imagery provides a consistent data record for tracking climate trends and identifying anomalous changes over the long term, making RS ideal for detecting the environmental determinants of health. Satellite data are also inputs into climate modeling and future projections, which offer a degree of support for future generations who are not yet able to advocate for themselves. The satellite's orbit provides global, intercomparable data on climate change, while the pixels that make up each image provide local data that may be used to assess potential health impacts of climate change. Since satellite data are collected remotely, human communities are observed, enumerated, and quantified in satellite imagery through the detection of villages, towns, and cities and associated land uses. Such superficial community characterizations give little insight on positive health needs or goals that may be affected by climate change.

The promise of RS to provide an impartial and accurate method for assessing climate change impacts on human health remains unfulfilled for several reasons. The spatial resolution of satellite imagery inherently privileges the collection of information about certain landscapes and features that are large enough or sufficiently homogenous to be resolved by a satellite's sensors. In humanized landscapes, the mixing of signals related to ecological and social processes is unavoidable, but the strongest signal within a given pixel dominates. This means that small-scale informal settlements, subsistence agricultural plots, and sparsely forested regions are more likely to be overlooked than cities, industrial farms, and dense forests.[24]

Marginalized communities most affected by climate injustice and health inequities are the least likely to be well-represented in satellite imagery, have access to it, or have the means to use it in ways determined by the community itself. For example, satellite nighttime lights-informed estimates of poverty tend to exclude rural, usually isolated households experiencing "extreme poverty" that do not generate a detectable amount of light.[25] Similarly, informal urbanization of Ho Chi Minh City was overlooked by traditional remote sensing approaches to detect urban growth that were not sensitive to small-scale dwellings and intermixed vegetation typical of informal settlements in the region.[26] The exclusion of socially vulnerable communities from an RS-based map of poverty or urbanization, for example, may also leave them out of decision-making processes or resource allocation needed to address the intersections of climate change and health. This double exclusion from detection and decision making may enable decisions that worsen health and deepen health inequities by actors who misrepresent or override the community's needs or concerns, assign blame where there is none, lead to delayed action or lack of action, or even indirectly contribute to state-sanctioned violence against a community through, for example, restrictive conservation plans, resource extraction, or dispossession.

CONSTRUCTING AND ANALYZING MULTIDIMENSIONAL JUSTICE

There is not just one way to "see" or pursue social justice owing to the multiple perspectives on what constitutes a just world in theory and practice. This section unfolds three interrelated dimensions—space, time, and relationships—of justice to look beyond concerns about the allocation of resources (i.e., distributive justice) to guide actions for justice-based futures.

Space

Weather variations caused by climate change yield spatially uneven impacts. The human climate niche of mean temperature of approximately 13 degrees Celsius is expected to shift in the next fifty years, with one-third of the global population (3.5 billion people) expected to be exposed to a mean annual temperature of greater than or equal to 29.0 degrees Celsius currently found only in 0.8 percent of the world's land area but projected to occur in 19 percent of land by 2070.[27] Warming temperatures will make some locations nonviable for human survival without constant cooling for many days, which is neither affordable nor supportive of livelihoods for the majority of the world's population.[28] Places face changes in the timing and amount of rainfall, with some experiencing more floods, others more droughts, and many experiencing both. These changes affect health differentially not only due to, for example, heat humidity and salinity stress affecting certain regions but also due to cascading impacts on access to nutrition, clean water, and a myriad of other place-specific resources necessary for health and well-being. Remote sensing contributes to all such mapping, while PAR can indicate the contextual impacts on health and salutogenic factors.

The concept of space goes beyond being a physical container. Space is shaped by social processes in the service of creating and sustaining structures of power, discrimination, and exploitation.[29] Spatialization—the construction of social space—can serve as a tool to manipulate populations and distract from their needs, and spatial power arrangements further determine where health-promoting and health-degrading activities take place. Principally and uncritically relying on detached top-down methods such as RS overlooks the linkages between space and power, whereas PAR if implemented properly can indicate how these processes occur and how to overcome them.

In line with Michael Glantz's conceptualization of winners and losers in human-caused climate change, global solutions that are positive in the aggregate can be damaging to certain people in certain places.[30] This aligns with how global development is pursued in different geographies, with development nearly identical to extraction and degradation in geographies with histories of marginalization and colonization, and thus inhibiting possibilities for agency and self-determined futures. For example, REDD+ (Reducing Emissions from Deforestation and forest Degradation), a global framework for reducing emissions from deforestation and degradation, has been criticized as identifying communities as the agents of forest loss rather than national policies that incentivize extraction and multinational actors that drive commercial extraction.[31] Under such frameworks, local people can thus be doubly victimized.

The spatial dimensions of justice note where people are situated in relation to resources and opportunities—where they are extracted, distributed, accessed, and owned—and how climate change may affect the resources that people depend on for their health and well-being. Participatory action research can help to identify necessary resources for positive health and how they are affected by climate change, drawing out discrete features, places, landmarks, boundaries, and use, along with how people understand the interconnections and interdependencies. Remote sensing , in contrast, can provide spatially continuous information on environmental conditions or change, usually measured within administrative regions defined by states or ecological boundaries such as biomes or watersheds, providing different insights into broader interconnections between actions and impacts on resources for positive health. It is not straightforward to establish spatial associations to specific communities between pixels in an image, much less individuals or households, without participatory or other sufficiently grounded information on spatial relationships between people and their access to and ownership of resources.

However, spatial justice implicates more than resource access. A singular focus on the distribution of resources can normalize and privilege structures and systems of power that oppress some and benefit others.[32] Spatial justice demands "greater control over how the spaces in which we live are socially produced wherever we may be located," highlighting the need for participatory processes and the ability to create and acknowledge overlapping senses of "place" in order to pursue justice.[33] This extends to ownership over the methods that provide insight into space. In 2019, the traditional owners and guardians of the Yankunytjatjara, Pitjantjatjara, and Anangu lands, via the Uluru-Kata Tjuta National Park board, successfully lobbied Google to remove satellite images of the sacred sandstone monolith Uluru from its maps service. Uluru is a place of enormous cultural and spiritual significance and has been a prominent focal point for Indigenous justice movements, including the current movement for constitutional recognition of Aboriginal and Torres Straight Islander people.[34] In 1985, the custodianship of Uluru was returned to the traditional owners, and in 2019 climbing Uluru was officially banned, including "virtually" climbing through satellite images.[35] Indigenous Law is an ancient, place-based worldview connecting the country and the land to its people;[36] within this understanding, human well-being is indistinguishable from culture, ancestral knowledge, relationships and kinship, and the health of the environment. The Tjukurrpa (as it is known by the Anangu people) recognizes how the land is imbued with ancestral spirits and knowledge that are the foundation of existence.[37] Protecting sacred sites is important not only from the perspective of land conservation or spirituality but also because the sites comprise the basis of life itself.[38] The decision to take Uluru off the grid not only helps preserve the local ecology and respect Indigenous culture but also restores to local Indigenous communities the right to chart their own pathways to positive health.

Despite notable exceptions such as this, consent or permission for landscape monitoring is rarely given a passing thought. Overconfidence in satellite-derived products and lack of on-the-ground verification can lead to biased assessments of poverty, population, and other parameters of interest without even the knowledge of the people within those landscapes. Similar pitfalls can be seen in PAR. While PAR is ideally initiated and led by local people, the entire process can be dominated by outside agendas and interests or misled by powerful local actors.[39] Moreover, due to the widespread displacement of Indigenous peoples, for example, sites of cultural and spiritual significance may be located at great distances and have contested ownership. Both methods thus have the potential to produce skewed or biased conclusions and to inform actions that augment rather than reduce deprivation, all while overlooking the spatial configurations and conceptualizations of salutogenic resources.

Spatial perspectives on justice point to the need to identify both the spatially discrete and continuous local resources that are important for health, where they are located, and how they are connected with other proximate and distal resources and processes. Local resources are affected by climate change generally but also within the context of place and expectations, which can obscure or illuminate potential solutions. Analysis must include resource ownership and how options for pursuing health are constrained by political and social boundaries, which may not follow environmental contours. Histories of colonization and modes of production explain how outside entities may own valuable local resources as well as how local entities may depend on outside resources in long production chains. Actions at the intersection of climate change and health must integrate understandings of how positive health and salutogenic factors are situated within space and take into consideration how actions spatially re/distribute power and ownership of potential futures.

Time

The effects of climate change on health can also be located in time, with connections between the past, present, and future. They are also experienced with different temporalities, with some effects being ever-present (e.g., sea level rise) and some being seasonal (e.g., more intense and less frequent cyclonic storms). Exceptional or extreme events tend to stand out in people's memories and hold greater sway over their decision making,[40] and RS can discern between anomalous outliers and evolving patterns through offering a record of the past to assess the normalcy of the present and model possible futures.

While RS's strengths lie in establishing a historical record to the present, PAR demonstrates *tempophilia*, which refers to "having an affinity for the present time, so highlighting and prioritizing what is happening now and focusing on immediate interests and needs" in the context of climate change even when connected to and thinking about one's children and further descendants.[41] Participatory action research shows how people tend to be more concerned with environmental changes that are readily perceived in the current moment, are contextualized with past experiences, and can be addressed through current and direct actions.[42] This perspective is not without merit. Because health scaffolds on itself, future health depends on meeting current needs; even within a human lifetime, a health outcome at one life stage may be a determinant of health in a future life stage. Families go on to influence the health of their descendants, and intergenerational health is determined by social inequities.[43] Tensions arise where actions that promote current health and well-being preclude healthy futures, including for future generations.

To understand the *now*, we must look beyond the present, to histories going back through millennia. We have extensive meteorological records in recorded history as well as detailed understanding of palaeo climates for all of Earth's history, a record that RS helps extend every day.[44] Perceiving changes over these extremely long time frames provides an understanding of how current changes to the climate compare to trends and patterns at multiple time scales, which can be contextualized by PAR. Projections of human social indicators provide insights into possible futures depending on our actions, which can be extrapolated to health outcomes. Unfortunately, the historical coverage of satellite imagery is not uniform over the earth, as geographic regions that were considered lower priority for data collection were monitored less frequently, with consistent, regular coverage beginning only in the 2010s. Participatory action research has the potential to fill in some of these gaps that occurred in the 1980s and 1990s, for example, by drawing on written and oral records and artistic expressions.

While time is universal, how we understand and experience it filters through culture, which can be mediated through PAR to define the temporal dimensions of climate change and health. Perspectives on time, which influence human behaviors and tendencies, vary considerably across cultures, including through dimensions of past, present, and future orientations; linear and cyclical time perspectives; and the pace or rhythm of life (i.e., how closely in time activities and experiences are situated).[45] Despite this diversity of lenses, Western perspectives of time dominate, depicting time in a linear line moving into the future and whose passage can be infinitely subdivided into units such as centuries, years, and seconds. This ever-ticking clock pressures action but falls short of capturing how humans live in, experience, and construct meaning through time. For instance, the lived present refers to an understanding of what is happening in a given moment within a structured experience. The lived present cannot be neatly plotted on a timeline, just as reading a phrase cannot be reduced to the letters that comprise it. Even more so, the lived experience "frames the now by placing it against a background of lived experience" by stringing together the moments that form a discrete experience or blur experiences together regardless of the units of time that pass between them.[46] Our sense of time and how we organize it into meaningful units depends on our perspectives and experiences, which are localized and so can be drawn out via PAR.[47] Each newly acquired satellite image, by contrast, can be thought of as a stand-alone anecdote, while the accumulation of images forms a timeline without human narrative or meaning.

Time is an essential resource for actions on climate change and health, though neither PAR nor RS position time as a salutogenic factor. Indeed, concerns abound that it may even be too late to realize environmental justice for Indigenous Peoples, for example, due to ecological and relational tipping points,[48] but PAR and RS together can provide glimpses into what justice could look like within these new paradigms. Health capabilities refer to what people have the potential to do given their assets, capacities, and opportunities but may or may not choose to pursue at a particular point in time.[49] For justice, health capabilities and opportunities should be nurtured over forcing specific health choices taken in specific timeframes, balanced against time-sensitive actions at critical junctures and points of "no return." For example, ground-based observations paired with RS led to the detection of the Antarctic ozone hole in 1985, supported identifying the causal role of chlorofluorocarbons, and revealed the possibility and necessity of addressing the problem through immediate, global action.[50]

Participatory action research can also offer insights on deeper dimensions of time as influenced by culture and human experience, but these perspectives are incomplete and limited. For example, the Gender Violence in the Amazon of Peru project used PAR to develop community-driven interventions for the prevention of gender-based violence in the Peruvian Amazon.[51] Using coproduced community timelines (a lifeline of the human life course, a seasonal calendar around agricultural and river activities, and a historical timeline of the inception and development of the community), participants could identify key moments across different time spans that affected gender-based violence. Of particular note was the impact of seasonal fluctuation and environmental unpredictability on agriculture and subsistence livelihoods, which became more precarious and stressful, translating into a number of health impacts, including violence. Using a participatory timeline approach enabled participants to identify environmental, health, and violence risk factors, and also, most importantly, develop locally led actions to tackle these issues.

Nonetheless, PAR insights may only loosely draw from histories that can be corroborated through other methodologies, such as RS, and are biased toward the present. Further, climate change can wield indirect effects on health through its social and environmental determinants, and these complex causal chains may not be well captured in the PAR process.[52] Participatory processes are necessary to pursue justice for health, but they may be very time-consuming for the people involved and take their time away from other collective actions on climate change and health as well as other productive or restorative activities with health consequences. However, RS provides opportunities to collect consistent and comparable data over time and provide counterfactuals to tease out different pathways of causation with key periods of distinction and influence. Yet, for both methods, the past is a poor predictor of future conditions and cannot take into consideration social-environmental transformations or transitions, with the potential for such changes to encompass those that are both harmful and helpful for health.

Relationships

Central to the question of justice for climate change and health is a consideration of relationships. People construct "imagined communities" around their perceptions of social belonging,[53] and these communities evolve through relational processes of creation and re-creation. Communities have fuzzy boundaries, and people belong to multiple overlapping communities organized around locations, interests, and conditions. Relationships between people and fundamentally pluralistic groups at all scales are not static, even when they may be represented, guarded, and divided by literal and figurative walls. Relational patterns may be teased out through PAR in order to understand how justice might be achieved for climate change and health. Remote sensing, however, struggles to assess relationships that are neither fixed in space over time nor directly detectable in imagery. Instead, relationships between pixels and people are modeled through complementary information on land tenure, administrative boundaries, or simple spatial proximity. Without complementary field or participatory data, fuzzy boundaries and multiple simultaneous identities tend to be overlooked in favor of simple one-to-one relationships.

While diversity is a natural and healthy part of human relationships, relationships between people are also marked by socially constructed differences, whereby the dominant group "others" people without power. These differences, which bear forcefully on the relationship between climate change and health, support structures of oppression, domination, and exploitation.[54] The quantitative data in a satellite image have the promise to be "read" in a consistent way across diverse communities regardless of cultural differences, and may offer a common reference point between groups to help align perspectives or nudge consensus building. However, othered groups often remain unseen and unheard in the interpretation of data, which is where PAR can be powerful in bringing forth their voices— or further marginalizing them if the techniques apply to only some people within a community. The diversity within these othered groups remains even more invisible, as people hold an outgroup homogeneity bias that assumes that members of a different group are more alike than is perceived in ingroups. For instance, Indigenous peoples are often essentialized by Western cultures as land protectors who hold a sacred harmony with the earth, which denies these groups the full range of options, preferences, and agency. As othered groups, they are forced to play specific roles at specific times in relationships (such as categorizing groups as victims, perpetrators, and agents of change, with little room for crossover), including in decision making for climate change and health.

Despite active difference making, which PAR can identify, there is paradoxically also a pervasive bias that people are fundamentally the same and require the same factors to support their positive health. Assuming that people share the same basic motivations, interests, and experiences demonstrates a lack of sensitivity to diversity and virtually always privileges the dominant group and implicitly expects everyone to adapt to dominant preferences, values, and goals. Indeed, the meaning and importance of passively collected satellite imagery must be extracted by researchers and practitioners using culturally dominant empirical approaches that may be orthogonal to the values of the community being imaged. Participatory action research is also not immune to this pitfall, as findings may be uncritically extrapolated to other groups.

Assuming universal concerns and imposing related solutions may be understood as cultural violence, or "those aspects of culture, the symbolic sphere of our existence—exemplified by religion and ideology, language and art, empirical science and formal science (logic, mathematics)—that can be used to justify or legitimize direct or structural violence" and deprive people of their identity and meaning needs through alienation.[55] Cultural violence seeks to persuade people to see certain forms of exploitation and repression as natural, or rather, not to see them at all. In the context of climate change and health, affected groups may use this as the starting point for PAR. Remote sensing can further document how marginalized people and their needs can become further invisibilized through such processes, including refugees being situated in places with disproportionate exposure to environmental hazards, not being accounted for in monitoring United Nations Sustainable Development Goal progress, and having unclear status as rights claimants for climate change and health.[56] Both RS and PAR must strive to recognize that different people are affected differently by climate change impacts across various relational divides.

Problems have been produced by playing a positive-sum game from a negative-sum perspective of exercising power over others and power over the environment, all while ignoring the interdependence of positive health. Solutions born from domination and oppression in terms of defining the problem, setting the agenda, and deciding the course of action are equally unlikely to lead to improvements. Power relationships must move away from power *over* to power *with* and power *to* in order to recognize fundamental interdependencies and arrive at the transformative solutions and partnerships that are needed for planetary and human health as well as justice. For both RS and PAR, there is the potential to explicitly work on relationships through the research process—between different communities of circumstance, practice, and interests—to develop compassionate mutual understanding and supportive actions for positive health.

CONCLUSIONS: THE JUSTICE JOURNEY

Justice is not an end destination, in part because there is no perfect way to understand, measure, or pursue justice. Rather, justice is an active process of learning and valuing beyond a single perspective. It is essential from a justice perspective to leverage multiple methods from multiple viewpoints to formulate understandings that are as complete as possible, while also acknowledging that any combination cannot provide everything needed and has drawbacks. Research on climate change and health should learn in this regard from many other fields, such as mapping volcanic hazards and reducing vulnerability for all genders and sexualities.[57] These knowledges include local, traditional, Indigenous, vernacular, and formal scientific approaches and so, by definition, involve PAR and RS.

As communities across the globe respond to interlocking crises and strive to pursue justice and well-being, there are opportunities to put into practice creative and inclusive approaches that knit together the spatial, temporal, and relational dimensions of justice for climate change and health. An example of a way forward is participatory three-dimensional mapping, which can apply to numerous topics—including climate change, health, and justice—since PAR assists in understanding local perspectives of mapping communities that RS cannot achieve, while RS can provide further layers, including of data not typically observable locally or determinable via PAR. Another example championed by the organization Stema uses participatory approaches to map salutogenic systems for positive health and from this bottom-up approach looks at ways to knit together novel datasets and technologies.[58] Starting data collection with community-led perspectives ensures that local knowledge takes a central place in shaping the contours of the evidence, while using additional technologically forward methods to boost the perceived value of the evidence with policymakers.

Together, PAR and RS can be used to show long time horizons of change across multiple regions and spatial scales across intersecting perspectives and relationships, illustrating ripple effects and the global interconnectedness of resources that contribute to health. But, existing platforms for knowledge and action based on RS should not integrate diverse actors merely through training them to be involved in data collection or usage with no intention of shifting the agenda. Instead, it is about bringing together all the knowledge forms to compare and contrast, and acknowledge and interpret similarities and differences in order to work through actions based on knowns, unknowns, certainties, and uncertainties. Shari Gearheard and colleagues reported a specific method for determining the local impacts of climate change on weather, which could also be applied to the local impacts of climate change on health.[59]

Shared solutions for climate change and health can be forged through partnerships that identify common interests below the surface of policy and action positions and embrace multiple ways of seeing and knowing. Solutions do not have to be universal or homogeneous; in fact, they cannot be and would be stronger drawing from a panoply of place-based and people-centered actions. The key is not to seek a perfect solution but rather to cultivate needed perspectives and partnerships to pursue justice for ever-changing societies on an ever-changing planet to apply ever-changing technologies and understandings without ever relying on a single one.

FUNDING

This research was supported through the Belmont Forum by the UK's Natural Environment Research Council (NERC) [grant number NE/T013656/1], the National Aeronautics and Space Administration (80NSSC18K0311), and a United States Institute of Peace grant on Environment, Conflict, and Peacebuilding.

NOTES

1 Jouni Paavola, "Health Impacts of Climate Change and Health and Social Inequalities in the UK," *Environmental Health* 16, no. 1 (2017): 61–68; Marina Romanello, Alice McGushin, Claudia Di Napoli, et al., "The 2021 Report of the Lancet Countdown on Health and Climate Change: Code Red for a Healthy Future," *Lancet* 398, no. 10311 (2021): 1619–62.

2 World Health Organization, "Ottawa Charter for Health Promotion," 1986, https://www.who.int/publications-detail-redirect/ottawa-charter-for-health -promotion.

3 Laura E. R. Peters, Geordan Shannon, Ilan Kelman, and Eija Meriläinen, "Toward Resourcefulness: Pathways for Community Positive Health," *Global Health Promotion* 29, no. 3 (September 2021): 5–13.

4 Sridhar Venkatapuram, *Health Justice: An Arguments from the Capability Approach* (Cambridge, MA: Polity Press, 2011).

5 Romanello et al., "2021 Report," 1619–62.

6 Jacqueline Middleton, Ashlee Cunsolo, Andria Jones-Bitton, Inez Shiwak, Michele Wood, Nathaniel Pollock, Charlie Flowers, and Sherilee L. Harper, "'We're People of the Snow': Weather, Climate Change, and Inuit Mental Wellness," *Social Science & Medicine* 262 (2020): 113137.

7 Florent Kohler and Eduardo S. Brondizio, "Considering the Needs of Indigenous and Local Populations in Conservation Programs," *Conservation Biology* 31, no. 2 (2017): 245–51.

8 Robert Chambers, "The Origins and Practice of Participatory Rural Appraisal," *World Development* 22, no. 7 (1994): 953–69.

9 Orlando Fals Borda, "Action Research in the Convergence of Disciplines," *International Journal of Action Research* 9, no. 2 (2013): 155.

10 Luciana Cordeiro, Cassia Baldini Soares, and Leslie Rittenmeyer, "Unscrambling Method and Methodology in Action Research Traditions: Theoretical Conceptualization of Praxis and Emancipation," *Qualitative Research* 17, no. 4 (2017): 395–407.

11 Borda, "Action Research," 155.

12 Orlando Fals Borda, "Participatory (Action) Research in Social Theory: Origins and Challenges," *Handbook of Action Research: Participative Inquiry and Practice* (2006): 27–37.

13 Mekou Youssoufa Bele, Denis Jean Sonwa, and Anne Marie Tiani, "Supporting Local Adaptive Capacity to Climate Change in the Congo Basin Forest of Cameroon: A Participatory Action Research Approach," *International Journal of Climate Change Strategies and Management* 5, no. 2 (2013).

14 On large institutional scales, see Nicholas Freudenberg and Emma Tsui, "Evidence, Power, and Policy Change in Community-Based Participatory Research," *American Journal of Public Health* 104, no. 1 (2014): 11–14.

15 Rachel Bezner Kerr, Hanson Nyantakyi-Frimpong, Laifolo Dakishoni, et al., "Knowledge Politics in Participatory Climate Change Adaptation Research on Agroecology in Malawi," *Renewable Agriculture and Food Systems* 33, no. 3 (2018): 238–51.

16 Kerr et al., "Knowledge Politics," 238–51.

17 Kerr et al., "Knowledge Politics," 238–51.

18 Cf. Borda, "Action Research," 155.

19 On researchers and funders, see Seth Shames, Krista Heiner, Martha Kapukha, et al., "Building Local Institutional Capacity to Implement Agricultural Carbon Projects: Participatory Action Research with Vi Agroforestry in Kenya and ECOTRUST in Uganda," *Agriculture & Food Security* 5, no. 1 (2016): 1–15. On local knowledge, see Max Liboiron, "Decolonizing Geoscience Requires More Than Equity and Inclusion," *Nature Geoscience* 14, no. 12 (2021): 876–77.

20 Eija Meriläinen, Ilan Kelman, Laura E. R. Peters, and Geordan Shannon, "Puppeteering as a Metaphor for Unpacking Power in Participatory Action Research on Climate Change and Health," *Climate and Development* 14, no. 5 (2021): 1–12.

21 Thomas Lillesand, Ralph W. Kiefer, and Jonathan Chipman, *Remote Sensing and Image Interpretation* (Hoboken, NJ: John Wiley & Sons, 2015).

22 Jacqueline Geoghegan, L. Pritchard, Yelena Ogneva-Himmelberger, et al., "'Socializing the Pixel' and 'Pixelizing the Social' in Land-Use and Land-Cover Change," in *People and Pixels* (Washington, DC: National Academy Press, 1998); Tracy A. Kugler, Kathryn Grace, David J. Wrathall, et al., "People and Pixels 20 Years Later: The Current Data Landscape and Research Trends Blending Population and Environmental Data," *Population and Environment* 41, no. 2 (2019): 209–34.

23 On application being highly subjective and interpretive, see Mia M. Bennett, Janice Kai Chen, Luis F. Alvarez Léon, and Colin J. Gleason, "The Politics of Pixels: A Review and Agenda for Critical Remote Sensing," *Progress in Human Geography* (2022): 03091325221074691. On political and cultural contextualization, see Kugler et al., "People and Pixels," 209–34. On relevant to justice and positive health, see Matthias Weigand, Michael Wurm, Stefan Dech, and Hannes Taubenböck, "Remote Sensing in Environmental Justice Tesearch—A Review," *ISPRS International Journal of Geo-Information* 8, no. 1 (2019): 20.

24 Jamon Van Den Hoek and Hannah K. Friedrich, "Satellite-Based Human Settlement Datasets Inadequately Detect Refugee Settlements: A Critical Assessment at Thirty Refugee Settlements in Uganda," *Remote Sensing* 13, no. 18 (2021): 3574.

25 Brock Smith and Samuel Wills, "Left in the Dark? Oil and Rural Poverty," *Journal of the Association of Environmental and Resource Economists* 5, no. 4 (2018): 865–904.

26 Arthur Acolin and Annette M. Kim, "Seeing Informal Settlements: The Policy Implications of Different Techniques to Identify Urban Growth Patterns from Satellite Imagery Using the Case of Informal Construction in Ho Chi Minh City," 2017, accessed from papes.ssrn.com (Social Science Research Network).

27 Chi Xu, Timothy A. Kohler, Timothy M. Lenton, Jens-Christian Svenning, and Marten Scheffer, "Future of the Human Climate Niche," *Proceedings of the National Academy of Sciences* 117, no. 21 (2020): 11350–55.

28 Romanello et al., "2021 Report," 1619–62.

29 Edward W. Soja, *Seeking Spatial Justice* (Minneapolis: University of Minnesota Press, 2010).

30 Michael H. Glantz, "Assessing the Impacts of Climate: The Issue of Winners and Losers in a Global Climate Change Context," *Studies in Environmental Science* 65 (1995): 41–54.

31 Margaret Skutsch and Esther Turnhout, "REDD+: If Communities Are the Solution, What Is the Problem?," *World Development* 130 (2020): 104942.

32 Anna Stanley, "Just Space or Spatial Justice? Difference, Discourse, and Environmental Justice," *Local Environment* 14, no. 10 (2009): 999–1014.

33 Soja, *Seeking Spatial Justice*, 7.

34 See, in particular, the "Uluru Statement from the Heart" calling for constitutional recognition for Australian Indigenous peoples at https://ulurustatement.org/.

35 Watarrka Foundation, "Uluru—A Sacred Aboriginal site," n.d., retrieved from https://www.watarrkafoundation.org.au/blog/uluru-a-sacred-aboriginal-site.

36 James Norman, "Why We Are Banning Tourists from Climbing Uluru?," *Conversation* 6 (2017).

37 Parks Australia, "Fact Sheet Tjukurpa," 2021, retrieved from https://parksaustralia.gov.au/uluru/pub/fs-tjukurpa.pdf.

38 Norman, "Banning Tourists."

39 Meriläinen et al., "Puppeteering as a Metaphor," 1–12.

40 Falk Lieder, Thomas L. Griffiths, and Ming Hsu, "Overrepresentation of Extreme Events in Decision Making Reflects Rational Use of Cognitive Resources," *Psychological Review* 125, no. 1 (2018): 1.

41 Ilan Kelman, Himani Upadhyay, Andrea C. Simonelli, et al., "Here and Now: Perceptions of Indian Ocean Islanders on the Climate Change and Migration Nexus," *Geografiska Annaler: Series B, Human Geography* 99, no. 3 (2017): 284–303, 287.

42 Kelman et al., "Here and Now," 284–303.

43 Robert S. Kahn, Kathryn Wilson, and Paul H. Wise, "Intergenerational Health Disparities: Socioeconomic Status, Women's Health Conditions, and Child Behavior Problems," *Public Health Reports* 120, no. 4 (2005): 399–408.

44 Jan Zalasiewicz and Mark Williams, *The Goldilocks Planet: The Four-Billion-Year Story of Earth's Climate* (Oxford: Oxford University Press, 2012).

45 C. Ashley Fulmer, Brandon Crosby, and Michele J. Gelfand, "Cross-Cultural Perspectives on Time," *Time and Work* 2 (2014): 53–75.

46 Bruce Ackerman, "Temporal Horizons of Justice," *Journal of Philosophy* 94, no. 6 (1997): 299–317, 302.

47 Ackerman, "Temporal Horizons of Justice," 299–317.

48 Kyle Whyte, "Too Late for Indigenous Climate Justice: Ecological and Relational Tipping Points," *Wiley Interdisciplinary Reviews: Climate Change* 11, no. 1 (2020): e603.

49 Laura E. R. Peters, Ilan Kelman, Geordan Shannon, and Des Tan, "Synthesising the Shifting Terminology of Community Health: A Critiquing Review of Agent-Based Approaches," *Global Public Health* 17, no. 8 (2021): 1–15.

50 Jonathan Shanklin, "Reflections on the Ozone Hole," *Nature* 465, no. 7294 (2010): 34–35.

51 The Gap Project: Preventing Gender Violence in the Amazon of Peru, last accessed September 6, 2023, https://www.gapprojectperu.com/.

52 Meriläinen et al., "Puppeteering as a Metaphor," 1–12.

53 Benedict Anderson, *Imagined Communities: Reflections on the Origin and Spread of Nationalism* (New York: Verso Books, 2006).

54 Iris Marion Young, *Justice and the Politics of Difference* (Princeton, NJ: Princeton University Press, 2011).

55 Johan Galtung, "Cultural Violence," *Journal of Peace Research* 27, no. 3 (1990): 291–305, 291.

56 On refugees' disproportionate exposure to environmental hazards, see Michael Owen, Andrew Kruczkiewicz, and Jamon Van Den Hoek, "Indexing Climatic and Environmental Exposure of Refugee Camps with a Case Study in East Africa," *Scientific Reports* 13, no. 1 (2023): 1–14. On refugees' not being accounted for, see Jamon Van Den Hoek, Hannah K. Friedrich, Anna Ballasiotes, Laura E. R. Peters, and David Wrathall, "Development after Displacement: Evaluating the Utility of OpenStreetMap Data for Monitoring Sustainable Development Goal Progress in Refugee Settlements," *ISPRS International Journal of Geo-Information* 10, no. 3 (2021): 153.

57 On mapping volcanic hazards, see Katharine Haynes, Jenni Barclay, and Nick Pidgeon, "Volcanic Hazard Communication Using Maps: An Evaluation of Their Effectiveness," *Bulletin of Volcanology* 70, no. 2 (2007): 123–38. On reducing vulnerability, see Jean-Christophe Gaillard, Andrew Gorman-Murray, and Maureen Fordham, "Sexual and Gender Minorities in Disaster," *Gender, Place & Culture* 24, no. 1 (2017): 18–26.

58 Jean-Christophe Gaillard and Emmanuel A. Maceda, "Participatory Three-Dimensional Mapping for Disaster Risk Reduction," *Participatory Learning and Action* 60, no. 1 (2009): 109–18. On Stema, see https://www.stema.org/.

59 Shari Gearheard, Matthew Pocernich, Ronald Stewart, et al., "Linking Inuit Knowledge and Meteorological Station Observations to Understand Changing Wind Patterns at Clyde River, Nunavut," *Climatic Change* 100, no. 2 (2010): 267–94.

CHAPTER 4

Nature-Based Solutions, Social and Health Inequalities, and Climate Justice in European Cities

CLAIR COOPER

In response to growing concerns about the relationship between urban biodiversity, health, and climate change, together with the challenges posed by the ongoing COVID-19 pandemic, the World Health Organization in partnership with the International Union for the Conservation of Nature and the Friends of Ecosystem-based Adaptation set up an expert working group on biodiversity, climate, One Health, and nature-based solutions. Their aim, to develop guidance and tools to support the deployment of nature-based solutions to improve human and ecosystem health and, in turn, the resilience of socioecological systems while also facilitating a healthy, just green recovery from COVID-19.[1] Nature-based solutions (NBS) are defined by the Internation Union for Conservation of Nature to be "actions to protect, sustainably manage and restore natural or modified ecosystems, which address societal challenges (e.g., climate change, food, and water security or natural disasters) effectively and adaptively, while simultaneously providing human well-being and biodiversity benefits."[2] Subsequent work by the European Commission to redefine the concept placed the central aims of NBS on the achievement of social and economic goals through the deployment of blue-green infrastructure projects.[3] Thus, despite the efforts of the working group, health and well-being benefits became secondary or an assumed by-product of work to support economic recovery despite evidence of the central role that green and blue space play an important role mediating ill health associated with climate change, especially in deprived communities that lack access to green space.[4] Research into the pathways that link NBS and health suggests that these interventions influence health by restoring mental health, building resilience through physical activity or improved fitness, buffering against the adverse health effects of air and noise pollution, and reducing the impact of the urban heat island effect or flooding.[5]

In contrast, Matilda Van den Bosch argues that these claims may be overemphasized since the NBS framework lacks consideration of the complex intimate relationship between environment and public health.[6] Adina Dumitru and colleagues argue this may be due to the scarcity of studies that examine the relationship between NBS and health or to uncertainty about the relationship due to lack of causal evidence.[7] Despite these uncertainties, scholars argue that these solution-orientated actions hold significant promise to harness the power of urban nature to protect, manage, and restore ecosystems and create bundles of ecosystem services to tackle interconnected societal challenges.[8] However, doubts remain over the potential of NBS to deliver health benefits due to the lack of clarity on how the concept relates to the complex relationship between health, social equity, and environmental justice.[9] Using evidence published in the *Urban Nature Atlas* that retrospectively classified different urban green and blue space projects as NBS, this chapter unpacks how the practice and implementation has evolved and how they interact with different social and economic determinants that influence health.[10] This chapter takes a systematic approach using quantitative relational techniques (such as multiple factor analysis) to unravel the complex and intertwined relationship between different features that influence and characterize the operationalization of NBS, social and economic determinants of health, and consequences for climate justice.

THE INTERSECTION BETWEEN HEALTH, NBS CLIMATE, AND ENVIRONMENTAL JUSTICE

This section sets out the theoretical framework for this chapter drawing on scholarship across climate and environmental justice, environmental inequity, and its relationship with the discourse of nature-based solutions.

Climate Justice, Environmental Justice, and Environmental Inequality

Nature-based solutions are typically framed as a public good that creates resources to help communities transition to sustainability and aid economic recovery, despite research that illustrates that groups who live in communities being rejuvenated by NBS are often excluded and made invisible by municipal actors responsible for promoting and enacting these interventions.[11] Hence, scholars question if the social arrangements and sociopolitical processes at play among actors involved in the governance and management of NBS can lead to climate and environmental justice.[12] Climate justice is a framework that confronts the intersection between climate change and how social inequalities are experienced as a form of social violence, and it is deeply intertwined and overlaps with concepts of social and environmental justice.[13]

Peter Newell and colleagues suggest there are four pillars of climate justice: procedural injustice, distributive, recognition, and intergenerational.[14] Procedural justice is concerned with the extent to which decisions made about responses to and impacts of climate change that influence the design, governance, and management of NBS are just (i.e., ethical and attentive to power dynamics). It also concerns the extent to which different groups across society are included in democratic processes or provided with opportunities to develop the capabilities to access and use resources that could aid climate adaptation;[15] procedural injustice overlaps with both recognition and distributional injustice. Recognition injustice concerns the recognition or respect given to different groups during decision-making processes and the extent to which the uneven capacity of some groups to participate is acknowledged.[16] Distributive climate justice relates to how the costs and benefits provided by NBS are spatially and temporarily distributed but also the extent to which the benefits can be accessed by society.[17] Nancy Fraser and Axel Honneth argue that the distributive justice paradigm also includes underlying processes that construct and influence distribution such as equity, recognition, and participation.[18] Scholars assert that intergenerational framing is also an important component of climate justice; this concerns the duties of the present-day society toward future generations as a means of holding different actors responsible for the decisions made about the equity of distribution of benefits of NBS to those currently in need, but also the needs of future generations.[19]

Relationship between NBS and Health Inequalities

Health is a complex, socially constructed concept that is open to different interpretations and definitions.[20] It is described as a state of physical, mental, and social well-being and a resource for everyday life that allows people to function and participate in activities as members of society.[21] Anthony Gatrell and Susan Elliot argue that the definition of health should account for the availability of personal and societal resources that allow us to cope with or manage our health or alter our environment.[22] This contrasts with the definition of inequity in health, which relates to unfair differences in health due to lack of resources and is an issue of social injustice[23] often associated with differences in socioeconomic position, ethnicity, and gender that are socially produced, unfair and unjust.[24] These systematic differences in health (or social determinants of health) are not randomly distributed but influenced by sociopolitical and socioeconomic processes that shape the circumstances in which people grow, live, work, and age and can worsen or improve differently when aspects of urban life interact and may be exacerbated by the effects of climate change.[25] Riley Dunlap and Robert Brulle suggest that those living in poverty are most vulnerable to the harmful effects of climate change on health due to their exposure to persistent, intersecting, and entrenched structural inequalities that cause inequalities in health.[26] However, there is growing evidence that suggests that access to green and blue space provided by NBS could help to alleviate some of the public health risks associated with climate change,[27] particularly in low-income and deprived communities where access to salutogenic resources are limited.[28]

DATA SOURCES AND METHODOLOGY

This section outlines the sources of data and methodology that was applied to the study.

Urban Nature Atlas

Naturvation or NATure-based URban innoVATION was a four-year project whose role was to investigate how NBS were being deployed across Europe and what could be achieved in cities by innovating with nature. Data collected by Naturvation was used to create an open platform of over one thousand NBS located across one hundred cities.[29] Figure 4.1 shows a map of the European cities that were selected for inclusion in the project by research team, including six urban regional innovation partnerships.

FIGURE 4.1. Map of cities selected for analysis by the Naturvation program, including urban regional innovation partnerships. Source: Naturvation, *Urban Nature Atlas*.

Using autonomous counting (or frequency counts of qualitative data),[30] Naturvation transformed qualitative data (gathered using discourse analysis, questionnaires, and structured interviews) for each case to create binary categorical variables representative of different urban conditions of NBS, ecosystems services produced, governance arrangements, stakeholders involved, forms of innovation, finance, and participatory method.[31]

Socioeconomic Position

To unpack how different factors that influence the implementation of NBS relate to different social and economic determinants that influence poor health, the study created an index of socioeconomic position (SEP) for cities. Socioeconomic position is an aggregate concept that is indicative of people's position within the social hierarchy of society, their likelihood of being exposed to harmful effects of urbanization or climate change, or their capability to access resources that might mediate these effects or enhance health.[32] Bruna Galobardes and colleagues define SEP as "socially derived economic factors that influence what positions individuals or groups within the multi-stratified structure of society have."[33]

Table 4.1 outlines the indicators and theoretical basis for the choice of indicators.

TABLE 4.1. SUMMARY OF THE THEORETICAL BASIS OF THE SOCIO-ECONOMIC POSITION INDICES

Indicators of Social Economic Position	Source
Education attainment	B. Galobardes, J. Lynch, and G. D. Smith, "Measuring Socioeconomic Position in Health Research," *British Medical Bulletin* 81 (2007); J. Lynch and G. Kaplan, "Socioeconomic Position," in *Social Epidemiology*, ed. L. F. Berkman and I. Kawachi (New York: Oxford University Press, 2000).
EU_SILC (Statistics on Income and Living Conditions) Occupation indicator	K. Purcell, "Work, Employment and Unemployment," *Key Variables in Social Investigation* 4 (2018): 1; Galobardes et al., "Measuring Socioeconomic Position in Health Research"; C. Salmond, J. Atkinson, and P. Crampston, *NZDep2013 Index of Deprivation*, 2014, https://www.otago.ac.nz/wellington/otago069936.pdf ; Bernd Wegener, "Job Mobility and Social Ties: Social Resources, Prior Job, and Status Attainment," *American Sociological Review* 56, no. 1 (1991): 60–71.
Average or median income	N. Krieger, D. R. Williams, and N. E. Moss, "Measuring Social Class in US Public Health Research: Concepts, Methodologies, and Guidelines," *Annual Review of Public Health* 18 (1997): 341–78; Salmond et al., *NZDep2013 Index of Deprivation*.
Poverty due to social transfers (social security payments) or low working hours	K. Purcell, "Work, Rmployment and Unemployment," 1; Salmond et al., *NZDep2013 Index of Deprivation*.

In addition to the concept of SEP, the study drew on the concept of social vulnerability that is influenced by such factors as social capital and networks, the age and fragility of individuals, and beliefs and customs,[34] and is often differentiated in terms of social stratification.[35] Thus, some of the determinants that influence social vulnerability also influence SEP and the determinants of poor health. The concept also overlaps with dimensions of environmental justice since it can also be influenced by access to resources, access to political power and representation, and the condition of urban infrastructure.[36] Table 4.2 summarizes the indicators used to represent different social, vulnerability, and health outcome conditions in cities.

TABLE 4. 2. SUMMARY OF SOCIAL, VULNERABILITY, HEALTH OUTCOME INDICATORS AND INDICA-
TORS REPRESENTING ENVIRONMENTAL CONDITIONS IN CITIES

Material Deprivation Indices	Social Vulnerability	Health Outcome Indicators
Education attainment	Lone pensioners	All-cause mortality
Average or Median income	Households with dependents	Mortality due to heart or respiratory disease
Poverty due to social transfers	Lone parents with dependants	Infant Mortality
Poverty due to low working hours	Foreign citizens born in EU country & non-EU country	All-cause mortality related to gender in citizens under 65 years
SILC Occupation indicators		

SOURCE: Eurostat, Cities (Urban Audit) Database, URL: Database—Cities (Urban Audit)—Eurostat
(europa.eu) Accessed: 01/10/2017.

To investigate how interactions between different characteristics of NBS
change over time and how the patterns that emerge relate to different social
and urban conditions that influence poor health, the study disaggregated
data about different urban green interventions retrospectively classified as
NBS by Naturvation into two multiblock datasets for analysis based on the
date of implementation of the NBS and the reference year of the urban audit
data. This allowed patterns in the relationships between NBS and different
social and urban conditions that influence poor health that have emerged
over time to be analyzed using quantitative relational techniques. Using
the FactoMineR package in R software, we compute a multifactor analysis visualizing the results in the FactoExtra package.[37] Multifactor analysis
is a geometric data analysis technique that represents structured datasets
as clouds of points in a multidimensional space.[38] However, unlike other
similar techniques (e.g., principal component analysis), multifactor analysis
simultaneously analyzes the strength of the relationships between different
observations described by multiple blocks, or sets of variables defined by the
same set of observations, within a structured data matrix as well as individual variables.[39]

RESULTS

This section presents the analysis of how the complex interrelationship
between the different characteristics of NBS and social and economic conditions that influence the social determinants of ill health has evolved across

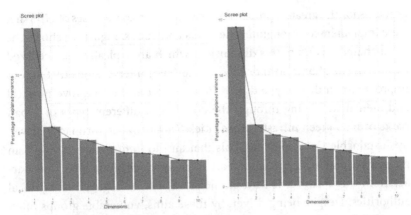

FIGURE 4.2. Scree plot showing the percentage variance represented by each dimension. Figure 4.2a NBS was implemented between 1990 and the 2000s; figure 4.2b NBS was implemented after 2010.

time and space. The sections that follow report on the characteristics of NBS that are most common across the cases implemented in the 1990s and 2000s (section 4.1) and those implemented after 2010 until 2017 (section 4.2). To identify the dimensions or groups of characteristics of NBS (also known as principal components) that are most important in each time period, the percentage variance represented by each dimension is plotted using a scree plot. Figures 4.2a and 4.2b show that the first five dimensions contribute the most variance across each of the dimensions in each time period. The sections that follow report on the principal characteristics of the group of NBS represented by each dimension and how these different variables that relate to SEP.

THE RELATIONSHIP BETWEEN NBS AND UNDERLYING STRUCTURAL CONDITIONS THROUGHOUT THE 1990S AND 2000S

This section describes how the groups of different characteristics of NBS and indicators of socioeconomic position, vulnerability, and health are grouped in the 1990s and 2000s.

Characteristics of NBS Implemented in the 1990s and 2000s

NBS for Urban Biodiversity, Recreation, and Health (Dimension 1)
Of those NBS implemented in the 1990s and 2000s, the first dimension

accounts for the greatest proportion of inertia across the cases of NBS that were implemented throughout the 1990s and 2000s. Figure 4.3 shows the main characteristics of this dimension, which are typically characterized by micro- (neighborhood) or meso-scale (city) interventions that aim to improve urban diversity and provide opportunities to improve physical and mental well-being through the creation of different types of urban parks or blue-green infrastructure cities. These NBS are primarily driven by Sustainable Development Goals that aim to improve access to urban greenspace and regenerate urban areas; governance of the NBS that characterizes this factor is led by municipal actors or cogoverned by local authorities. Despite being driven by these aims, vulnerable groups often excluded from climate decisions, such as the disabled or marginal communities such as families with dependents or the elderly, are unrelated to this dimension. Consequently, opportunities to reconnect citizens, particularly marginalized groups, with ecological infrastructure that could influence the upstream determinants of ill health are missed.

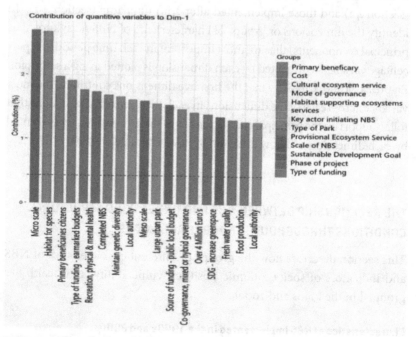

FIGURE 4.3. Contribution of different characteristics of NBS, health, social, economic, and vulnerability indicators on the total variance of the first dimension (or principal component).

NBS for Social and Health Inequality (Dimension 2)

The second dimension is the second-largest contributing group of NBS that, unlike the first dimension, is strongly influenced by the indices of SEP for cities, followed by gender-related all-cause mortality (see figure 4.3). Unlike the first and largest group of NBS, the NBS that characterize the second-largest dimension in the 1990s and 2000s aim to justly distribute the benefits of NBS by creating community growing spaces or rewilding derelict spaces for the benefit of vulnerable social groups (families with dependents, lone parents, and pensioners) (see figure 4.4). These interventions aim to provide salutogenic resources that could alleviate the risk of ill-health effects of climate change and urbanization. This finding is also reflected in the contribution that social indicators such as social housing, private rental, and lack of amenities make to the loading of this dimension.

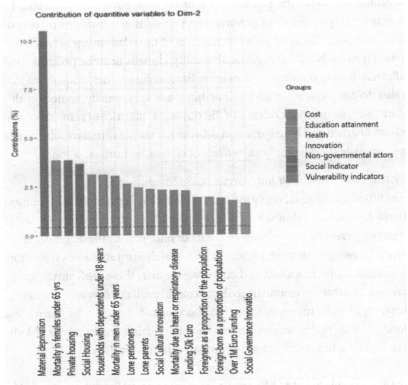

FIGURE 4.4. Contribution of different characteristics of NBS, health, social, economic, and vulnerability indicators on the total variance of the second dimension (or principal component).

Buffering Socioeconomic Distress (Dimension 3)

The third dimension that characterizes NBS deployed in the 1990s and 2000s is also strongly influenced by SEP. The findings suggest that NBS are being deployed to help cities transition toward urban sustainability by adopting an inclusive cross-sectoral approach to the governance of NBS involving research, state and municipal government, and community actors. Typically financed through donations or crowdfunding, these solutions are co-governance or governed and managed by nongovernmental actors to create opportunities for urban growing to benefit marginal groups.

Blue-Green NBS to Regulate Climate, Air Quality, and Flooding (Dimension 4)

The study also suggests that NBS that characterize the fourth dimension are deployed on a street scale or mesoscale, governed by nongovernmental organizations, from the private sector or regional government enabled by corporate responsibility or greenwashing agendas and financed by corporate investment. Despite growing evidence of the relationship between different types of NBS that regulate air quality, climate, or noise pollution and alleviate health inequality in marginalized communities, indicators that relate to SEP, mortality, and vulnerability are very weakly related to this dimension suggesting evidence of distributional injustice of resources (provision of green and blue space and ecosystems services) that could alleviate and mediate the adverse health effects of climate and urbanization.

System Innovation with Blue-Green NBS (Dimension 5)

The fifth dimension is largely influenced by the total cost of the intervention (from five hundred thousand euros to two million euros) to improve water quality by creating NBS along sea coasts or blue-green infrastructure. Governed by nongovernmental actors from research institutions or civic actors and financed by European and corporate finance, these NBS aim to influence sustainable development goals for coastal protection by adopting a systems approach to reconfigure socioecological systems by providing food-growing opportunities, improving water quality, creating urban biodiversity, and stimulating economic growth through tourism.

The Relationship between NBS and Underlying Structural Conditions after 2010

This section describes how the groups of different characteristics of NBS and indicators of socioeconomic position, vulnerability, and health are grouped after 2010.

NBS for Biodiversity, Recreation, and Health (Dimension 1)

Of the NBS deployed after 2010, just under half of the variance is attributed to the first dimension, which typifies NBS implemented on a neighborhood or street scale. The analysis also suggests that governance arrangements have transitioned from those that are government-led or co-governed by municipal actors to those being cogoverned or governed by nongovernmental actors involving community groups or private companies (see figure 4.5). However, closer examination of the modes of participation suggests that while cities strive to recognize the role of different actors, engaging them in coplanning, consultation, and information dissemination processes, there is little evidence of the empowerment of citizens as central actors.

The NBS that typify this dimension also aim to create multifunctional ecosystems that provide services for flood regulation, recreation, food production, and aesthetic appearance for the benefit of communities. However,

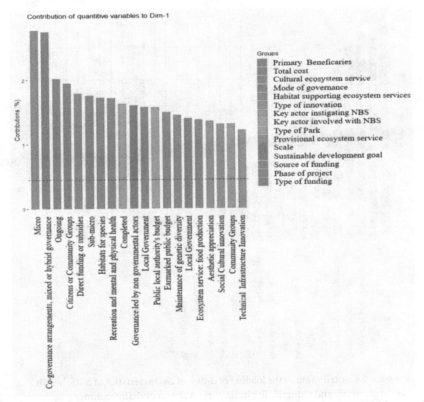

FIGURE 4.5. Contribution of the loading of different characteristics of NBS, health, social, economic, and vulnerability indicators on the first dimension.

the distribution of these services does not appear to be related to social groups that are most vulnerable to inequalities in health and the adverse health effects of climate change. This is evident from the weak contribution that marginal groups and structural conditions make to the dimension.

Multifunctional Blue-Green NBS Enabled by Regional Partnerships (Dimension 2)

The second dimension that typifies NBS deployed after 2010. Groups of community actors or citizens are led or jointly govern implementation of NBS in collaboration with national government and the private sector (see figure 4.6). While these NBS are governed by coalitions of community, private sector, and state actors, analysis suggests that modes of participation are limited to crowdfunding, coplanning, and consultation.

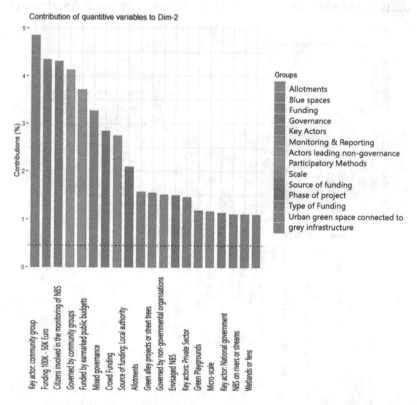

FIGURE 4.6. Contribution of the loading of different characteristics of NBS, health, social, economic, and vulnerability indicators on the second dimension.

Other approaches that empower citizens to influence institutional processes and climate decisions, or develop capabilities and skills in the management of NBS, are missing from this dimension.

Multifunctional NBS Aim to Support Disadvantaged Communities (Dimension 3)

The third dimension that characterizes NBS deployed after 2010 is strongly
influenced by SEP; variables indicative of vulnerability (such as lone parents) and all-cause mortality related to gender are more weakly associated
with this dimension. As with the second dimension, collaboration is
important, involving community groups or governmental actors with
support from local government, business, and civic society, but not disadvantaged groups. The findings suggest that these interventions are primarily for the benefit of local government and not communities and
disadvantaged groups.

Multifunctional NBS for Climate Adaptation Enabled by European or National Legislation or Strategy for NBS (Dimension 4)

This dimension is strongly characterized by a wide range of ecological
domains (such as green roofs, rewilding derelict spaces, and pocket parks)
implemented on a city or street scale. While these interventions aim to
cocreate ecosystem services that could regulate air quality and improve
biodiversity and water quality, opportunities to encourage agency and
help people reconnect with nature are missed. There is also evidence of
participatory and recognitional injustice as actors from civic society or
private companies involved in the governance of NBS exclude citizens and
marginalized groups from institutional processes.

NBS for Deprived Coastal Communities (Dimension 5)

This dimension is strongly influenced by government-led interventions
involving regional governments and nongovernmental organizations. Socioeconomic position is also strongly influential, suggesting that these NBS aim
to regenerate disadvantaged coastal communities by providing equitably
distributed ecosystem services for recreation opportunities, stimulating
growth through tourism, and providing coastal protection. Cities achieve
this through cocreating multifunctional green spaces (such as green corri-
dors, walls, or parking lots), though these interventions lack other types of
NBS that could positively influence health inequality, such as community

gardens or urban parks. To ensure just participatory processes and encourage agency, different modes of public participation include the involvement of citizens in data collection and citizen oversight.

DISCUSSION

This chapter examines how the different characteristics that influence the implementation of NBS have evolved and how they relate to different social and economic determinants that influence poor health in European cities. The findings show that the health and well-being of urban inhabitants has been a key driver for the enactment and deployment of NBS since the early 1990s. However, analysis of the cases retrospectively classified as NBS that were implemented between the 1990s and the 2000s suggests that the distribution of the benefits of NBS that could alleviate the impact of climate change are largely unrelated to SEP, social vulnerability, or mortality except for two dimensions (2 and 3). Furthermore, across each of these dimensions, there is repeated evidence of procedural and recognition injustice as communities, particularly marginal groups, are denied the opportunity to develop capabilities that could help them transition to sustainability. After 2010, the analysis shows that institutional processes between cities and urban inhabitants become more collaborative but are largely inequitably distributed with the exception of two dimensions that are characterized by NBS whose distribution relate to social and health inequality.[40] While the core principles that form the concept advocate fair and equitable distribution of societal benefits of NBS and broad participation, it is evident that inattention to historical and structural injustices and uneven power relations embedded in governance processes has led to the use of NBS as a strategy to stimulate economic growth under the guise of a climate management strategy.[41] Despite this, a small cluster of community-led NBS have emerged that have great potential to deliver just, multifunctional NBS that could help to alleviate the social determinants of ill health and improve public health outcomes of those vulnerable to climate change due to historical and preexisting structural conditions. Supported by investment from local government, grassroot actors adopt inclusive participatory methods from crowd-sourcing funding for citizen management, monitoring, and evaluation of NBS to deliver just transitions to sustainability in their communities.

NBS Implemented between the 1990s and 2000s Create the Issue of Climate Injustice

Retrospective classification of NBS deployed before 2010 reveals that many NBS were used to transform decaying urban infrastructure. Through community–public sector collaboration, large-scale blue-green infrastructure projects were created to increase greenspace for recreation and regulate flooding, especially in coastal areas. These findings are consistent with research by Richard Marshall, who suggested that investment in large-scale blue infrastructure projects was often used by postindustrial cities to transform and rebrand "decaying" cities to attract investment for growth.[42] While these NBS are often collaborative and claim to support improvement in health and well-being through the creation of passive forms of welfare, this study finds that these NBS are mainly driven by place-based marketing or greenwashing to support economic recovery.[43] In doing so, powerful actors deploying these solutions are blind or indifferent to the politics of the maldistribution of environmental benefits of NBS, leading to distributional injustice and increasing the risk of maladaptation.[44] Here inattention to the nexus between social and economic determinants of ill health, the influence of urban conditions, and availability of environmental goods not only results in distributive injustice but also reduces the capacity of those vulnerable to the ill-health effects of urbanization and climate to adapt and risks reinforcing existing inequalities exacerbating their susceptibility to ill health.

A similar pattern can be seen in the third cluster of NBS implemented before 2010 whereby NBS are being used to secure foreign investment to regenerate shrinking cities influenced by the early legacy of the fall of socialism and the deregulation of urban spatial planning.[45] Under the guise of climate urbanism and social justice imperative, powerful actors utilize lucrative funding opportunities associated with green growth and climate adaptation to support the rejuvenation of deteriorated infrastructure.[46] These cities, particularly postsocialist and postindustrial cities, aim to rebrand and transform areas perceived as wastelands (e.g., community gardens) or relics of the socialist era to help attract a talented young and skilled workforce.[47] However, indifference to the politics of maldistribution and inattention to preexisting structural conditions associated with severe deprivation lead to uneven distribution of resources that could improve the upstream determinants of ill health. Participatory parity leads to the exclusion of citizens (including disadvantaged groups) in democratic processes, leading to procedural and recognitional injustice as they are denied their right to be included in processes that could help develop capabilities that could aid in climate adaptation but also improve quality of life.[48]

Issues of Climate Injustice Resolved after 2010?

The findings suggest that following 2010, institutional conditions that influence the governance and management of NBS have been more collaborative and transition from municipal to nongovernmental actors, but exclusion of urban inhabitants' particularly marginal groups continues and is still evident in four of the five dimensions. However, there is also emerging evidence that the distribution of resources that NBS creates, such as different types of blue or green space, is beginning to be related to structural conditions that contribute to poor health, but only two of the five dimensions are designed to improve access and develop capabilities to engage with infrastructure that could improve well-being. Furthermore, there are few NBS specifically designed to meet the needs of the marginal groups, especially the young and elderly, suggesting that issues of intergenerational injustice could be being created by NBS.

Except for the fifth dimension, which adopts a just approach to encourage disadvantaged groups to engage with NBS, the findings suggest that institutional processes continue to deprive these groups of an opportunity to participate in decision-making processes. Influenced by European or national policy or subsidies for NBS, coalitions of private and public sector actors increasingly play a role in the governance and management of large-scale multifunctional blue-green infrastructure projects with an emphasis on technological innovation to transform socioecological systems. Helen Toxopeus and colleagues suggest that hybrid governance strengthens justice outcomes of NBS by explicitly including citizen involvement and participatory governance in NBS.[49] However, these results suggest that while these interventions aim to transform socioecological systems, they fall short of doing so since they lack consideration of adaptation of the socio-technical system; moreover, participatory processes are socio-spatially selective. Consequently, citizens and marginalized groups continue to be largely unrecognized as central actors with different modes of participation bounded by opportunities for coplanning and consultation leading to procedural and recognition injustice.

In contrast to the first two dimensions that characterize NBS after 2010, the presence of a city network or regional partnerships characterize the third-dimension which aims to cocreate multifunctional NBS that aim to reduce the risk of flooding and improve biodiversity and opportunities for food growing for citizens, including disadvantaged groups. However, there is little evidence that these NBS relate to structural conditions that influence ill health, but they are instead related to a higher proportion of citizens with higher education attainment. This could be indicative that investment in NBS is being used to attract a young, highly educated creative class to growing cities.[50] Despite this, nongovernmental actors that govern and manage these interventions aspire to be inclusive by employing methods such as citizen monitoring and task force involvement. However, techoeconomic narratives still dominate, leading to the reproduction of inequalities in the maldistribution of resources that could improve health or aid climate adaptation.

CONCLUSIONS

Comparison of the relationship between different characteristics that influence the implementation of NBS has evolved considerably since the emergence of the concept in the 2000s. Evidence presented in this chapter suggests that actors governing these NBS implemented in the 1990s and 2000s frequently missed opportunities to design interventions that could influence well-being and encourage agency through this infrastructure, and thereby achieve the potential of public health benefits that NBS could provide. Furthermore, repeated exclusion of community and disadvantaged groups from institutional and participatory processes not only led to procedural and recognitional forms of climate injustice but also missed opportunities to distribute resources in areas where they could mediate the adverse effects of social and health inequality and support adaptation to climate change.

Since 2010, NBS are being governed and jointly managed by coalitions of actors to deliver "idealized" benefits to disadvantaged groups, but in reality, they are primarily driven by processes of marketization and regeneration in cities continuing the trend of procedural and recognition-al injustice. One explanation for this is that civic and community actors imagine the delivery of benefits of these interventions through a lens of "politics of possibility."[51] This leads to these actors inadvertently focusing on synthetic appreciation of the urban social-nature dialect. Consequently, engagement with civic society may focus on powerful urban actors at the expense of marginalized individuals or groups, creating issues of distributional injustice.[52] Despite this, there is evidence that more just NBS are beginning to emerge that adopt equitable governance processes. These NBS have great potential to mediate the ill-health effects of climate and urbanization, but in light of the extent of evidence of different forms of climate justice highlighted by this analysis, there needs to be a fundamental shift in the priorities that shape national and local policy instruments that influence the design, governance, management, and monitoring of NBS. Only by doing so will we ensure that the allocation of the benefits of NBS is not only equitable and just but also avoids reproducing structural conditions that can worsen vulnerability to the ill-health effects of climate change and urbanization.

ACKNOWLEDGMENTS

This research has been funded by the European Commission's Horizon 2020 research and innovation program under grant agreement no. 730243 and participating partners in the Naturvation project.

NOTES

1 World Health Organization, "New WHO-IUCN Expert Working Group on Biodiversity, Climate, One Health and Nature-Based Solutions," March 30, 2021, https://www.who.int/news/item/30-03-2021-who-iucn-expert-working-group -biodiversity.

2 Emmanuelle Cohen-Shacham, Gretchen Walters, C. Janzen, and S. Maginnis, eds., *Nature-Based Solutions to Address Global Societal Challenges* (Gland, Switzerland: IUCN, 2016), xiii.

3 Thomas. C. Wild, Freitas Tiago, and Sofie Vandewoestijne, eds., *Nature-Based Solutions: State of the Art in EU-Funded Projects* (Luxembourg City, Luxembourg: Publications Office of the European Union, 2020).

4 Richard Mitchell and Frank Popham, "Greenspace, Urbanity and Health: Relationships in England," *Journal of Epidemiology & Community Health* 61 (2008): 681–83; Richard J. Mitchell, Elizabeth J. Richardson, Niamh K. Shortt, and Jamie R. Pearce, "Neighbourhood Environments and Socioeconomic Inequalities in Mental Well-Being," *American Journal of Preventative Medicine* 49 (2015): 80–84; Jenny J. Roe, Catherine W. Thompson, P. A. Aspinall, et al., "Green Space and Stress: Evidence from Cortisol Measures in Deprived Urban Communities," *International Journal of Environmental Research and Public Health* 10 (2013): 4086–4137.

5 Terry Hartig, Richard. Mitchell, Sperp de Vries, and Howard Frumkin, "Nature and Health," *Annual Review of Public Health* 35 (2014): 207–28; Iana Markevych, Julia Schoierer, Terry Hartig, et al., *Exploring Pathways Linking Greenspace to Health: Theoretical and Methodological Guidance, Environmental Research* 158 (2017): 301–17; Alice. M. Dalton, Andy P. Jones, Stephe. J. Sharp, et al., "Residential Neighbourhood Greenspace Is Associated with Reduced Risk of Incident Diabetes in Older People: A Prospective Cohort Study," *BMC Public Health* 16 (2003): 1171.

6 Matilda Van den Bosch and Åsa O. Sang, "Urban Natural Environments as Nature-Based Solutions for Improved Public Health—A Systematic Review of Reviews," *Environmental Research* 158 (2017): 373–84.

7 Adina Dumitru, Niki Frantzeskaki, and Marcus Collier, "Identifying Principles for the Design of Robust Impact Evaluation Frameworks for Nature-Based Solutions in Cities," *Environmental Science and Policy* 112 (2020): 107–16. On the scarcity of studies, see Magdalena van den Berg, Wanda Wendel-Vos, Mireille van Poppel, et al., "Health Benefits of Green Spaces in the Living Environment: A Systematic Review of Epidemiological Studies," *Urban Forestry & Urban Greening* 14 (2015): 806–16.

8 Cohen-Shacham et al., *Nature-Based Solutions*; Niki Frantzeskaki, "Seven Lessons for Planning Nature-Based Solutions in Cities," *Environmental Science & Policy* 93 (2019) 101–11; Chris M. Raymond, Niki Frantzeskaki, Nadja Kabisch, et al., "A Framework for Assessing and Implementing the Co-benefits of Nature-Based Solutions in Urban Areas," *Environmental Science and Policy* 77 (2017): 15–24.

9 Clair Cooper, Niall Cunningham, and Louise J. Bracken, "Exploring Different Framings of Nature-Based Solutions with Respect to Governance, and Citizen Participation, Beneficiaries, and Quality of Life Outcomes," *Environmental Planning and Policy, Environmental Science & Policy* 150 (2023): 103592.

10 Isabelle Anguelovski, "Tactical Developments for Achieving Just and Sustainable Neighborhoods: The Role of Community-Based Coalitions and Bottom-to-Bottom Networks in Street, Technical, and Funder Activism," *Environment and Planning C: Government and Policy* 33 (2015): 703–25; Riley Dunlap and Robert J. Brulle, eds., *Climate Change and Society: Sociological Perspectives* (Oxford: Oxford University Press, 2015).

11 Pedro H. C. Torres, Danielle de Souza, Sandra Momm, et al., "Just Cities and Nature-Based Solutions in the Global South: A Diagnostic Approach to Move beyond Panaceas in Brazil," *Environmental Policy & Planning* 150 (2023): 24.

12 Madison Powers and Ruth Faden, *Structural Injustice: Power, Advantage, and Human Rights* (New York: Oxford University Press, 2019); Helen Toxopeus, Panagiota Kotsila, Marta Conde, et al., "How 'Just' Is Hybrid Governance of Urban Nature-Based Solutions?," *Cities* 105 (2020): 102839.

13 Peter Newell, Srivastava Shilpi, Lars O. Naess, et al., "Toward Transformative Climate Justice: An Emerging Research Agenda," *Wiley Interdisciplinary Reviews: Climate Change* 12, no. 6 (2021): e733.

14 Newell et al., "Toward Transformative Climate Justice."

15 Martha C. Nussbuam and Amartya Sen, *The Quality of Life* (Oxford: Oxford University Press, 1993).

16 David Schlosberg and Lisette B. Collins, "From Environmental to Climate Justice: Climate Change and the Discourse of Environmental Justice," *Wiley Interdisciplinary Reviews: Climate Change* 5, no. 3 (2014): 359–74; David Schlosberg, *Defining Environmental Justice: Theories, Movements, and Nature* (Oxford: Oxford University Press, 2007); N. Fraser and A. Honneth, *Redistribution or Recognition? A Political-Philosophical Exchange* (London: Verso, 2003).

17 David Schlosberg, *Defining Environmental Justice*; J. A. Rawls, *Theory of Justice* (Oxford: Clarendon Press, 1972); Newell et al., "Toward Transformative Climate Justice."

18 Fraser and Honneth, *Redsitribution or Recognition?*

19 Newell et al., "Toward Transformative Climate Justice"; L. H. Meyer, ed., *Intergenerational Justice* (Oxford: Routledge, 2016).

20 Anthony C. Gatrell and Susan J. Elliott, *Geographies of Health: An Introduction* (Hoboken, NJ: John Wiley & Sons, 2014).

21 On health as a resource, see the World Health Organization, "The World Health Organization Quality of Life Assessment (WHOQOL): Development and General Psychometric Properties," *Social Science & Medicine* 46 (1998): 1569–85. Gerry McCartney, Frank Popham, Robert McMaster, and Adam Cumbers, "Defining Health and Health Inequalities," *Public Health* 172 (2019): 22–30.

22 Gatrell and Elliott, *Geographies of Health*.

23 Katherine E. Smith, Clare Bambra, and Sarah Hill, eds., *Health Inequalities: Critical Perspectives* (Oxford: Oxford University Press, 2016).

24 Margaret Whitehead, "A Typology of Actions to Tackle Social Inequalities in Health," *Journal of Epidemiology & Community Health* 61 (2007): 473–78.

25 Commission on Social Determinants of Health (CSDH), *Closing the Gap in a Generation: Health Equity through Action on the Social Determinants of Health: Final Report* (Geneva: World Health Organization, 2008); Orielle Solar and Alec Irwin, "A Conceptual Framework for Action on the Social Determinants of Health," *Social Determinants of Health Discussion Paper 2 (Policy and Practice)*, WHO Document Production Services, Geneva, Switzerland, 2010; Riley J. Dunlap and Robert J. Brulle, eds., *Climate Change and Society: Sociological Perspectives* (Oxford: Oxford University Press, 2015).

26 Dunlap and Brulle, *Climate Change and Society*.

27 Nadja Kabisch and Matilda van den Bosch, "Urban Green Spaces and the Potential for Health Improvement and Environmental Justice in a Changing Climate," in *Nature-Based Solutions to Climate Change Adaptation in Urban Areas: Theory and Practice of Urban Sustainability Transitions*, ed. Nadja Kabisch, Horst Korn, Jutta Stadler, et al. (Cham, UK: Springer, 2017); Elizabeth A. Richardson, Jamie Pearce, Helena Tunstall, et al., "Particulate Air Pollution and Health Inequalities: A Europe-Wide Ecological Analysis," *International Journal of Health Geographies* 12 (2013): 34.

28 Mitchell and Popham, "Greenspace, Urbanity and Health"; Mitchell et al., "Neighbourhood Environments"; Viniece Jennings, Cassandra J. Gaither, and Richard Schulterbrandt. Gragg, "Promoting Environmental Justice through Urban Green Space Access: A Synopsis," *Environmental Justice* 5 (2012): 2–5.

29 Naturvation, *Urban Nature Atlas*, url: https://www.naturvation.eu, accessed September, 2017. See also https://www.una.city.

30 David R. Hannah and Brenda Lautsch, "Counting in Qualitative Research: Why to Conduct It, When to Avoid It, and When to Closet It," *Journal of Management Inquiry* 20 (2011): 14–22.

31 Naturvation, *Urban Nature Atlas*.

32 Michel Marmot, *Fair Society, Healthy Lives: The Marmot Review: Strategic Review of Health Inequalities in England Post-2010* (2010).

33 Bruna Galobardes, John Lynch, and George D. Smith, "Measuring Socioeconomic Position in Health Research," *British Medical Bulletin* 81 (2007): 21.

34 Susan L. Cutter, Bryan J. Boruff, and W. Lynn Shirley, "Social Vulnerability to Environmental Hazards," *Social Science Quarterly* 84 (2003): 242–61.

35 Kerstin Krellenberg, Florian Koch, and Sigrun Kabisch, "Urban Sustainability Transformations in Lights of Resource Efficiency and Resilient City Concepts," *Current Opinion in Environmental Sustainability* 22 (2016): 51–56.

36 Cutter, Boruff, and Shirley, "Social Vulnerability," 242–61.

37 Brigitte Le Roux and H. Rouanet, *Geometric Data Analysis: From Correspondence Analysis to Structured Data* (Amsterdam: Kluwer Academic Publishers,

2004); R Core Team A. and R Core Team, *A Language and Environment for Statistical Computing: R Foundation for Statistical Computing*, 2022, https://www.R-project.org/.

38 Brigitte Le Roux, Solene Bienaise, and Jean-Luc Durand, *Combinatorial Inference in Geometric Data Analysis* (Boca Raton, FL: CRC Press, 2019).

39 B. Escofier and J. Pagès, "Multiple Factor Analysis (AFMULT package)," *Computational Statistics & Data Analysis* 18 (1994): 121–40.

40 Cohen-Shacham et al., *Nature-Based Solutions.*

41 On historical and structural injustices, see Alasia Nuti, *Injustice and the Reproduction of History: Structural Inequalities, Gender, and Redress* (Cambridge: Cambridge University Press, 2019).

42 Richard Marshall, *Waterfronts in Post-industrial Cities* (London: Spon Press, 2001).

43 On place marketing and greenwashing, see Thorsten Schuetze, Lorenzo Chelleri, and June-Hyung Je, "Measuring Urban Redevelopment Sustainability: Exploring Challenges from Downtown Seoul," *Sustainability* 9, no. 1 (2016): 40.

44 Adriana A. Zuniga-Teran and Andrea K. Gerlak, "Multidisciplinary Approach to Analyzing Questions of Justice Issues in Urban Greenspace," *Sustainability* 11 (2019): 3055; G. Johnson and M. Loralea, eds., *Political Responsibility Refocused: Thinking Justice after Iris Marion Young* (Toronto, Canada: University of Toronto Press, 2013); Iris Marion Young, *Justice and the Politics of Difference* (Princeton, NJ: Princeton University Press, 2011).

45 Jacub Kronenberg, Annegrat Haase, Edyta Łaszkiewicz, Attila Antal, et al., "Environmental Justice in the Context of Urban Green Space Availability, Accessibility, and Attractiveness in Postsocialist Cities," *Cities* 106 (2016): 102862; Dagma Haase, Annegrat Haase, Dieter Rink, and Justus Quanz, "Shrinking Cities and Ecosystem Services: Opportunities, Planning," in *Atlas of Ecosystem Services: Drivers, Risks, and Societal Responses*, ed. M. Schröter, A. Bonn, S. Klotz, et al. (Cham, Switzerland: Springer, 2019).

46 On climate urbanism, see Joshua Long, and Jennifer L. Rice, "From Sustainable Urbanism to Climate Urbanism," *Urban Studies* 56 (2019): 992–1008.

47 Kronenberg et al., "Environmental Justice."

48 On participatory nonparity, see Fraser and Honneth, *Redsitribution or Recognition?*

49 Toxopeus et al., "Hybrid Governance" 102839.

50 Richard Florida, "The Creative Class and Economic Development," *Economic Development Quarterly* 28 (2014): 196–205.

51 Chiara Tornaghi and Barbara Van Dyck, "Research-Informed Gardening Activism—Steering the Public Food and Land Agenda," *Local Environment* 20 (2014): 1247–64.

52 On marginalized individuals or groups, see Nik Heynen, "The Scalar Production of Injustice within the Urban Forest," *Antipode* 35 (2003): 980–98.

Part 2
Implementing and Assessing Interventions

CHAPTER 5

Climate Change and Evolving Roles for Nurses

BARBARA SATTLER

The American Public Health Association was one of the first health associations to host sessions about global warming with nurses among the first presenters.[1] That was more than twenty years ago. Then Hurricane Katrina happened in 2005 and nursing organizations awakened to a new landscape of extreme weather events, not quite yet connecting them to climate change but definitely aware that there were shifts occurring with weather patterns.[2] As scientists became increasingly concerned about the global temperature rising, some health professionals, including some nurses, began to connect the dots and recognized the importance of heeding early warnings.

The Alliance of Nurses for Healthy Environments (ANHE), established in 2008, is an international organization whose mission is to integrate environmental health (including climate change) into nursing education, practice, research, and advocacy, and it was an early leader in nursing and climate change.[3] Active members of ANHE helped to write state, national, and international nursing associations' Climate Change Position Statements, including the Climate Change Position for the American Nurses Association and the International Council of Nurses (ICN).[4] In 2017, in Barcelona, Spain, ICN featured a plenary on climate change followed by a packed breakout room in which nurses from all over the world shared how climate change was "showing up" in their respective countries—droughts, crop failures, changes in fishing patterns, and other worrisome conditions.

An international *Nursing Collaborative on Climate Change and Health* was convened and many major nursing organizations are now starting to incorporate climate change into their policy lexicon. Nurses are active at diverse levels in climate change, including executive sustainability positions in health systems, directing university-level research and educational centers, and advocating in legislatures and regulatory agencies.

This chapter highlights some of the leaders in the nursing and climate change landscape. It is by no means exhaustive in scope but meant to provide both information and inspiration about how individual nurses and their nursing organizations can be agents for positive change at the neighborhood level, within our health care institutions, at the state and national level, and even at the international level. This author has had the pleasure of knowing the nurses whose work is described in this chapter and has had the honor of working side by side with many of them. My approach to this chapter has been experiential rather than academic.

Some history . . .

In the mid-1990s, Holly Shaner was a nurse at New Hampshire's Dartmouth Hitchcock Hospital. At home she was a habitual recycler, even dividing her recycled glass by color. When she was at work, she became increasingly distraught that almost all the hospital waste was being placed in "red bags" whose final disposition was a medical waste incinerator. She knew that red bag waste should have been limited to infectious waste. At the time that Shaner began to lead the charge to reduce "red bag" waste, which was disposed of by incineration, there were thousands of medical waste incinerators in the United States.

Collectively, the medical waste incinerators were responsible for almost 50 percent of the mercury pollution in the air in the United States because mercury was used in so many health care products—think fever thermometers and sphygmomanometers. In addition, the inappropriate incineration of hospital plastics was one of the major air pollution sources of dioxins, a highly toxic family of chemicals known to cause cancer, damage the immune and reproductive systems, affect childhood development, and cause endocrine disruption.[5]

Like so many other nurses, who are compulsive "fix it" people, Shaner launched a successful campaign to have her hospital segregate its waste and ensure that paper, glass, plastic, and potentially infectious and hazardous chemical waste would all wind up in the appropriate waste stream, thus significantly reducing the need for incineration, reducing her hospital's waste costs, and actualizing health professionals' maxim of "first do no harm." Ideally, paper, glass, and plastics should be separated and recycled. Food waste should be composted, and hazardous chemical waste and pharmaceutical waste should be sent to a hazardous waste site.[6]

During the time that Shaner was discovering the impacts of hospital waste, the environmentalist community was also seeing how harmful medical waste incineration was to the air. In 1996, an unprecedented meeting of national environmentalists and health professionals was convened, and a new and lasting collaboration was born called Health Care Without Harm. The nurses in the room included Shaner; Susan Wilburn, who codirected the American Nurses Association's Center for Occupational and Environmental Health (which has since been disbanded); Karen Ballard from the New York State Nurses Association; Charlotte Brody, who went on to codirect the national Health Care Without Harm organization; and Barbara Sattler, who at the time was a professor at the University of Maryland, and the author of this chapter.

Health Care Without Harm's initial and successful campaigns in the United States were to stop medical waste incineration by promoting the use of safer alternative technologies (such as steam sterilization) and to halt the use of mercury-containing devices in health care. The number of medical waste incinerators in the United States was ultimately reduced from almost three thousand to less than one hundred. There is still much work to do in developing countries, where incineration is still both a common way to dispose of waste and an emitter of greenhouse gases. Health Care Without Harm went on to engage hospitals and health care systems, develop sustainable food (farm-to-hospital) programs, promote environmentally preferable purchasing campaigns, and ultimately became involved in climate change and created a Health Care Climate Challenge. The Health Care Climate Challenge, which is international in scope, has over 350 participants representing more than twenty-six thousand hospitals and health centers in forty-three countries and is based on the following three pillars:

- *Mitigation:* Reduce health care's own carbon footprint.
- *Resilience:* Prepare for the impacts of extreme weather and the shifting burden of disease.
- *Leadership:* Educate staff and the public about climate and health and promote policies to protect public health from climate change.

Shanda Demorest is a nurse who works for Practice Green Health, the hospital and health system organization allied with Health Care Without

Harm. Her focus is specifically on climate change and she has developed a Nurses Climate Challenge that has already reached over fifty thousand nurses. She worked closely with the Alliance of Nurses for Healthy Environments to create an associated Nurses Climate Challenge: Schools of Nursing Commitment by which over sixty major nursing schools have made commitments to integrate climate change into their curriculum in order to better prepare our future nursing workforce to understand and respond to climate change and to advocate for climate-healthy policies.[7] Demorest started with nurses who were active in Health Care Without Harm's Nurses Work Group and networked with other nurses all over the country who were participating in climate change–related conferences and/or publishing papers about climate change. She used social media to expand her reach and recruit interested faculty. This set the stage for the cocreated curriculum materials that are now being used in nursing schools.[8]

Demorest has worked closely with another nursing climate champion, Elizabeth Schenk, who is the executive director of Environmental Stewardship at Providence Health Care System, which is made up of fifty-two hospitals and over one thousand clinics. Schenk has been a national pioneer in promoting hospital-based climate solutions. She is also active at her local level, in her community of Missoula, Montana, where she is working on grassroots approaches to climate mitigation and adaptation through her town's Climate Ready Missoula.[9] Schenk also developed the Climate, Health, and Nursing Tool (CHANT), a ten-minute survey asking respondents about awareness, motivation, and behaviors related to climate change and health that has been continuously tracking data since 2017.

NURSES AND THE FIGHT AGAINST PLASTICS

The other thing that we learned early in the Health Care Without Harm campaign days was about plastics. First of all, there are a lot of plastics in health care—we are awash in them. Secondly, many of them are polyvinyl chloride plastics and, during the process of manufacturing PVC plastics, dioxins are released into the air. Dioxins are a family of chemicals that are highly toxic. They are one of the most carcinogenic chemicals known. We learned that by burning PVC plastics, dioxins are also emitted. So our success in limiting medical waste incineration meant that we were limiting the creation of dioxins at the end of the plastics' lifecycle, which was good, but that did not stop the dioxins from being emitted during the manufacturing process.

The third problem is that plastics don't ever really go "away." Globally, our beaches and oceans are awash in them and they are winding up in our air, drinking water, and food. In the environment, they disintegrate into smaller and smaller particles (microplastics) until they are so small that they can enter the human body and have been discovered in human lungs, livers, spleens, kidneys, and brains as well as in placentas and newborns' meconiums. There is growing evidence that plastics may be interfering with cellular, tissue, and organ function.[10]

The fourth and most important thing we know about plastic is how it's related to climate change. Plastics are derived from fossil fuels, which we know to be one of the prime culprits with regard to climate change because of the emission of greenhouse gases. In the 1970s when I was starting out as a nurse, bedpans, emesis basins, and urinals were all made out of stainless steel. After use, bedpans were cleaned and sterilized in a device colloquially called a "hopper," which was a bedpan cleaner that was plumbed into the walls outside of patient areas throughout the hospital. After a bedpan completed its cleaning and sterilization process, it could be used by the same or even another patient. Central supply cleaned the emesis basins and urinals. No patient was sent home with these supplies. These stainless supplies lasted forever—they were sustainable.

Today, part of the ubiquitous presence of plastics in our hospitals is in the form of plastic bedpans, urinals, and emesis basins. After patient use, they are either tossed out or given to the patient to take home. Once home, they are tossed out. They are not recyclable. While some of the plastics in the health care setting will be hard to avoid, returning to sustainable, reusable products whenever possible is part of the shift that we must make. Nurses can help to carry the torch on this.

There is even a fifth problem with plastics that can cause direct health effects. The most common "plasticizer" additive, DEHP, is used to make PVC plastics, such as IV bags and tubing, more malleable. However, it is both toxic (it is an endocrine-disrupting chemical) and can leach out of the plastic and into the IV fluid such that the patient will be unwittingly receiving a dose of DEHP along with their requisite fluids and medications. The National Toxicology Program issued an alert with a recommendation that

DEHP-free IV bags and tubing be used in neonatal intensive care units, particularly when caring for baby boys, where it has the potential to be harmful to their reproductive organs.[11]

All five of the problems with plastics are being described to illustrate that our climate and environmental health issues are intimately related and are simultaneously creating extraordinary health risks. It is also true that by reducing the use of harmful, fossil fuel–derived products we can address a number of ecological and human health risks and at the same time reduce a primary contributor to climate change.

CREATING CHANGE AT THE LOCAL LEVEL

There are a great many nurses who are working on the local level. Kenna Lee, a hospice nurse in California, a location that has been repeatedly affected by devastating wildfires, is actively involved in creating and realizing the four goals of her small town's climate action plan:

- Reduce citywide greenhouse gas emissions to net zero by 2030.
- Sequester carbon from the atmosphere using nature-based solutions.
- Prepare for climate impacts that cannot be avoided.
- Center equity and community engagement in the city's climate actions.

In addition to being involved in local climate action plans, nurses are joining and managing Medical Reserve Corps and Red Cross nursing units by training and planning and, when necessary, deploying into communities where fires, floods, and extreme heat events place whole communities at risk. They help to identify the needs of our most vulnerable community members; what they have learned is that there are many distinct vulnerabilities, including the very young and the very old, people with physical disabilities and cognitive disabilities, people in addiction, pregnant women, women who have just had babies, and many more populations. Nurses in climate disasters learn a new lexicon of vulnerabilities and often see how race, poverty, and immigration status multiply problems in the immediate disaster and hinder recovery.

At a neighborhood level, Lisa Hartmayer, who was also active in her hospital's climate change programs, decided that she wanted to help prepare people on the street where she lived. She was guided by a wonderful program called the Cool Block—The Cool City Challenge. This program literally operates at the block level, engaging neighbors in building community as they learn to assess their personal climate impacts. One participant described how Cool Block "has changed things in our neighborhood. Not only have we lowered our carbon footprints and prepared for any emergency, but we actually talk to each other, wave to each other, and get together regularly even after the program was over."[12] Mapping our neighborhoods, knowing where vulnerable populations live, what languages they may speak, and other important information can be a lifesaver when disasters hit. Given that climate change is, in fact, on us, building local community resiliency is an important step in limiting its impacts on a local level.

Shortly after Health Care Without Harm was created, a Nurses Work Group was established that catalyzed nurses' efforts around the country, often resulting in the formation of hospital-based "green teams." Sometimes these green teams were just nurses and other times they were multidisciplinary formations. In the early 2000s, Lisa Hartmayer, who worked at the University of California San Francisco Medical Center, helped to convene a green team and started with a daylong program on climate change. She worked on a range of issues and spurred on one of the first hospital-based "Meatless Mondays" traditions based on the fact that meat animal production creates considerably more greenhouse gases than growing vegetables (and that plant-based diets are healthier). Hartmayer helped make this dietary service change by working with the hospital's food services director, who came to her climate program and was sufficiently inspired to help make the changes.

Around the same time as Hartmayer's activities on the West Coast, on the East Coast, Denise Choiniere was enrolled in an environmental health nursing graduate program. A week after she listened to a lecture on heavy metal toxicity, she was at her job in the telemetry unit putting a Holter monitor on a patient. As was protocol, she put a new battery in the monitor and then, as she was about to toss the used battery in a red bag (the accepted practice at the time), she realized that the lithium in the battery (and any other heavy metals that are in batteries) would be incinerated and would be released into the air from the incinerator smokestack. This outcome was no longer acceptable to her.

Choiniere got busy finding a company that would recycle small batteries and initiated a battery recycling program on her unit. When she described this success in my class, where she had learned about metal toxicity, and I asked her how many small batteries her workplace, the 650-bed University of Maryland Medical Center, purchased every year, she said she didn't know. But she returned to class the following week with the answer: ninety-seven thousand. Ninety-seven thousand small batteries were incinerated annually at her hospital, which we both found appalling. With appreciation from the administration, Choiniere went back to the battery recycling company and helped get small battery containers installed all over the hospital. These containers were then collected by the battery recycling company and made into new batteries. This was the beginning of a sustainability culture shift at her hospital and it also changed the trajectory of Choiniere's career.[13]

FROM THE INDIVIDUAL TO THE SYSTEMIC

Choiniere had heard of the idea of green teams and asked the director of nursing at her hospital if she could help start one. The director was pessimistic about there being interest in such a thing among her nurses, but Denise was persistent. Before convening a meeting, Choiniere, being the good organizer that she was, went around the hospital talking to nurses about a prospective green team and finding out about their interest. Convening nurses is harder than herding cats, given that their 24-7 schedules are extremely challenging. Nevertheless, over sixty nurses showed up at Choiniere's inaugural meeting of the green team. Her director, who peeked her head in during the meeting, assumed she was in the wrong room because she could not imagine sixty nurses being interested in sustainability and environmental health. She was wrong. Many nurses are interested, but it usually takes a leader (or two) to help build the scaffolding on which this work can be supported, whether it's in a hospital or a professional nursing organization.

The hospital recognized that Choiniere was one of those leaders who could help build the scaffolding. They asked her to leave bedside nursing and do the sustainability work full-time. Addressing hospital waste has been a seminal activity for many nurses who champion sustainability. Red bag waste is one of the most expensive waste streams for hospitals (only potentially infectious waste should be placed in red bags); most of the rest of hospital waste that cannot be recycled should be placed in "regular" waste containers that are handled much the same as our household trash. Choiniere and Shaner both helped save their hospitals considerable costs, which is prized by the hospital leadership in the C-suite. There was also a developing cachet in being a "green" hospital, as all things "green" were being marketed and valued. Ultimately, Choiniere became the director of materials management and sustainability for her hospital and responsible for, among other things, all hospital purchasing decisions.

When green teams were first being convened in the United States, in hospitals and universities and other institutions, the focus was on reducing waste and energy use because of their obvious effect on the environment, all of which still holds true. But these days, the lens for sustainability efforts and the chief motivation is also mitigating climate change and reducing greenhouse gases. The health care sector is responsible for 10 percent of US greenhouse gases. Many hospitals have installed solar panels, switched to electric vehicles, and made efforts to reduce their overall energy usage. The national organization of health care engineers, made up of those people who are primarily responsible for the hospital's physical plant, created an Energy to Care Program specifically to address climate change. The program started with 239 facilities in 2014 and now has more than 3,650. Since the inception of this program hospitals have collectively saved more than $550 million.[14]

NURSES FOCUS ON FOOD PRODUCTION

In addition to energy decisions, food choices and agricultural practices have a huge impact on human and ecological health, and this was another area in which Choiniere was an early pioneer. Her university hospital employed and served many people who lived in food deserts. She knew that the average food product travels fifteen hundred miles from farm to table, with its associated fifteen hundred miles of tailpipe pollution and greenhouse gases. Choiniere was able to start a farmer's market in the square directly across from her hospital by which employees, community members, and patients' families gained access to fresh and local produce. Thereafter, several hospitals in Baltimore followed suit.

Another pioneer is pediatric nurse Atiya Wells, who saw a vacant lot tucked in between brick rowhouses in Baltimore and dreamed of a community garden. Atiya has cultivated an impressive presence now known as BLISS Meadows (BLISS standing for Baltimore Living in Sustainable Simplicity). Atiya is truly a visionary—someone who turned her dream into reality and is cultivating minds along with cultivating plants. She has incorporated educational elements about African Americans' agricultural histories into her work in BLISS Meadow's predominantly African American community, created a safe place for children to experience nature, and provided the community with healthy food at the local market. BLISS Meadows expanded from its initial two acres to ten and is a positive force in the neighborhood. It is guided by a goal of climate justice and equitable access to nature (and food) and integrates a set of objectives known as its "five pillars": animal husbandry, conservation, community greenspace, environmental education, and food access.[15]

While Atiya Wells was growing food, nurses in California were concerned about food production on a larger scale and about the hundreds of thousands of California farmworkers who were growing food for the whole country. Nurses who are members of the Food, Agriculture, and Farmworker Committee of the California Nurses for Environmental Health and Justice have three areas of focus, all of which are climate-change related. One of these is addressing farmworker health and safety, specifically during extreme weather events and fires. By allying themselves with farmworker advocacy groups they have been able to use their trusted nurses' voices to promote better occupational health standards for one of the most vulnerable communities in the United States. Farmworkers fall short on every metric for social determinants of health, making them extraordinarily vulnerable to the effects of climate change. And, notably, farmworkers are responsible for almost everything we eat three times a day.

While concern about farmworkers is a common thread, each of the initial members of the Food, Agriculture, and Farmworker Committee came with a particular history and interest. Mechelle Perea-Ryan, from California's Central Valley, is a nurse practitioner at a community clinic where she observes many of the agricultural risks that are experienced by farmworkers.

She has become particularly involved in a western state's initiative to develop better and more standardized protective regulations for farmworkers during extreme heat events and when wildfire smoke is at dangerous levels. Perea-Ryan helped the committee to better understand this issue and to develop advocacy strategies that included writing a set of recommendations to the federal Occupational Safety and Health Agency. She also trained a multidisciplinary team of primary care providers who work at a rural network of clinics about farmworkers' specific vulnerabilities to extreme heat and smoke exposure, now common climate-related events.[16]

Meghan Adelman, another nurse member of the committee, has had a long history of engagement with nutrition, including graduate work in nutrition and public health. She and her husband are also owners of a well-known health food company. As a member of the committee, Adelman has helped to organize multiple educational programs for nurses throughout California on the health benefits of plant-based diets for both humans and the planet. Promoting plant-based diets is contextualized in the programs while teaching nurses about healthy soils, the unhealthy impacts of pesticides, and the relationship between climate change and agricultural practices and food choices. These programs have included a preconference workshop for school nurses during their annual conference and a daylong program at an organic farm with a farm tour as part of the educational program.[17]

Erika Alfaro is also a leader in the committee and president of her local National Association of Hispanic Nurses chapter. She is a public health nurse in a community that includes a robust agricultural sector. In addition to specific concerns about farmworkers' health and safety, Alfaro is concerned about maternal and child health, adequate housing, and immigration issues in farming communities. Together with Katie Bolin, another nurse and committee member, and the Center for Farmworker Families, Alfaro helped to organize a farmworker tour for nurses to visit with farmworkers in the Salinas Valley. Nurses met with women farmworkers at a strawberry farm to talk about occupational and immigration challenges and then were invited to visit their government-provided farmworker housing.

There were many important insights from this tour, including the fact that farmworker families were forced to leave in November and were not allowed to return until April, even though the housing would remain vacant during that interim period. Particularly for those who were undocumented, interim housing sometimes meant living in vehicles and garages. The revelation of this farmworker challenge motivated the nurses to seek policy solutions to farmworker housing difficulties. They are now working with food and farm coalitions to address farmworker housing along with a comprehensive policy that includes climate-healthy agriculture.[18]

The same nursing committee is working with the California Climate and Agriculture Network (CalCAN) promoting climate-resilient approaches to agriculture. California produces 46 percent of the fruits and nuts in the United States, and 25 percent of its produce. It's the biggest agriculture state, and the climate and health impacts of the predominantly industrial model of producing food are contributing to air pollution, depleting potable water aquifers, and releasing carbon into the atmosphere (from tilling the soil for annual crops). In 2023, CalCAN convened the Food and Farm Resilience Coalition for a state legislative initiative that has four pillars: sustainable agriculture, farmworker health and well-being, climate-friendly access to food, and regional food system infrastructure (that will better support small and middle-sized farms). Barbara Sattler, the author of this chapter, was asked to represent the California Nurses for Environmental Health and Justice on the steering committee for the coalition, thus bringing a nursing voice to the effort and keeping the nurses' committee apprised of and engaged in legislative advocacy opportunities.[19]

Nurses have begun to realize that they can continue to take care of patients with asthma and other lung diseases and help communities wrestle with the agricultural sector for access to clean water, but they can also start to understand and advocate for more sustainable agricultural practices that will keep our patients and communities healthy. By doing daylong workshops around the state, the California Nurses for Environmental Health and Justice has helped to develop nurses' knowledge and advocacy abilities to promote farmworker health and safety, as well as climate-safe and sustainable agriculture.

All of the members of the Food, Agriculture and Farmworker Committee are interested in pesticides, and Vanessa Forsythe, a retired school nurse, is representing nurses within another statewide coalition, Californians for Pesticide Reform. Pesticides constitute a category of chemicals that are designed to harm a living organism, such as a plant, an insect, or a microbe. Many pesticides are made with petrochemicals—meaning that many of the chemicals that make up pesticides, including the inerts, are derived from fossil fuels. And, regrettably, many pesticides can be harmful to humans; some are what we call "forever chemicals," because they stay intact, do not break down, and continue to be toxic whether they are in the air, water, soil, our food, or our bodies.

The range of potential human health risks from the wide array of commonly used agricultural pesticides is expansive, including neurological disorders such as autism, reproductive health risks, cancers, and even mental health diagnoses such as depression. Reducing pesticide use will serve multiple functions, all of which will improve human, ecological, and climate health. Nurses are adding their voices to public health-focused campaigns to reduce agricultural pesticide use and to improve "right to know" laws so that individuals and communities have better access to information about when and where agricultural pesticides are being used and their associated harmful risks. Fosythe is also active in a group called Safe Agriculture, Safe Schools that is specifically calling for notification when harmful agricultural chemicals are being sprayed in fields that are adjacent to schools and daycare centers.

NURSES FOCUS ON FOSSIL FUELS

During an annual leadership retreat of the Alliance of Nurses for Healthy Environments a handful of nurses started to talk about a new activity in their communities in New York, Maryland, Ohio, and Pennsylvania. They had bits and pieces of information but didn't have the big picture. Quietly, the gas and oil industry was beginning to buy and lease land in areas where there was shale under the ground, from which natural gas could be extracted, a practice known as unconventional hydraulic fracking, later simply termed "fracking." These curious nurses, including Ruth McDermott-Levy from Pennsylvania, Kathy Curtis from New York, and Sattler (the author of this chapter) convened an ANHE Fracking Committee that worked for years on a wide range of educational programs and policy initiatives. In New York and Maryland, largely because of public health concerns, fracking was banned before it got started, but Ohio and western Pennsylvania became hotbeds for fracking. In addition to working on state policies, McDermott-

Levy helped to write the position statement for the Pennsylvania Nurses Association calling for a ban on fracking that was later adopted by the American Nurses Association.[20]

Lisa Campbell, a public health nurse who worked in a poor Latino community in Texas, discovered that oil companies were sending their waste products to these poor communities, where they were disposing of them by drilling holes and pouring the toxic waste liquids down into the earth. Campbell, who was also affiliated with a university, worked with university faculty from the law school to begin to address this environmental injustice. They advocated for policies to protect communities from becoming dumpsites for the very powerful oil industry in Texas.

Many nurses continue to work in communities that are plagued by human and ecological effects of fracking. And they continue to work to prevent more pipelines from being developed and yet another outgrowth of natural gas development, "ethane crackers," plants that convert the ethane from natural gas into plastic products. Nurses are working on the whole life cycle of gas and oil, including helping to reduce plastic use in health care.

ADVOCACY AND ACTIVISM IN NURSING

Nurses' unions have been visibly active in addressing climate change and its many contributors. Historically, members of the California Nurses Association turned out in droves to demonstrate against "big oil" during the Keystone XL Pipeline controversy; oppose crude oil trains from coming through environmental justice communities in Oakland, California; and bear witness at the Standing Rock Sioux Reservation in North Dakota.[21] They have been active proponents of the Green New Deal, a comprehensive climate change proposal that calls for an end to our fossil fuel economy, a national investment in renewable energy, and good union jobs in the new clean-energy economy.

Other nursing unions, such as Service Employees International Union's Nurse Alliance, have been equally engaged and are providing workshops for nurses in California on how to respond to local climate-related disasters, including how to set up medical shelters when people with injuries and illnesses are displaced because of floods and fires. Through their unions, many nurses have been involved in disaster response both where they live and in other parts of the country and world where emergency services are needed. Some nursing unions have sent contingencies to the international scientific meeting on climate change known as the Conference of Parties, often referred to by its acronym COP, during which time they have declared that we are in a climate emergency and in need of swift and comprehensive climate smart policies and practices by all nations of the world.

Nurses are involved in campaigns to get their hospitals and health systems to divest their investment funds away from fossil fuel corporations, arguing that if our first motto is "do no harm," we cannot be supporting corporate activities that are harmful to people and the planet. Their stand is an important signal to the fossil fuel industry; however, many hospitals still have their funds in fossil fuels. On this and many issues, we still have a long way to go.

Laura Anderko, while a nursing professor at Georgetown University, was in 2013 recognized by the Obama White House as a "Champion of Change" for her tireless work in educating nurses and other health professionals about the health risks associated with climate change, with a particular emphasis on children. Anderko was the first nurse in the country to direct a federally funded Pediatric Environmental Health Specialty Unit in the mid-Atlantic region. Focusing on prevention, she advocated in partnership with organizations such as the American Lung Association to promote an understanding of "health in all policy," including how the Clean Air Act affects climate change and the associated health impacts. Anderko's teaching and advocacy recognized the need for improvement in the social determinants of health and the disproportionate race-related climate impacts in communities of color.[22]

At the national level nurses have been involved in many policy initiatives. With Katie Huffling as the executive director, the Alliance of Nurses for Healthy Environments has taken the lead on most of the national nursing climate-related efforts. They have organized lobby days, including flying in nurses from all over the country. They have worked diligently on climate-related regulations, including methane regulations (methane being a powerful greenhouse gas). The Alliance also organized a White House meeting with president Barack Obama to talk about the ways in which nurses could be both involved in and supported by federal climate change initiatives. The Alliance also helped mobilize nurses to join the

"Fire Drill Fridays" in Washington, DC. These events were organized by Jane Fonda, a well-known actress and climate activist. During these events, nurses marched in the streets, and Annabell Castro Thompson, who was then the president of the National Association of Hispanic Nurses, spoke on the podium with Fonda about climate justice. Several nurses, including this author, joined Fonda in a sit-in in the Senate building, which constituted committing civil disobedience, and we, along with Fonda, were consequently arrested.

Nurses across the country and the world have been involved in climate change actions and advocacy. As noted, there is much work left to do, and clearly an urgency in getting it done. Nurses have not yet realized their full power to press for the changes that need to occur. Only a small percent of nurses actually belong to any nursing organization. This lack of participation diminishes our collective potential and our organizations' abilities. First, nursing organizations that have robust memberships can afford to hire professional policy and advocacy staff, which is essential when working with state or federal legislation and regulations.[23] It is membership dues that help to staff nursing organizations. Second, if nurses do not belong to an organization, they will have less power to advocate for the kinds of policy changes we urgently need. Faculty must help prelicensure nurses to understand that our "trusted voices" have value only if we use them, and if we use them strategically they will have even more value.

Working individually, nurses have accomplished remarkable things; working with the full power of our nursing organizations and coalitions of nursing organizations will help nurses to exert more force in the policy arena. Now is the time to work in our neighborhoods, in our cities and counties, in our states and capitals, and at the international level. If we work strategically with our efforts, we can be an awesome force for climate and human health.

In the realm of climate change, it is easy to become the bearer of depressing news. The nurses whose efforts have been described in this chapter recognize the dire situation that we are in but have remained buoyant and hopeful. They have learned that focusing on solutions and finding your "tribe," the people who share your values, can help build one's resilience in this sometimes difficult work. They have been visionary and have created work that other nurses have wanted to join in. They are the nursing leaders we have been waiting for.

NOTES

1   In 1981, the American Public Health Association's national conference had as its theme "Energy, Health, and the Environment." In 1993, the theme was "Building Healthy Environments: Physical, Economic, Social and Political Challenges," accessed April 24, 2023, https://www.apha.org//media/Files/PDF/meetings/Annual/Past_Annual_Meeting_Locations_Dates.ashx.

2   National Nurses United, "A Decade after Katrina, RNs Reflect on What We've Learned and Refused to Learn," National Nurses United, August 26, 2015, https://www.nationalnursesunited.org/blog/decade-after-katrina-rns-reflect-what-weaeutmve-learned-and-refuse-learn.

3   Alliance of Nurses for Healthy Environments, https://envirn.org/about/.

4   International Council of Nurses Climate Change position statement, "Nurses, Climate Change, and Health," accessed July 3, 2023, chrome extension://efaidnbmnnnibpcajpcglclefindmkaj/https://www.icn.ch/sites/default/files/inline-files/ICN%20PS%20Nurses%252c%20climate%20change%20and%20health%20FINAL%20.pdf.

5   For problems with medical waste incineration, see Healthcare Environmental Resource Center, "Pollution Control and Compliance Assistance Information for the Healthcare Industry," accessed July 3, 2023, https://www.hercenter.org/facilitiesandgrounds/incinerators.php.

6   Ian McMillan, "Global Warning: News Analysis," *Nursing Standard* 12, no. 52 (September 16–22, 1998): 16.

7   Cara Cook, Shanda Demorest, and Elizabeth Schenk, "Nurses and Climate Action," *American Journal of Nursing* 119, no. 4 (April 2019): 54–60.

8   Shanda L. Demorest and Teddie M. Potter, "Climate Change and Health: An Interdisciplinary Exemplar," https://open.lib.umn.edu/changeagents/chapter/climate-change-and-health/, accessed April 25, 2023; see Nurses Climate Challenge website at https://nursesclimatechallenge.org/school-of-nursing-commitment.

9   Elizabeth Schenk's 2013 PhD dissertation is titled "Development of the Nurses' Environmental Awareness Tool." See Schenk's survey, accessed July 3, 2023, https://envirn.org/nurses-climate-survey/.

10 Anabel González-Acedo et al., "Evidence from In-Vitro and In-Vivo Studies on the Potential Health Repercussions of Micro- and Nanoplastics," *Chemosphere* 280 (October 2021): https://doi.org/10.1016/j.chemosphere.2021.130826.

11 National Toxicology Program's report by the Center for Evaluation of Risks to Human Reproduction, 2006, accessed July 3, 2023, https://ntp.niehs.nih.gov/ntp/ohat/phthalates/dehp/dehp-monograph.pdf.

12 Private conversation with Lisa Hartmayer.

13 For an interview with Denise Choiniere, see Barbara Sattler, "The Greening of a Major Medical Center," *American Journal of Nursing* 111, no. 4 (April 2011): 60–62, doi:10.1097/01.NAJ.0000396559.09620.94.

14 See the announcement of the American Society for Health Care Engineering's 2022 Energy Star Partner award, accessed April 25, 2023, https://www.ashe.org /ashe-receives-2022-energy-star-partner-year-sustained-excellence-award.

15 Timothy B. Wheeler, "Baltimore Naturalist Transforms Neglected Land into BLISS," *Bay Journal,* November 9, 2020, https://www.bayjournal.com/ news/people/baltimore-naturalist-transforms-neglected-land-into-bliss/ article_6ce1ac02-2294-11eb-ab46-a38b1772ed3d.html.

16 Mechelle Perea-Ryan is a professor of Nursing at California State University, Stanislaus. She is part of the leadership of the California Nurses for Environmental Health and Justice. Listen to Perea-Ryan on the webinar "Farmworker Exposure to Extreme Heat and the Importance of Cancer Screenings," June 24, 2021, https://www.calnursesforehj.org/past-recordings/.

17 Meghan Adelman, "The Healing Power of a Whole Foods, Plant-Based Diet: A Nutrition Prescription for Cancer," University of San Francisco Master's Projects and Capstones, 2017, accessed July 3, 2023, https://repository.usfca.edu/cgi/view content.cgi?article=1536&context=capstone.

18 See Alfaro's bio at CA Nurses for Environmental Health and Justice, https://www .calnursesforehj.org/leadership/.

19 See the CalCAN website at https://calclimateag.org/. The climate focus of the organization is boldly featured.

20 Ruth McDermott-Levy, Nina Kaktins, and Barbara Sattler, "Fracking, the Environment, and Health," *American Journal of Nursing* 113, no. 6 (June 2013): 45–51; see also Ruth McDermott-Levy and Victoria Garcia, "Health Concern of Northeastern Pennsylvania Residents Living in an Unconventional Oil and Gas Development Development County," *Public Health Nursing* 33, no. 6.

21 National Nurses United, press release, "Nurses: Keystone XL Pipeline Still Significant Threat to Public Health and Climate Change," March 24, 2017, https:// www.nationalnursesunited.org/press/nurses-keystone-xl-pipeline-still-signif icant-threat-public-health-and-climate-change.

22 Anderko's personal website, the "Mid-Atlantic Center for Children's Health and the Environment," can be accessed at https://www1.villanova.edu/university /nursing/macche.html.

23 For example, the California Nurses for Environmental Health and Justice Leadership Council is made up of nurses from around the state who meet biweekly to help coordinate nurses' environmental health engagement in California. See accessed July 3, 2023, https://www.calnursesforehj.org/leadership/.

CHAPTER 6

Maya Agriculture, Green Land Grabs, and Climate Justice in Quintana Roo, Mexico

JOSÉ MARTÍNEZ-REYES AND CAMILLE C. MARTINEZ

In recent decades, the Máasewal Maya of central Quintana Roo, Mexico, and the forest and multispecies beings that they have co-lived with for generations are experiencing profound transformations. Tourism expansion, mega development projects, changes in land tenure, privatization of communal lands (*ejidos*), and increasing periods of drought and extreme rain associated with climate change are putting enormous pressure on the Máasewal's ability to continue traditional gardens that have been the quintessential activity to provide food security and diverse nutrition for their communities. Traditional gardens have enabled the Maya to survive and thrive since the collapse of the Classic Period (c. 250–900 CE).[1] However, climate change is exacerbating a decline in traditional agriculture that has important implications for climate justice and health, including a growing dependence on a processed, less-nutritious diet, and the health consequences associated with food insecurity. Thomas L. Leatherman, Alan H. Goodman, and J. Tobias Stillman documented an increase in chronic diseases such as diabetes and obesity among the Maya in the Yucatán Peninsula due to changes in diet.[2] Moreover, in addition to climate change challenges there are challenges from biodiversity conservation. How do the Maya explain the changes in food security? How do they adapt to these new circumstances that challenge their relationship to the natural world?

Based on interviews conducted in the summers of 2015 and 2018 with Maya farmers from the villages of X-Hazil and Tres Reyes in the Yucatán Peninsula and ethnographic observations, this chapter focuses on Maya cultural, ecological, and religious explanations of the current climate crisis and the crisis of traditional agriculture, which are linked to colonial and postcolonial history. The challenges to the Maya cultural practices resulting from land privatization and significant climate alterations are treated here as fundamental. These challenges must be addressed if the Maya have any hope of achieving climate justice and improving their health and nutritional well-being. Next, we document communal responses to climate injustice. The two villages engaged in measures to prevent the expansion of green land grabbing and reinforced their traditional ecological knowledge and adaptive strategies to maintain their moral ecology with the forest, enabling them to negotiate the unfolding events that are causing local leaders to question what lies ahead for the future of the Maya's forest, health, and well-being.

In this chapter, we set out to explain the current predicament of the Máasewal Maya's traditional farming, showing that inequality and climate change are intricately linked to health and well-being. We begin with the historical perspective of the events that explain who the Maya are and the kinds of ethnic conflicts they've had to endure to protect their culture, land, and resources. The second section describes the rise of tourism, particularly the building of Cancún, and how this event started a boom that transformed Maya villages and affected public health by altering the diets of a significant part of the local population. The third section shows how a scheme to privatize nature with the excuse of protecting land as a carbon-offsetting mechanism threatens land tenure and climate justice for the Maya. Finally, we discuss the moral ecology of the milpa and how Maya farmers explain and adapt to climate change to avoid further injustices.

HISTORICAL PERSPECTIVE: A HISTORY OF CONFLICT

To provide a comprehensive examination of climate change, public health, and inequality, historical understandings of race, ethnic, and class conflict must take center stage. The Maya who dwell in today's Zona Maya of central Quintana Roo are, for the most part, descendants of the rebels who took up arms against Yucatec Whites and Ladinos (people of mixed White and Indigenous descent) in one of the most important events in Mexican history, the Caste War of Yucatán, which officially lasted from 1847 to 1901.[3] This event has been regarded as "the most successful Indian revolt in New World history" because of the Máasewal Maya's success in establishing complete autonomy from the Yucatec and Mexican governments until 1901, when Mexican troops arrived in the region's political and religious capital, known as Noj Kaj Santa Cruz Balam Ná to the Maya, or Chan Santa Cruz to Mexicans.[4]

The war was over Mexico's attempt to control and subdue the Maya to exploit the zone's rich natural resources, as well as to expand and consolidate its frontier toward Belize. Once Mexicans regained control of the region, timber and *chicle* (gum) became important commodities for the expansion of capitalism.[5] Yucatecans also declared that the export of these products would bring progress and development to a region otherwise labeled as "backward" and "barbaric."[6]

The Máasewal had a reputation of being fierce warriors that kept Mexican capitalist exploration and extraction companies away for many decades with a few exceptions. They allowed the extraction of timber to the British from Belize to secure then necessary weapons and supplies to pay for the war effort.[7] However, after the war, the Mexican government started a pacification campaign that included the establishment of Spanish-language schools for Maya children. Additionally, a boom in the chicle industry opened the Maya Zone to forest exploitation by outsiders. Moreover, the results of the Mexican Revolution began to be felt in the region after 1934, when president Lázaro Cárdenas implemented the agrarian reform in Quintana Roo. New *ejidos* (communal lands with restrictions on sales) were seen as one of the most radical measures of social justice because they gave land titles to landless peasants. The Mexican Revolution's land redistribution program has been one of the most profound in Latin America.[8] Redistributions continued to be organized in Quintana Roo throughout the 1950s. This had mixed results among the Mexican population but effectively usurped communal lands from the Maya. Unlike a great number of places in Mexico where issues of land tenure were a major problem and the root of inequality, the Máasewal had occupied and engaged the Eastern peninsula and were not bound to *ejidos*. Ultimately, a portion of the region was designated as *ejido* land, restricting the Máasawal's land tenure rather than enlarging it. However, the designation made them legal landowners in the eyes of the state, which would ward off land developers and Ladinos in the following years. Years later, mass development projects, such as the creation of Cancún and the expansion of the tourism industry, as well as conservation projects newly threatened Maya land tenure.

THE HEALTH RISKS OF THE TOURIST AND BIODIVERSITY CONSERVATION PANACEAS

In the 1970s, Mexico promoted tourism as a tool for economic growth, particularly in Quintana Roo.[9] The creation of Cancún spearheaded this growth. It turned an isolated fishing village into a massive hotel complex. For the past thirty years, Quintana Roo has been one of the fastest-growing states in Mexico, economically and demographically. The population doubled during the period from 1990 to 2000, due to migration from other Mexican states. This massive tourism industry profoundly transformed the ninety kilometers of coast between Cancún and Tulum, creating new tourist spaces along virtually every beach. The coastal area is referred to as the Riviera Maya by the tourism industry for promotional purposes. A plethora of hotels have been built, from megaresorts to small-scale "eco-hotels."

Tourism has had enormous repercussions on the Maya communities of the Yucatán. Many Maya men and women were recruited to work in the lowest-paying jobs, primarily in construction and service jobs in the tourist towns.[10] Back in their respective communities, farmers eventually abandoned their milpas, which prompted changes in the quality of their nutrition and in their relationship with the environment. Dietary surveys conducted in Yucatán and Quintana Roo showed a growth in consumption of "junk foods" and drinks.[11] Results from long-term dietary surveys in Yalcobá, Yucatán, and Cobá, Quintan Roo showed what is referred to as the "double-edged sword of malnutrition": childhood undernutrition and adult obesity.[12] While these transformations were taking place, biodiversity conservation also began to significantly alter the landscape, especially after the creation of the Sian Ka'an Biosphere Reserve in 1986.[13] In principle, nature reserves are a positive response to the rapid decline of biodiversity by protecting areas rich in flora and fauna that are considered important as extractive capitalism continues to expand globally. However, their implementation has been at the expense or to the detriment of local Indigenous communities.[14]

Communities whose *ejido* lands surround the biosphere reserve have been pressured or convinced to participate in "alternative development projects" with the promise of generating alternative sources of income.

With the expanding role of UNESCO (United Nations Educational, Scientific, and Cultural Organization) to meet biodiversity and climate change challenges, nongovernmental organizations (NGOs) proliferated and were able to secure funds to promote "sustainable use" projects within the parameters of UNESCO definitions of sustainable use of reserve buffer zones. Examples of such projects in the last two decades range from organic agriculture to raising parrots for the pet market to ecotourism trails to rustic furniture from specific small underbrush trees to making arts and crafts for the tourist market. Communities began to feel jaded because development projects never took off and they never saw the benefits of the considerable time and effort invested in those initiatives. One example was when the women in the community developed intricate, time-consuming artisan crafts (silhouettes of forest animals made from butterfly wings) but had no plans or resources for marketing them. The intention of such projects was to push communities to rely more on "alternative development" and less on their traditional shifting agriculture that requires burning, which is seen as an environmental threat to forest managers and environmentalists. Ultimately, the projects generated little or no income yet took critical time away from traditional farming, hunting, and other domestic activities that contribute to the Maya's well-being.

GREEN LAND GRABS AND CLIMATE INJUSTICE

A direct consequence of climate change adaptation is the exacerbation of turning land into a commodity. Climate injustice in Quintana Roo is being perpetuated by privatization schemes that threaten traditional land tenure. Land privatization is creating increased tensions within Maya communities between people that want to continue the current system of communal land tenure and those that are pressured to sell their land rights through a recent biodiversity conservation scheme taking place in the communities around the Sian Ka'an Biosphere Reserve. Roberto Hernández, a former Banamex bank owner and one of Mexico's wealthiest individuals, started the scheme of buying *ejido* land rights from individual farmers living in towns from Tulum and extending southward to the Sian Ka'an Biosphere Reserve in what is known as the Zona Maya.[15] Hernández, along with his wife, created a conservation trust called Fundación

Claudia y Roberto Hernández with the purported intention of "protecting the Maya Forest."[16] They created the Alianza Selva Maya (Maya Forest Alliance) in 2010 to develop conservation projects to "preserve the natural abundance of the Yucatan peninsula" by bringing together environmental NGOs, government, private capital, foundations, and *ejido* representatives to develop strategies for conservation, including promoting ecosystem services (carbon capture credits).[17] These types of programs and policies are being supported by the United Nations' REDD+ Program (Reducing Emissions from Deforestation and Forest Degradation).[18] Ultimately, Hernández's conservation trust scheme is a de facto privatization of nature, or a green land grab, as they bought up swathes of land. Local Maya farmers conveyed to us that by using third party groups, Hernández disingenuously convinced local folks to sell their land rights by promising thousands of pesos. Community members mentioned that Hernández also simultaneously lures agrarian government officials into approving such transactions. These land grabs have also been documented by other researchers in the area.[19]

The concerns of regional Maya leaders are coming to fruition in the Tres Reyes *ejido*. We learned that one person representing Hernández bought eight *ejido* rights. Later at an *ejido* meeting in Tres Reyes, the buyer confirmed that he worked for Hernández. Each *ejido* member in Tres Reyes has access to approximately four hundred hectares of land. While it is a large amount of land, most of the *ejido* is forest cover, and the traditional agricultural practice amounts to only a few hectares in production each year. However, if one person has eight *ejido* rights, they will have the right of use to approximately 3,200 hectares. The foundation's intention is to accumulate land and pronounce that it is all protected forest to take advantage of the economic programs for ecosystem services. If this trend continues, Hernández's foundation would benefit enormously while disenfranchising the Maya despite going against the intent of REDD+, who claim that "REDD+ will require the full engagement and respect for the rights of Indigenous Peoples and other forest dependent communities."[20]

To make matters worse, Hernández has already bought other *ejido* lands. The local press found Hernández using family member's names to buy what amounts to the majority of the *ejido* rights in the Pino Suarez *ejido*, just south of Tulum.[21] Maya leaders worry that land that is supposed to be for use by Maya farmers, "land that God has given us," as they attest, will soon be gone.[22] Hernández is taking advantage of the dire economic situation (in particular, the extreme difficulty of growing enough food in their cornfields because of climate change) and enticed Maya farmers with thousands of pesos, which they saw as a panacea to take care of their immediate needs. However, by selling their *ejido* rights, they essentially gave away their rights to land for a small sum. In the *ejido* of Tres Reyes, the prices paid varied. The first person sold his for seventy thousand pesos (approximately six thousand US dollars). Others requested more money, seeing that the buyer was willing to pay more. Nonetheless, they all now find themselves in a position in which they have no land to work.

A substantial number of Maya from this region have experienced working in the tourist industry at some point in their lives. Paradoxically, the tourism industry both provoked and offers an alternative source of income to food insecurity. Many Maya shared stories of how cutbacks due to a low tourism season made it impossible to find jobs, forcing them to return to the area. As they explained, having access to land is crucial. If other options failed, having access to land became their safety net. If privatization continues to become more and more prevalent, it will lead to more marginalization for the Máasewal. Further research should examine whether the COVID-19 pandemic, which severely affected tourism, hastened the return of Mayas who had left their communities.

Despite these land sales, during early speculation some community members saw how quickly they spent what they viewed as a lot of money, and how quickly their sense of economic security disappeared. It led them to see the writing on the wall, so to speak. As one community member expressed, "I will never give up my rights. My *tatich* [grandfather] said that he who gives up his land gives up his future. This is land for my grandchildren. Why would you give up the land of your grandchildren? I will not give up my land for any money in the world."[23] Those who manage to survive without having to sell find a measure of security and satisfaction in being on their land. Some that had migrated and returned expressed that life in the forest, albeit hard, continues to be a source of pleasure and means for continuing traditions connected to their land as opposed to the grueling life of the service and construction industries. Indigenous advocacy groups in Quintana Roo condemned the scheme of allowing the privatization of *ejido* rights because they recognize that it is detrimental to the future of their communities. Moreover, the usurping of Maya land makes a mockery of REDD+'s alleged mission of respecting the rights of Indigenous populations. REDD+ should be held accountable for turning a blind eye to green land grabbing because their program undermines the struggle for climate justice.

MILPA ADAPTATION OR CULTURAL DECLINE? HOW THE MAYA PERCEIVE CLIMATE CHANGE

As forest dwellers, the Maya have a profound engagement with their environment and have pragmatic views of performing their work. Their attitude toward the environment stems from the everyday struggle to render the resources of their land as productive as they can for their survival while also fostering a forest ethic. For ages, traditional gardens have provided a diverse, nutritious diet. Leatherman, Goodman, and Stillman documented an increase in chronic diseases such as diabetes and obesity among the Maya in the Yucatán Peninsula due to changes in diet because of decreased access to traditional foods.[24] While the Maya have changed their engagement with the land to confront different challenges, traditional agriculture has been an ongoing adaptation and learning process that they have practiced for a long time. Working in the milpa, household gardens, and hunting, although hard work, are both practical and rewarding to them. In this section, we provide a general description of traditional gardening, particularly the milpa, the traditional ecological knowledge involved, and how the Yucatec Maya perceive and explain the changes associated with a warming planet.

While men predominantly work the milpa, women also engage with it. Sometimes they plant and harvest, but mostly they influence it by selecting seeds. Maya women have a much more prominent role in cultivating the *pet pach* (the Maya name for household gardens) and tending domestic plants and animals in their *solares*, the two other modes of traditional farming that contribute to Maya subsistence and well-being and are integral to their moral ecology. Both practices diversify their diet and complement the milpa. The *pet pach* are small to medium-sized garden units in the immediate surroundings of the village. *Solares* are smaller gardens located near the household. Both men and women work the *pet pach*. Women predominantly manage the *solares*. Milpas are approximately one to eight kilometers from the village.

The *pet pach* and *solares* are cultivated with a variety of crops primarily for household consumption and occasionally for sale when there is a surplus. The plantings include root vegetables, tomatoes, peppers, leafy greens, and fruit trees (especially papaya, orange, lime, and mango). The gardens surveyed in households in the villages of Tres Reyes and X-Hazil included plants that were ornamental, medicinal, and gastronomic. Some of the crops are planted in raised beds known as *ka'anche'* in the *solares* to keep household animals such as pigs, hens, and turkeys from eating or damaging them before being transplanted. This process allows for close protection of species that need care, especially during the first few weeks or months of their existence. Maya women begin to engage with their environment at an early age, especially in the *solar*. Through this engagement, they perceive the environment, and create attachment to place, practicing the Maya moral ecology. This practice of working in the solar only begins to scratch the surface of the ways in which Maya women contribute to the moral ecology, and further research should be undertaken.

Life in the Zona Maya revolves around the annual cycle of the milpa. All of the Maya's activities center on the processes, activities, and rituals associated with this agricultural practice. To outside observers, which include environmentalists, NGO personnel, state managers, and even some anthropologists, the process of the milpa appears simple: the Maya farmers select a plot of land then "cut the forest, burn it, plant it, weed it and harvest it."[25] The simplicity of the model belies the complexity and integrality of the milpa in the Maya moral ecology. By "moral ecology" we mean that a critical mass of the native inhabitants of the Maya forest have an intimate relationship with the forest that permeates their everyday engagement with the environment. It blends history, identity, spiritual beliefs, communion with other species, and making a livelihood with the milpa lifeworld, their socioecological model of traditional gardening and hunting practice.[26] On closer scrutiny, the milpa process is much more complex than it appears. While we cannot fully capture the complexity here, the following is a sketch of the milpa process and its place within the moral ecology.

The first step in creating a milpa is *ximbal k'aax* (site selection), in which the person takes a walk into the forest to select the site for the new

field. This is usually a daylong activity. The farmer walks around, survey-
ing his *ejido* land, and then selects the potential area for a milpa based on
the size and quality of the vegetation. The presence of specific plants and
trees indicate to the Maya whether the land has good soil. For example,
they look for the *cheechem* (black poisonwood), *ox* (ramón), and *ya'* (chi-
cozapote). They also look at the quality of the soil and whether there is a
substantial amount of rocks or limestone. The current soil classification
falls into two categories: red (*chak lu'um*) and black (*ek* or *box lu'um*); the
latter is considered the most fertile soil and is thus preferred. The Maya
also look for the presence of water sources, such as natural wells and sink-
holes (which are found sparingly throughout the forest). These long-held
practices generate a traditional ecological knowledge that includes changes in the land as a result of climate change.

The next step is the *jol ch'ak* (to clear), which is the process of clearing the
site with a machete to form a small path around the selected area to mark
the clearing and to facilitate its measuring. Before the Maya start the clearing, they make sure they place themselves in the rising sun's path and pro-
ceed to cut from east to west (moving in the direction of the sun) and then
from south to north. This process takes place in February and March, well
before the rainy season begins in May. With the uncertainty that climate
change has introduced, there is concern about the proper times to appor-
tion different stages of the work in preparation for the planting season.

The third step, *we p'is k'aax* (forest measure), is the process of measuring the size of the milpa by *mecates* or *p'isik'aan* (rope measure), which is
done by two people with a rope approximately twenty meters long. One
mecate will measure twenty square meters. The number of mecates will
give a rough idea of how much time will be needed to prepare the site for
the burning.

The fourth step is the *kol*, or felling and cutting the vegetation inside
the selected area. It takes place at various times of the year depending on
the amount of vegetation, that is, on whether the milpa is in its first, second, or third year. This is because a milpa cornfield is used for three years
and is maintained differently each of those years. Not all the vegetation
is felled and burned; the process is strategic rather than a mindless act of
clearcutting. Several species are preserved for medicinal, practical (i.e., for
construction or cooking), or religious uses. Other species are cut down to
about three feet tall so that they can flourish again once the milpa is left to

begin the fallow phase, or the time of regeneration. For example, on species that is preserved is the ya' (zapote), whose fruit is consumed by forest animals like deer. The ramón is another species that is protected for its use in religious ceremonies. Its leaves are also consumed by deer and other species. The huano (*sabal japa*) is preserved for its value for construction. After felling the vegetation and spreading it so that it burns evenly (*p'uyk'am che'*), the next step is the *mis pach kol* (firebreak, or *guardarraya* in Spanish). Maya farmers take great care to prevent the spread of the fire to other areas of the forest by clearing an area one to two meters wide around the milpa prior to the burning. The next step is the *tok*, or burning of the milpa. This takes place in April and May, just before the rainy season, when warm winds are blowing in the area. The fire is set in a corner where the wind favors it so that it spreads toward the center of the milpa.

The ashes will help to provide the topsoil with nutrients; once they have settled, it confirms that conditions are right for planting (*pak'al*). For a few days before this, the community gathers the seeds that they have accumulated throughout the year and engages in a final selection of corn seeds. The women in the household actively take part in this task by contributing the knowledge about selection that they have accumulated for years. Children are also taught how to select the best seeds for planting. Once the Maya have selected the seeds, they mix the corn with the beans, lima beans, and pumpkin seeds. These will be put together in a gourd (*lek*) or bag that an individual planter will have with him or her. Some women help in the planting of the milpa, although not to the degree that they plant in their home gardens. To plant the seeds, the farmer pokes a hole with the planting stick (*xuul*) carried in one hand. He uses his other hand to take several seeds (maize, beans, lima beans, and pumpkin) from the gourd or pouch strapped around his shoulder. It is by sense of touch that they select the number and kind of seeds that they will plant. Then they lean a little bit forward, and with the seeds held by the tips of their fingers, they let them drop with such precision that none fall outside the hole. The hole is then covered with soil using the *xuul*. The Maya did this in such a graceful manner that it appeared effortless.

Once the corn grows its cobs, the Maya perform the *wats'*, the process in which the stems of the corn are bent to protect the cobs so that when it rains water drops out and does not get trapped inside the corn. This usually takes place three or four months after the planting. *Wats'* also speeds up the drying of the corn by cutting the water and nutrient flow from the stem. It provides protection from birds as well. The Maya must be careful not to bend the stems too low because animals might eat them or the moisture from the ground might rot the crop.

Once the *wats'* has taken place and the corn dries, it is time for the *jooch nal* (harvest). In October we see the first harvesting of corn, the most important harvest. Harvesting crops in the milpa will continue through the following March, April, and May when the *ib* (lima beans) and the ik (squash) are harvested. The corn is stored after the cobs are divided between the *i'nal*, which are the best of the harvest, and the *alnal*, which are of lesser quality. The *alnal* will be consumed first and will be used for feeding animals in the solar. The harvest for other crops will last up to five years (during what is sometimes called the fallow stage), including the root crops such as *camotes* (yams) and *yucca* (cassava).

In interviews in Tres Reyes and X-Hazil, farmers spoke of one of the biggest problems associated with the milpa: the declining output of corn. "It doesn't produce as before. A long time ago, between one and three hectares would last all year; now we plant five or more and it doesn't produce enough for the whole year," says a Maya farmer Josué Ek Balam.[27] The community has recently noticed a decline in rain, and the size of the forest canopy is smaller than before. Younger generations are particularly discouraged by the need for more effort to produces a lower-yielding corn, and see that it becomes increasingly difficult to eke out a living in agriculture, even if combined with other productive activities.

Although living in the forest is never easy, most community members attribute the fluctuations of crop yield to climate change. In addition to the lack of rain, Ek Balam said that these changes can be noted in the character of the soil and in the trees. In another community, we had the opportunity to walk in the forest with Florentino Can Ek, who, in addition to being a farmer is also an apiculturist. Once we got to his beehives to supply water, he explained how the bees are producing less and dying faster than before. He also attributed this to climate change.

A main concern expressed throughout our interviews was the changing rain pattern. The Maya are facing increased periods of drought associated with climate change. These changes are putting enormous pressure on their resources—the forest, forest wildlife, their traditional agriculture—and make them more dependent on government subsidies. The hardships also encourage emigration to find the relatively few jobs the tourism industry provides. Facing this array of difficulties, local leaders, including the Maya dignitaries associated with the Church of the Talking Cross, continue to question the prospects for future generations. As a people that have endured profound struggles, including invasion and war, they often respond with an apocalyptic sentiment.

One of the farmers complained that it's getting more and more difficult to get permits for burning land, which is an essential part of preparation for planting in shifting agriculture. Changing rain patterns are making things more unpredictable. "Planting corn usually took place in May; now we have to plant some in May and some in June," says Eulogio Cab of X-Hazil.[28] He explains further: "Rain is not falling when it's due. In recent years rain has been delayed. We had a practice called in Maya *xtik imuk*, which is planting corn, beans, and squash in dry soil just before it rains. When it rains all the corn grows tall and level. Just beautiful. But now we can't plant because rain has been delayed more and more. This means that the practice of *xtik imuk* is being lost. Last year we had to wait until rain came." He says that the corn will still grow when planted after the first rains, but it is different. Yet he has also noticed the opposite pattern: "There are times in which the rain is starting earlier than usual. Two years ago, the rain came in April, and I had not burned, nor completely cleared my plot. I had to plant corn without burning. So, the climate seems to have changed tremendously." Heat waves are another big concern. "The sun is hotter than before," Cab said. "We say in Maya "*Jach* choko' k'in" (The sun is *really* hot). It feels as if it is moving closer to us." He emphasized "*Jach*," which means "very," and the phrase is used to refer to climate change generally. Another farmer made a connection to increased pests and climate change, remarking that "the milpa has been invaded by more pests. It's noticeable when planting one hectare of chili peppers." Peppers are grown to sell in the market, and he noticed that the quantity and quality were more difficult to achieve because of pest infestations, which, he lamented, compelled them to use pesticides. He noted that the *pet pach*, or house gardens, including tomatoes, watermelons, plantains, and bananas, do well without pesticides.

Don Eulogio also observes that climate change is affecting the forest vegetation and its wildlife. "I've noticed how animals are not finding enough food in the forest and they are attacking our root crops and fruit trees, not the fruit of the trees they usually eat from. Now they are depending more on our labor." Another example of problems with wildlife is the rise of close encounters with jaguars approaching the village. Farmers say that predators like the jaguar used to keep their distance because they had enough food in the forest.

Climate change has also intensified hurricanes and storms, and this was on the Maya's list of concerns. Don Eulogio told us that there had long been a fear of hurricanes, but the fear was generally contained to a shorter period than today. "When I was a child, our elders always, always feared the month of September. That month was the one that hurricanes could hit." They used to say that if they reached the end of September without a big storm, they would be done for the season. But now, he says, you have hurricanes more frequently as early as June and as late as November. The increase in storms has the potential to completely destroy a year's harvest. Elogio recounted that a friend planted ten acres and had a successful harvest. The following year, just as the corn was tall and about to pollinate, a storm hit and knocked down all the corn stalks and he lost everything.

Don Eulogio expressed with certainty that the climate has been altered. When asked what he attributes the warming weather to, he responded that humans are responsible. Particularly, he specified, the ones that continue with deforestation and all the factories in big cities. For this reason, he argues, "We can never give up on reforestations." This observation points to the injustice of climate change. Don Eulogio saw clearly that things being done on the outside are affecting the local environment and community. Likewise, he saw that climate change was caused by industrialization and outsiders. Don Eulogio mentioned "reforestations" because he is constantly planting trees. He sees this as one way to help the forest and, ultimately, the community both locally and more broadly.

Although the causal argument for climate change based on burning fossil fuels is very commonly heard these days, the Yucatec Maya have another explanation for the profound changes, one that was prophesized by the Máasewal elders. "Our grandparents were sages," says Pedro Canul.[29] "I remember when I was a little boy, my grandmother telling me that the *huaches* [a term used for White foreigners] were going to build a four-lane road and that would be the beginning of the end of the world." He recounted this as they were expanding the Tulum-Felipe Carrillo Puerto highway from two to four lanes, which opened the gates for privatizing *ejido* land. Some of the land has been parceled out, and *propiedad privada* (private property) signs have begun surfacing along the highways. Other tensions are putting pressure on the land and community. The new Tren Maya project has divided the Maya community between those who don't want to see any more fragmentation crosscut the forest and others who trust that President Obrador will bring prosperity to the region.

Threats to the forest and, above all, loss of access are always a concern for the Maya. We once posed a hypothetical question to various people in the community: What would happen if for some reason you lost access to the forest? The most common answer was, "We die," and, as someone said, "I will work on the milpa until I have no more energy left in my body." These responses say much about the importance of the forest to the Maya. When we asked about the significance of the forest, the most common response was that it meant "survival" and was the "source of life." When the Maya assert that "el monte da vida, pero hay que trabajarlo" (the forest gives life, but you must work it), they recognize that nature is not a category but a relationship from which a particular moral ecology of a forest is constituted. The forest exists, and by dwelling in it and engaging with it, it gives life and good health, and the Maya ecology comes into being.

CONCLUSION

Today, the Máasewal Maya find themselves at a crossroads in an era of growing awareness of biodiversity loss and climate change. In negotiating how to maintain their livelihoods and connection to the land they have implemented a wide range of developmentalist, Western-devised "conservation" projects. These conservation projects imposed from the outside continuously promised to "improve" their lives while restricting their access to the land. Although communities on the border of the reserve were motivated to join green development projects in the hope that they would provide opportunities and an alternate livelihood, they were largely unsuccessful because these opportunities were imposed upon them rather than valuing and listening to the communities' traditional ecological knowledge, thus contributing to food insecurity.

Yet the Máasewal have earned a bad reputation among environmental managers because shifting agriculture involves burning sections of the forest, and many environmentalists and biologists equate this practice with destruction of the environment, even though in many instances it helps biodiversity. This is precisely what specialists of the milpa argue: fire briefly disturbs the natural environment, but the Maya work *with* the forest once a milpa is planted and managed.

"Moral ecologies" are bodies of knowledge about the environment. They include what anthropologists refer to as traditional ecological knowledge and the ethos of peoples' relationship with their environment that ground and guide who they are and how they act within their environment. Milperos recognize that their work in the forest, although difficult, usually results in better and more stable benefits for the community. However, hopes of autonomy have begun to falter as two interwoven challenges that had been simmering slowly on the back burner have taken central prominence. The first challenge was the de facto green land-grab scheme that further reduced their access to land. The second challenge is climate change, both its ecological impact *and* the conservation strategies being implemented to allegedly confront it.

The communities understand that environmental damage is a threat. Thus, they would likely continue to support conservation efforts if they perceived a fit for them. Most everyone agrees that protecting the forest is beneficial because animals take refuge and reproduce in the protected space. However, they also know that there is inequality in the relationship between themselves and those in charge of the reserve. Conservationists and government bureaucrats who work with them must recognize their dependence on the forest, not only for their livelihood but also for their cultural reproduction, and must integrate this understanding into conservation strategies or the resulting efforts are likely to fail and leave lasting negative impacts on local communities. Building on local knowledge is where we see the groundwork for communities that want to build sustainable, healthy communities and not be subordinated to a continuation of colonial relations. The Sustainable Solutions Lab at University of Massachusetts Boston also offers an example. In April 2022, we hosted a panel of four Maya farmers (men and women) so that we could listen to their traditional ecological knowledge and the actions they are taking to confront climate change and address food sovereignty. The panelists reiterated some of the above sentiments and provided new insights about the challenges for climate justice. Listening to and learning from Indigenous communities is a decolonial practice necessary to achieve climate justice. Our hope is that the Máasewal can find a path to autonomy, climate justice, health, and claiming commons without the ills of conflict, repression, and marginalization.

NOTES

1 Anabel Ford and Ronald Nigh, *The Maya Forest Garden: Eight Millennia of Sustainable Cultivation of the Tropical Woodlands* (Walnut Creek, CA: Left Coast Press, 2015).

2 Thomas Leatherman, Alan H. Goodman, and J. Tobias Stillman, "A Critical Biocultural Perspective on Tourism and the Nutrition Transition in the Yucatan," in *Culture, Environment and Health in the Yucatan Peninsula: A Human Ecology Perspective*, ed. Hugo Azcorra and Federico Dickinson (Walnut Creek, CA: Springer, 2020): 97–120.

3 Nelson A. Reed, *The Caste War of Yucatan: Revised Edition* (Stanford, CA: Stanford University Press, 2001).

4 Victoria Reifler Bricker, *The Indian Christ, the Indian King* (Austin: University of Texas Press, 2010), 87.

5 Herman W. Konrad, "Capitalism on the Tropical-Forest Frontier: Quintana Roo, 1880s to 1930," in *Land, Labor, and Capital in Modern Yucatan*, ed. Jeffery T. Brannon and Gilbert M. Joseph (Tuscaloosa: University of Alabama Press, 1991).

6 Marie Lapointe, *Los mayas rebeldes de Yucatán* (Merida, Mexico: Maldonado Editores, 1997).

7 Reed, *Caste War of Yucatan*.

8 Fernando Benítez, *Lázaro Cárdenas y la Revolución Mexicana, III: El Cardenismo* (Mexico City: Fondo de Cultura Económica, 2015).

9 Magali Daltabuit and Oriol Pi-Sunyer, "Tourism Development in Quintana Roo, Mexico," *Cultural Survival Quarterly* 14, no. 1 (1990): 9–13.

10 María Bianet Castellanos, *A Return to Servitude: Maya Migration and the Tourist Trade in Cancún* (Minneapolis: University of Minnesota Press: 2010).

11 Thomas L. Leatherman and Alan Goodman, "Coca-colonization of Diets in the Yucatan," *Social Science & Medicine* 61, no. 4 (2005): 833–46.

12 Thomas L. Leatherman, Morgan K. Hoke, and Alan H. Goodman, "Local Nutrition in Global Contexts: Critical Biocultural Perspectives on the Nutrition Transition in Mexico," in *New Directions in Biocultural Anthropology*, ed. Molly K. Zuckerman and Debra L. Martin (Hoboken, NJ: John Wiley & Sons, 2016): 49–65.

13 Sian Kaan, like much of the Yucatán's tourism area, is named in Mayan and means "origin of the sky." It is over 1.3 million acres and one of the largest protected areas in Mexico. These were the traditional lands of the Maya. Today, no Maya are included in the management.

14 José Martínez-Reyes, *Moral Ecology of A Forest: The Nature Industry and Maya Post-Conservation* (Tucson: University of Arizona Press, 2016).

15 Gabriela Torres-Mazuera, Sergio Y. Raul Benet Madrid, and R. Benet Keil, "Tres décadas de privatización y despojo de la propiedad social en la Península de Yucatán," *Ciudad de México: Consejo Civil Mexicano para la Silvicultura Sostenible, AC (CCMSS)*, 2021, https://www. ccmss. org. mx/wp-content /uploads/2020_22_TresDecadasPrivatizacion.pdf.

16 Martínez-Reyes, *Moral Ecology of A Forest.*

17 Alianza Selva Maya, n.d., accessed February 2012, https://www.alianzaselvamaya .com.

18 REDD+ was established in 2005 by the United Nations Framework Convention on Climate Change (UNFCCC) to reverse deforestation and loss of carbon sequestration in developing nations through financial incentives for maintaining forests to forest land holders.

19 Torres-Masuera, Madrid, and Keil, "Tres décadas."

20 "Climate Change REDD+," UNEP [United Nations Environment Program], accessed May 14, 2004, http://www.unep.org/climatechange/reddplus/Introduc tion/tabid/29525/Default.aspx.

21 Gustavo Marín Guardado, "Turismo, ejidatarios y 'mafias agrarias' en Tulum, Quintana Roo, México: El caso del ejido José María Pino Suárez," in *Sin tierras no hay paraíso: Turismo, organizaciones agrarias, y apropiación territorial en México,* ed. G. Guardado (Tenerife, Spain: Pasos, Revista de Turismo y Patrimonio Cultural, 2015), 91–113.

22 Prior to this recent threat, privatizing *ejido* land was made much easier because of the North American Free Trade Agreement and the changes to the agrarian reform law of the Mexican constitution (Article 27), making it possible to sell *ejido* land to Mexican capitalist developers and paving the way for dispossessing farmers and indigenous groups from their land.

23 Personal interview, June 2016.

24 Leatherman, Goodman, and Stillman, "Critical Biocultural Perspective," 97–120.

25 Ellen R. Kintz, *Life under the Tropical Canopy: Tradition and Change among the Yucatec Maya* (New York: Holt, Rinehart and Winston, 1990), 120.

26 Martínez-Reyes, *Moral Ecology of a Forest.*

27 Josué Ek Balam, personal communication.

28 Eulogio Cab of X-Hazil, personal communication, June 2018.

29 Pedro Canul, personal interview.

CHAPTER 7

Structural Inequalities and Extreme Heat in the Boston Region

SAJANI KANDEL AND ANTONIO RACITI

EXTREME HEAT AND UNDERSERVED COMMUNITIES IN DORCHESTER

The latest National Climate Assessment reports an increase in the annual average temperature in every region in the United States.[1] The yearly average temperature over the contiguous United States has increased by 1.2 degrees Fahrenheit (0.7 degrees Celsius) over the last few decades and by 1.8 degrees Fahrenheit (1 degrees Celsius) in the previous century.[2] Additionally, an increase of about 2.5 degrees Fahrenheit (1.4 degrees Celsius) in the annual average temperature is expected over the next few decades regardless of future emissions.[3] Given projected future population increase, extreme heat will continue to be a substantial public health policy issue in the United States for decades to come.[4]

The impact of extreme heat is evident in medium- to large-size cities worldwide.[5] Central city temperatures are higher than their nearby suburban and ex-urban areas due to the differences between their environments, a well-documented phenomenon known as the urban heat island effect (UHI).[6] The UHI effect is caused by high concentrations of dark, heat-retaining surfaces such as asphalt and concrete on paved roads, parking lots, tall buildings, anthropogenic heat waste, land cover, vegetation, and other morphological features commonly present in urban environments.[7] The temperatures in urban areas due to UHI can be 0.9 to 7.2 degrees Fahrenheit (0.5 to 4 degrees Celsius) higher during the day and 1.8 to 4.5 degrees Fahrenheit (1 to 2.5 degrees Celsius) higher during the night depending on the climate, city size, population density, urban forms, and method of measurement.[8] The combined effect of urban heat island and increasing extreme heat events due to climate change significantly increases heat exposure in cities.

Considerable research evidence worldwide has identified the disproportionate impact of extreme heat exposure on residents of low-income communities. Tirthankar Chakraborty and colleagues found that urban heat exposure in 88 percent of US cities correlates negatively with income.[9] Another study on 175 metropolitan areas in the United States found that over 70 percent of people with incomes below the poverty line have a higher heat exposure than those with incomes twice above the poverty line.[10] The relationship between extreme heat and poverty also involves a more complex relationship with racial identity. While people of color tend to have lower incomes than White populations in the United States, it is hard to isolate economic factors to explain the unequal distribution of urban heat island intensity exposure. Angel Hsu and colleagues found that the average person of color lives in a census tract with higher summer daytime surface urban heat island intensity than non-Hispanic White in all but six of the 175 largest urbanized areas.[11] Susanne Benz and Jennifer Burney also found that neighborhoods with higher Black, Hispanic, and Asian populations shares are hotter than the more White, non-Hispanic areas in 1,056 US counties, even when controlling for income.[12] Recent evidence identifies the effect of historic urban planning policies and practices that promoted segregation in the distribution of heat across communities. For example, previously redlined neighborhoods categorized as "D" areas are now an average of 2.6 degrees Celsius warmer than "A"-coded residential areas in 108 studied cities in the United States.[13] Seventy-four percent of the neighborhoods graded as "high-risk" or "hazardous" eight decades ago are low-to-moderate income today, and 64 percent are currently minority neighborhoods.[14]

This chapter explores how extreme heat has become a concern for Boston communities and what types of planning initiatives have been implemented to address this issue. We are interested in heat planning initiatives undertaken in historically disenfranchised communities. Our aim is to understand how these communities respond to ongoing heat planning initiatives and what type of work still needs to be done to strengthen, expand, or reinvent current efforts. We focus on the Dorchester neighborhood in Boston, which is home to many socially vulnerable residents and where a series of heat planning initiatives have been implemented. The chapter begins with an introduction to the problem of extreme heat in Boston, paying particular attention to disenfranchised groups. Drawing on interviews with Dorchester residents and community leaders, the chapter highlights the strengths and limitations of the city's current efforts from the residents' perspective. The chapter concludes with a few reflections regarding the nature of these initiatives and insights for enhancing the impact of the city's extreme heat initiatives.

PLANNING PRACTICES ADDRESSING PUBLIC HEALTH CONCERNS

Heat is the leading cause of death across all-natural disasters in the United States.[15] Between 1986 and 2020, severe heat events have claimed 127 lives per year in the United States (see figure 7.1).[16] According to the US Centers for Disease Control and Prevention, an average of 702 heat-related deaths (415 with heat as the underlying cause, and 287 as a contributing cause) occurred in the United States annually between 2004 and 2018, and 90 percent of these fatalities occurred between May and September.[17]

The effects of extreme heat can range from dizziness, muscle cramps, fainting, and heatstroke and, if untreated, can lead to death. Elderly and very young residents, those homebound or confined to bed or unable to care for themselves, socially isolated individuals, those lacking access

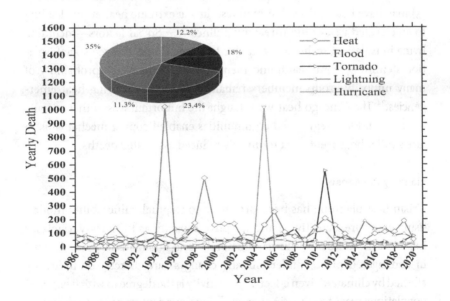

FIGURE 7.1. Annual deaths of different disaster events between 1986 and 2020 (line graph). The pie chart shows the percentage of fatalities between 1986 and 2020 in each disaster category. Data source: (NOAA, 2021). Graphic re-elaboration by the authors.

to air conditioning, outdoor workers, low-income residents, homeless individuals, and those with preexisting illness or medical conditions are more vulnerable to extreme heat and its associated health impacts.[18] Older adults, young people, and individuals with disabilities do not adjust to sudden changes in temperature and are more likely to have chronic conditions that undermine normal body responses to heat.[19]

Extreme heat morbidity and mortality are often much higher among racial minorities and low-income groups.[20] During the 1995 Chicago heat wave, 49 percent of decedents were Black; 46 percent were White; and 5 percent were from other racial/ethnic groups.[21] Between 2004 and 2018, among all race/ethnicity groups, non-Hispanic American Indian/Alaska Natives (0.3 perone hundred thousand) and non-Hispanic Blacks had the highest death rate (0.3 perone hundred thousand) of heat-related deaths, respectively. The disparities in heat-related impacts across racial and ethnic groups are associated with social vulnerability. For instance, low-income populations (1) often do not have the financial means to adequately cool their living space using air conditioners,[22] (2) may be less likely to participate in help-seeking behaviors during heat events, such as making health-related calls, or (3) may have health risks that manifest during extreme heat events leading to an elevated risk of life-threatening illnesses.[23] Social factors—including living in isolation, secluded living due to fear of crime, or even social isolation stemming from health and mental illness increase the probability of many minority group members remaining unnoticed during heat emergencies.[24] The Chicago heat wave taught the important lesson that socially cohesive and less segregated communities enabled coping mechanisms to face similar heat events that ultimately reduced associated deaths.

Planning Response

Urban heat planning has been driven by conceptual vulnerability models using exposure, sensitivity, and adaptive capacity as primary variables.[25] Exposure indicates how "humans, natural assets, and material goods located in endangered places under climatic changes" are directly or indirectly affected by climate-driven effects.[26] Sensitivity is the degree to which exposed populations react to climatic changes.[27] Finally, adaptive capacity describes a population's ability to handle adversity through anticipatory and preventive actions.[28] While these three dimensions can be measured and analyzed separately, their interconnection defines the community's resiliency to heat within most vulnerability models. Most heat preparedness plans and vulnerability assessments involve data-driven geographic analysis using Census-based demographic and socioeconomic characteristics to understand vulnerability to heat.[29] This conceptual model of vulnerability has shaped municipal actions to address extreme heat issues by encouraging efforts to physically reduce the intensity and duration of heat exposure both during and after periods of extreme heat.[30]

While many communities have implemented heat action plans to reduce the health impacts of extreme heat, assessing the role of municipal heat-related interventions remains largely unexplored.[31] Literature suggests that effective heat action plans reduce health risks if these include education and awareness strategies;[32] surveillance and impacts monitoring;[33] access to cooling centers; green spaces; and design and modifications of existing infrastructures.[34] Some studies have found that public cooling centers tend to be underutilized in communities experiencing heat events.[35] This underutilization was attributed to socioeconomic barriers for those most in need of such centers, lack of knowledge of facilities, poor transportation access, fear of crime at and near these centers, the inability of vulnerable populations to leave home or travel, and negative associations with cooling centers.[36] Additionally, even though urban greening projects decrease air temperatures by lowering solar radiation and increasing evapotranspiration of warm air,[37] it remains less clear how evenly distributed these positive impacts are across populations within marginalized communities.[38]

Heat Planning in Boston

In the northeast region of the United States, the average temperature increased by almost 2 degrees Fahrenheit (0.16 degrees Fahrenheit per decade) between 1895 and 2011.[39] While the average summer temperature in Boston from 1981 to 2010 was 69 degrees Fahrenheit, it may rise as high as 76 degrees Fahrenheit by 2050, and 84 degrees Fahrenheit by 2100.[40] Based on city projections, there were eleven days per year over 90 degrees Fahrenheit between 1971 to 2000, but the may be as many as forty days over 90 degrees Fahrenheit by 2030, and ninety days over 90 degrees Fahrenheit by 2070.[41] The Climate Central Report ranks Boston as the sixth-worst heat island among the 159 cities in the United States based on its UHI Index.[42] In Boston, "high-risk"–graded neighborhoods during redlining now have longer heat event duration than other parts of the city (see figure 7.2). The "high-risk" or category "D" neighborhoods are 7.5 degrees Fahrenheit hotter in the day, 3.6 degrees Fahrenheit hotter at night, and have 20 percent less parkland and 40 percent less tree canopy than highest-graded areas "A."[43]

In Boston, preparing for extreme temperatures related to climate change can be traced back to 2009. At that time, the Commonwealth of Massachusetts' Executive Office of Energy and Environmental Affairs organized the Climate Change Adaptation Advisory Committee to assess the impacts and mitigation strategies needed to combat climate change. One year later, the Commonwealth's Department of Environment partnered with Tufts University to publish a report titled "Preparing for Heatwaves in Boston" as part of an integrated plan that outlined necessary actions to reduce the risk of predicted effects of climate change.[44]

Climate Ready Boston (CRB), a primary document for climate change adaptation and preparedness, was finalized in 2016 to operationalize the city's high-level climate preparedness goals identified in its first comprehensive plan after fifty years, Imagine Boston 2030. *Climate Ready Boston* incorporated a climate vulnerability assessment for extreme heat events, identified potential health impacts, and extrapolated future local mortality rates based on climate projections.[45] It includes a small chapter that features the most recommended urban heat management strategies, such as green spaces, street tree canopies, and access to open spaces. The 2021 Natural Hazard Mitigation Plan 2021 (a updated version of the 2015 plan) is a plan by the Office of Emergency Management to identify the risks and

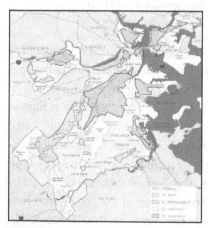

FIGURE 7.2. On the left: redlined neighborhoods in Boston. Source: City of Boston, city-wide analysis (2021). On the right: Daytime summer air temperature distribution. Source: City of Boston. Heat vulnerability analysis (2021).

vulnerabilities associated with natural disasters for home, business, and critical infrastructure and develops strategies to mitigate damages. It is not a part of CRB but a coordinated effort that includes climate change projection and strategies for extreme weather events. The Open Space Plan 2020–2025 and the twenty-year Urban Forestry Plan present recommendations for increasing green spaces and tree canopies for urban heat reduction, with a high focus on a marginalized neighborhood that lacks such amenities.

The City of Boston's Department of Environment currently provides fifteen pages of informational materials and guidelines for minimizing the environmental impacts of proposed projects.[46] Some recommended heat mitigation strategies include green roofs or eco-roofs, cool roofs, and reflective pavements as well as information on improved insulation techniques, weatherization programs, and sustainable building products and design.[47]

To advance the city's climate resiliency and carbon reduction goals, Boston Zoning Code Article 37[48] and the Climate Resiliency Review Policy ensure that all major building projects must be planned, designed, constructed, and managed to minimize adverse environmental impacts.[49] Article 37 also requires all major projects to achieve a minimum "certifiable" level of the US Green Building Council Leadership in Environmental and Energy Design (LEED) Rating System. There are no regulations to enhance the extreme heat performance of existing residential and nonresidential buildings.

Regarding Boston's response to heatwaves, the Boston Public Health Commission's Department of Emergency Management sends out emergency temperature alerts to preregistered residents through its Alert Boston System.[50] In addition, local television channels, radio stations, and emergency alert system provide information regarding the high temperature forecasts. The City of Boston's Emergency Management website lists all the cooling and emergency centers, along with multilingual education materials on personal heat safety tips.[51]

EXPLORING (EXTREME) HEAT PLANNING IN DORCHESTER

Climate Ready Boston identifies Dorchester as one of the communities most vulnerable to extreme heat events due to the large share of socially vulnerable individuals, chronically ill residents, and older individuals. Dorchester is the largest and one of the most diverse neighborhoods in

Boston, and comprises 18 percent of Boston's population.[52] As of 2019, 56.2 percent of Dorchester residents were foreign-born, of which 40 percent are from the Caribbean, 23 percent are Asian, and 21 percent African. In Dorchester 33.5 percent of residents are non–English speakers. Seventy-two percent identify as people of color (compared to 52 percent city-wide);[53] 35 percent identify as Black; Dorchester's median income is 23 percent lower than the city average; and 18.2 percent of the community's population makes less than fifteen thousand dollars.[54] In Dorchester 23.7 percent of residents live in poverty, giving this neighborhood the highest poverty rate of any Boston neighborhood.[55]

Dorchester has an aging housing stock and many heat-trapping surfaces, including highways, arteries, and parking lots that add to the community's urban heat footprint.[56] The modeled surface temperature collected by the Wicked Hot Boston Project in 2019 reported a high difference in ambient air temperatures within different subareas of the neighborhood. The Ashmont T station in Dorchester recorded one of the most elevated temperatures, 102.6 degrees Fahrenheit , in Boston.[57] In addition, the neighborhood lacks quality green spaces.[58] It has less than 10 percent tree cover, which is much less than the city average of 27 percent.[59] Understandably, the area has many vulnerable individuals and does not have any amenities to provide cooling benefits to reduce the urban heat island effect that amplifies heat exposure.

In the following section, we draw on interviews with residents and leaders to illuminate their perceptions of extreme heat and the city's recent heat planning interventions. We interviewed members and leaders of local community associations, nonprofit organizations, senior centers, churches, and other social service institutions working in the neighborhood. One of the authors also participated in participatory observation of community meetings, nonprofit town halls, civic associations, and tenant organization forums in the summer of 2021. We sought to understand how local leaders report residents' extreme heat experiences, coping mechanisms during heat events, and opportunities to access local services. Neighborhood residents were briefed about this study and invited to participate in a follow-up research phase, which is not discussed in this chapter. This chapter's reflections and highlights are based on ten interviews. While we do not claim the number of interviews to be exhaustive, we believe they are representative of the views of most Dorchester community leaders regarding local heat planning.[60] All interviewees identified themselves as BIPOC (Black, Indigenous, People of Color) individuals and are residents of the two neighborhoods shown in Figure 7.3.

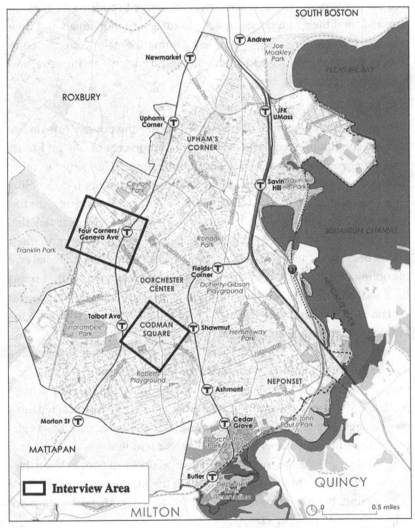

FIGURE 7.3. Dorchester District, Boston. The boxes identify Four Corners and Codman Square Neighborhoods, where the authors conducted resident interviews, participant observation, and community meetings. The area is home to most of Dorchester's African American residents. Map source: Climate Ready Explorer (2021).

Flags on Heat Strategies

The interview campaign findings highlighted extreme heat as an increasingly serious concern for Dorchester residents. Social, economic, and environmental factors exacerbate underserved residents' heat exposure and limit their adaptative capacity to safeguard individuals and their families. The listed "red flags" concisely summarize major concerns community members shared and highlight useful insights to improve heat planning practices in Boston.

Cost of Cooling

Nationwide, about one fifth of households below the poverty line do not have air conditioning equipment. However, 80 percent of Boston homes have either room or window air-conditioning units.[61] Compared to other midsized cities, a higher percentage of Boston residents have cooling units; however, there is little information on how many residents use this equipment. By participating in community meetings we learned that many residents could not take advantage of air conditioners in their homes because of the high electric service costs. Overall, the financial cost of cooling units is the barrier most community members face that limits their usage during the summer months. A woman living in Dorchester explained, "Our winter bills are high . . . very high. We must stick with the heat for winter, and in summer, we try to save as much as possible."[62]

According to the American Council Energy-Efficient Economy, a quarter of low-income households in Boston spend more than 19 percent of their income on energy bills, making them significantly energy burdened.[63] During community meetings, interviewees described how they struggle to pay winter heating bills and prefer to decrease their electric charges in the summer months. One resident living in a traditional Dorchester triple-decker mentioned, "I live on the third floor of the triple-decker. My apartment is overheated during the summer months. I have a window unit, but it's not enough to cool down. Using AC all day does not help. It feels like living in a sauna."[64]

To address such concerns, the federally funded Low-Income Home Energy Assistance Program seeks to help low-income households cover heating and cooling costs. This funding is primarily designated for the winter months to help with heating costs in northern cities. However, the program does not aid low-income families facing the cooling charges associated with the summer months. Limited programs by local community-based organizations and nonprofits, in partnership with the city and private foundations, have been providing aid to purchase and install air conditioning units for low-income residents with documented medical issues. On this point, one of the local nonprofit representatives commented, "There is a need for more innovative and longer-term solutions for energy and cooling. I wish the city would put more on its own 'skin in the game.' The city needs to commit long-term investments in this realm."[65]

The City of Boston provides a weatherization program via Action for Boston Community Development in cooperation with National Grid and Eversource to help low-income residents become more energy-efficient and comfortable while lowering their energy costs and helping the environment. Both renters and homeowners are provided with assistance through this program. However, some have concerns regarding the limited capacity of this initiative to address an increasingly chronic structural problem. One of the interviewees commented, "Boston currently helps residents by offering no-cost home energy audits and recommendations for home energy improvements. Many absentee landlords in the area and low-income renters cannot benefit from these programs. Unless these programs become regulated, poor residents, rent-burdened tenants, immigrants, etc., will not benefit. And these are the people who are most impacted by climate change, including heat."[66]

The Boston Housing Authority provides residents with energy assistance and manages cooling units and services for public housing tenants. Massachusetts state regulations require property owners to provide heat from September 15 to June 15 in all residential buildings. Historically, July is the hottest month in Boston. However, in recent years, the metro area has been getting early high-temperature days. One of the public housing residents reported, "All these windows were shut airtight and couldn't be opened if we wanted to install a window unit. I live on the top floor of the housing; with heat on, it is unsafe and unhealthy conditions for us."[67]

This issue was reported several times in community meetings not only in public housing complexes but also in private residential neighborhoods.

The state sanitation code requires property owners to make heat available until June 15. Some landlords opt to keep it on, making apartments unbearable to live in for the residents, especially for seniors. The state sanitary code is silent on air conditioners. This void often prevents housing managers from switching on central air conditioning before June 15. The state issued an additional guideline (105 Code of Massachusetts Regulations CMR 410.201) in 2018 to clarify that the code does not prevent any property owner (including local housing authorities) from turning on air conditioners before June 15, if the minimum temperature is maintained in each dwelling. However, it is hard to implement these guidelines effectively. For example, many buildings are not equipped with cooling systems; others with complex heating, ventilation, and air conditioning systems need several days to switch between heating and cooling. Additionally, window-only air conditioning units are not necessarily efficient in lowering the apartment's temperature when temperatures are very high outside. With changing climate, advocacy groups have been pushing to change the decade-old state sanitary code and revise the end of heat month to May 30. However, substantial changes have yet to happen.

How Useful Are Cooling Centers?

Cooling centers are primarily opened to provide relief to families without air conditioning during heatwaves across many communities in the United States.[68] In Boston, for instance, Boston Centers for Youth and Families opens their community centers from 9 a.m. to 5 p.m. to help residents during heat emergencies. These are often not helpful for residents living in urban heat islands facing prolonged heat exposure. Community meetings and interviews have repeatedly raised the issue of lack of permanent infrastructure investments to reduce heat peaks. One resident explained, "We do not have any community cooling centers nearby—the nearest one closes at 5 p.m. There are summer days when the apartment is unbearable, and it does not have to be a heatwave. Also, this past year there were many high-temperature days in early May, and most cooling centers opened only after June 15th. These centers were not helpful even though we wanted to use them. I think about elderly and disabled people . . . what would they do? With so many early heat waves, perhaps these centers should revise their timings and schedule."[69]

Transportation was raised as both a resource and a barrier for people trying to travel to cooler places during heat events. Cooling centers are not easily accessible to Dorchester residents whose travel to these facilities requires walking to nearby bus or train facilities and subsequently waiting at an overheated bus stop or on a subway platform during high-temperature days. This may limit access to these centers during heat events for residents, older adults, and people with disabilities. While some community members positively referenced transportation programs for seniors or paratransit for individuals with disabilities, participants reported either not having awareness of these centers or not being comfortable visiting the centers for heat relief.[70]

Tree Equity and Green Gentrification

Several planning documents have frequently referenced the positive effects of tree canopies on reducing urban heat island footprint.[71] Currently, 28 percent of Boston's land area is covered by trees. This percentage is significantly less than other urban areas in Massachusetts, which have about 65 percent of land covered by trees.[72] Neighborhoods such as Chinatown and Dorchester have a significantly lower proportion of tree canopies. Boston has made little progress in increasing tree canopies as recommended in its most recent Open Space Plan.[73] The number of roads, sidewalks, and other impermeable surfaces and the demand for more housing reduce the space available to plant trees. One resident commented, "There is not a single tree in our area. Walking around the neighborhood and to the transit is hard during the summer months. The streets feel boiling!"[74]

While there are significant disparities in shaded green spaces and tree cover in the areas experiencing increasing heat issues, the fear of green gentrification is also tangible within the community. "Green infrastructure" has been used as a primary urban planning intervention for mitigating climate change impacts (such as flooding, heat, etc.) in Boston.[75] However, there is growing anxiety among residents who associate green spaces with gentrification. Some community leaders fear that green spaces designed to combat different forms of climate change, including extreme heat, may push out the poor and working-class residents. Interviewees emphasized public decisions that had disadvantaged their communities over many generations, from redlining that led to years of disinvestment to interstate highway projects that demolished several Black neighborhoods. One resident expressed their fears in the following way: "Many residents in the area are already concerned about 'gentrification' in the neighborhood. Rents are rising, and many of our friends' families are displaced. Plans to create parks and green spaces without consideration around the area may foster the problem. There is widespread fear among communities of being pushed out."[76]

The theory of green gentrification is an emerging concept in peer-reviewed literature as well as in popular discourse. It is based on the idea that adaptation projects through green infrastructures contribute to differences in property value and thus contribute to the displacement of former residents searching for more affordable housing and neighborhood options elsewhere. Some of the recent scholarly work highlights the strong association between green spaces and displacement in cities such as Atlanta (Fourth Ward), Austin, and Boston (East Boston).[77] In particular, East Boston has become one of the most gentrifying areas of the city largely experiencing this phenomenon.[78] While the literature on green gentrification suggests looking at just planning to help prevent green gentrification, not many development projects have been inspired by those lessons.[79]

DISCUSSION

Over the years, urban heat research has affected two central areas of planning: heat risk response and urban heat management. Heat risk response refers to an emergency dimension of planning, emphasizing municipal strategies focusing on short-term strategies during extreme heat events.[80] Examples of this type of planning include improved forecasting, early heat warning system, emergency shelters, and cooling stations. In contrast, urban heat management refers to municipal heat management strategies designed to reduce heat exposure, intensity, and duration during and after periods of extreme heat events.[81] Examples include urban greening and albedo modification using reflective and lighter-color materials in urban infrastructures such as streets and parking lots.[82]

From the onset of heat planning in the 1970s, heat has been understood as an "emergency" issue. This is the primary reason extreme heat issues are typically categorized under public health and emergency management issues.[83] Scholars in disaster management believe that heat should not be recognized only as an "emergency" issue but should also have a "chronic" challenge to be managed. For some scholars, defining heat as an emergency issue reflects a reactive approach that fails to consider the extreme vulnerability of marginalized populations to this aspect of climate change.[84] However, scholars in the natural hazard mitigation planning paradigm seek to integrate both approaches by defining urban heat management as "local planning actions to lessen the exposure to the risk through mitigation and recovery action."[85] Boston's extreme heat planning approach is ahead of state-wide adaptation planning efforts because it is crafting planning, regulatory, and management measures explicitly dealing with expected heat waves. However, the effectiveness of these efforts may be undermined by the limited effort the city has made to engage those directly affected by extreme heat and heatwaves in the planning process.

Boston's extreme heat planning approach's effectiveness may also be limited by the lack of a comprehensive extreme heat mitigation and management plan. This issue is even more relevant to environmental justice communities such as Dorchester, where the disconnection between residents' needs and concerns and provided programs appears problematic. While the piecemeal approach versus a more comprehensive approach has been assessed as problematic in this research, it also is an issue that the city is currently trying to address. As we are writing this chapter, the City of Boston published its first "heat resiliency study." The analysis of this recent study was not part of the research scope presented in this chapter, and additional work is needed to understand its impacts in local communities such as Dorchester. Our preliminary set of community interviews and participant observations in Dorchester surfaced a few constructive recommendations designed to strengthen heat planning practices in Dorchester:

- **RECONSIDER PRIORITIES.** The first recommendation acknowledges the need to bring heat planning to the forefront of climate change policy making. Since Boston is considered one of the cities most vulnerable to sea level rise and coastal and inland flooding, it is challenging to prioritize extreme heat.[86] At the moment, extreme heat planning advocates must compete for limited time and resources to secure the attention of important public and private sector development actors who are focused on other climate change priorities. Technical assessments determining what should be considered a climate change priority need to be revisited and reevaluated to include issues such as extreme heat often overlooked within the mainstream climate change agenda. Most importantly, community needs, struggles, values, and experiences need to be studied and prioritized in decision making to create socially sensible adaptation actions.

- **LEARN THROUGH PRACTICE.** The second recommendation is to use the heat planning process as an immediate testing ground for true inclusive and equitable processes. Heat can be an inconvenience for some individuals and communities, but it can also be lethal for the vulnerable and marginalized. Our research shows that increasing heat exposure deeply impacts the daily lives of poor and working-class communities. It shows that the current heat management strategies are driven by approaches grounded in knowledge generated through expert-driven vulnerability assessments and are often translated from contexts where climate, weather, and microclimatic variation are different. Additionally, questions on the wide applicability of concrete strategies for cooling emerged. These are related, for instance, to the accessibility of cooling centers, the high energy burdens that residents must sustain for air conditioners, and the creeping fear of being displaced in the name of landscape upgrades for resilience planning. Those flags push for a renovated research agenda focused on reevaluating the understanding of heat issues among different demographics to generate new pathways to make heat planning more effective and equitable.

- **LOOK BACK TO LOOK FORWARD.** A third recommendation is to look critically at the existing old infrastructural system. Like many cities in New England, Boston has a very old infrastructural system that is often taken for granted in adaptation planning. In the context of our research, this topic is relevant because many of the neighborhoods historically neglected from public investments have suffered from a lack of infrastructure upkeep. In these neighborhoods, a careful and detailed restoration planning and design approach should be undertaken to upgrade existing and deteriorated public infrastructures to address climate change concerns. In recentering the focus of adaptation to forms of rehabilitation of the built environment in disenfranchised communities, it is crucial to experiment with new forms of community-led planning processes. Such processes would guarantee an upgrade of those infrastructures so that communities have power in the decisions affecting their living environment and control the management of newly established projects/programs. On the one side, this approach will foster enhanced initiatives for heat planning and bring much-needed public attention to long-forgotten areas of the city; on the other, it will advance concrete examples of community-led development that would foster social cohesion in those neighborhoods. Such efforts are long overdue in these contexts where skepticism toward government-led initiatives on resilience keeps ramping up, and genuinely community-sensitive approaches are needed to address serious climate concerns.

CONCLUSIONS: WHAT FORM OF HEAT PLANNING SHOULD BE FOSTERED?

In Boston, the overall planning response to extreme heat impacts across the city has relied on efforts that parallel the national trend. On the one hand, there is a stark separation in approaches between "emergency" and "chronic" planning objectives and strategies; on the other hand, there is little overall clarity about who is responsible for specific kinds of initiatives. This latter issue results in a scattered approach inconsistent with a potential comprehensive planning process. The shortcomings of our existing approaches to extreme heat planning are currently emerging from the voices of those who experience firsthand adaptation management and emergency measures for heat and call for alternative adaptation planning strategies.

The previous reflections on preliminary participant observation and resident interview data presented in this chapter offer a cautionary tale for climate change planning in Dorchester and, more generally, in Boston. These data suggest that engagement processes focused on climate change planning need to take a two-pronged approach. They must combine collective reflections on the meaning of technology and technical solutions to climate change with research focused on local concerns not directly related to climate. In other words, there is an urgent need to have planning processes in place that, while continuing to look at technical innovations, need deeper engagement with communities to reflect on the effect of those innovations and collectively decide what actions need to be prioritized and fostered.

As researchers within the Department of Urban Planning and Community Development (UPCD) at University of Massachusetts Boston, we seek to make concrete, more inclusive approaches to planning for resilience. The Summer Immersion in Community Resiliency Planning Program has been developed by UPCD in partnership with the Boston Planning and Development Agency (BPDA), Boston Public Schools, and Madison Park Technical Vocational High School. The pilot program in summer 2022 establishes a four-week experiential class for fifteen high school students of Madison Park Vocational Technical School in collaborative research with residents and leaders to determine the nature, extent, and impact of extreme heat on current residents. This experience constitutes an innovative community/university partnership experiment to collectively reflect on dealing with a critical climate-related issue affecting one of the most underserved communities and then explore community-led urban planning, policy, and design mitigation strategies. While its results are not included in this chapter, it is relevant to share its launch as a concrete example in the effort to implement new forms of collaboration between local communities and long-term established institutions (such as the BPDA) to shape better approaches to heat planning.[87]

NOTES

1 David Reidmiller, Christopher W. Avery, David R. Easterling, Kenneth E. Kunkel, et al., Impacts, Risks, and Adaptation in the United States: Fourth National Climate Assessment, vol. 2 (Washington DC: US Global Change Research Program, 2018), https://nca2018.globalchange.gov/.

2 Katharine Hayhoe, Donald J. Wuebbles, David R. Easterling, et al., *Our Changing Climate: Impacts, Risks, and Adaptation in the United States: Fourth National Climate Assessment*, ed. David Reidmiller, Christopher W. Avery, David R.

Easterling, Kenneth E. Kunkel, et al., vol. 2 (Washington DC: US Global Change Research Program, 2018), 72–144 72–144, doi: 10.7930/NCA4.2018.CH2.

3 Hayhoe et al., *Our Changing Climate*.

4 Scott C. Sheridan and Michael J. Allen, "Temporal Trends in Human Vulnerability to Excessive Heat," *Environmental Research Letters* 13, no. 4 (2018): 043001.

5 The Centers for Disease of Control define "extreme heat" as summertime temperatures that are much hotter and/or humid than average. The definition of "extreme heat" varies based on many different factors, such as location, weather conditions (temperature, humidity, and cloud cover), and the time of year. For example, a May temperature of 92 degrees Fahrenheit in Boston is extreme heat, whereas a May temperature in Phoenix would have to reach more than 100 degrees Fahrenheit to be considered extreme.

6 Timothy. R. Oke, "City Size and the Urban Heat Island," *Atmospheric Environment* 7, no. 8 (1973): 769–79.

7 Oke, "City Size," 769–79; Christopher Coutts and Micah Hahn, "Green Infrastructure, Ecosystem Services, and Human Health," *International Journal of Environmental Research and Public Health* 12, no. 8 (2015): 9768–98.

8 James Milner, Colin Harpham, Jonathon Taylor, et al., "The Challenge of Urban Heat Exposure under Climate Change: An Analysis of Cities in the Sustainable Healthy Urban Environments (SHUE) Database," *Climate* 5, no. 4 (2017).

9 Tirthankar Chakraborty, Angel Hsu, Diego Manya, et al., "Disproportionately Higher Exposure to Urban Heat in Lower-Income Neighborhoods: A Multi-City Perspective," *Environmental Research Letters* 14, no. 10 (2019); Colleen R. Reid, Marie S. O'Neill, Carina J. Gronlund, et al., "Mapping Community Determinants of Heat Vulnerability," *Environmental Health Perspectives* 117, no. 11 (2009): 1730–36; Susanne Amelie Benz and Jennifer Anne Burney, "Widespread Race and Class Disparities in Surface Urban Heat Extreme across the United States," *Earth's Future* 9, no. 7 (2021): 1–14.

10 Angel Hsu, Glenn Sheriff, Tirthankar Chakraborty, and Diego Manya, "Disproportionate Exposure to Urban Heat Island Intensity across Major US Cities," *Nature Communications* 12, no. 1 (2021): 1–11.

11 Hsu et al., "Disproportionate Exposure," 1–11.

12 Benz and Burney, "Widespread Race and Class Disparities."

13 Jeremy Hoffman, Vivek Shandas, and Nicholas Pendleton, "The Effects of Historical Housing Policies on Resident Exposure to Intra-Urban Heat: A Study of 108 US Urban Areas," *Climate* 8, no. 1 (2020): 1–15. "Redlining" refers to the historical practice of denying home loans or insurance based on an area's racial composition. Created by the Home Owners' Loan Corporation in 1938 and later used by the Federal Housing Administration, the colored map used the race of inhabitants as a key criterion when designating risk for mortgage lenders in neighborhoods, giving rise to the term "redlining." The categories were A (best), B (still desirable), C (declining), and D (hazardous). Anywhere Blacks

lived and anywhere Blacks lived nearby were colored red to indicate that these neighborhoods were too risky for mortgages.

14 Bruce Mitchell and Juan Franco, *HOLC "Redlining" Maps: The Persistent Structure of Segregation and Economic Inequality*, National Community Reinvestment Coalition technical report, March 2018.

15 David Hondular, Rovert C. Balling, Jennifer K Vanos, and Matei Georgescu, "Rising Temperatures, Human Health and the Role of Adaptation," *Current Climate Change Report* 1, no. 10; NOAA, *Weather Fatalities 2020: 80 Years of Severe Weather Fatalities*, 2020, accessed May 10, 2021, https://www.weather.gov/hazstat/.

16 NOAA, *Weather Fatalities 2020*.

17 Ambarish Vaidyanathan, Josephine Malilay, Paul Schramm, and Shubhayu Saha, "Heat-Related Deaths United States, 2004–2018" *Centers for Disease Control Morbidity and Mortality Weekly Report* 69, no. 24 (2020): 729–34, https://doi.org/10.15585/mmwr.mm6924a1.

18 Melanie Boeckmann and Hajo Zeeb, "Justice and Equity Implications of Climate Change Adaptation: A Theoretical Evaluation Framework," *Healthcare* 4, no. 3 (2016): 65; Elisabeth Anne-Sophie Mayrhuber, Michel L. A. Dückers, Peter Wallner, et al., "Vulnerability to Heatwaves and Implications for Public Health Interventions—A Scoping Review," *Environmental Research* 166, no. 10 (2018): 42–54.

19 US Centers for Disease Control and Prevention, *Heat-Related Deaths— Philadelphia and United States, 1993–1994*, 1994, accessed May 10, 2021, https://www.cdc.gov/mmwr/preview/mmwrhtml/00031773.htm.

20 Kristie L. Ebi, "Effective Heat Action Plans: Research to Interventions," in *Environmental Research Letters* (Bristol, UK: Institute of Physics Publishing, 2019).

21 Eric Klinenberg, *Heat Wave: A Social Autopsy of Disaster in Chicago*, 2nd ed. (Chicago: University of Chicago Press, 2015).

22 Natalie R. Sampson, Carina J. Gronlund, Miatta A. Buxton, et al., "Staying Cool in a Changing Climate: Reaching Vulnerable Populations during Heat Events," *Global Environmental Change* 23, no. 2 (2013): 475–84.

23 Joyce Klein Rosenthal, Patrick L. Kinney, and Kristina B. Metzger, "Intra-Urban Vulnerability to Heat-Related Mortality in New York City, 1997–2006," *Health and Place* 30 (November 2014): 45–60.

24 Carina J. Gronlund, "Racial and Socioeconomic Disparities in Heat-Related Health Effects and Their Mechanisms: A Review," *Current Epidemiology Reports* 1, no. 3 (2014): 165–73; Klinenberg, *Heat Wave*.

25 Olga V. Wilhelmi and Mary H. Hayden, "Connecting People and Place: A New Framework for Reducing Urban Vulnerability to Extreme Heat," *Environmental Research Letters* 5, no. 1 (2010): 70–59; B. L. Turner, Roger E. Kasperson, Pamela A. Matsone, et al., "A Framework for Vulnerability Analysis in Sustainability Science," *Proceedings of the National Academy of Sciences of the United States of America* 100, no. 14 (2003): 8074–79.

26 Joern Birkmann, Omar-Dario Cardona, Martha L. Carreño, et al., "Framing Vulnerability, Risk and Societal Responses: The MOVE Framework," *Natural Hazards* 67, no. 2 (2013): 193–211.

27 Wilhelmi and Hayden, "Connecting People and Place," 70–59.

28 Wilhelmi and Hayden, "Connecting People and Place," 70–59.

29 Reid et al., "Mapping Community Determinants," 1730–36; Daniel P. Johnson, Jeffrey S. Wilson, and George C. Luber, "Socioeconomic Indicators of Heat-Related Health Risk Supplemented with Remotely Sensed Data," *International Journal of Health Geographics* 8, no. 57 (October 2009): 57–70.

30 Brian Stone, Kevin Lanza, Evan Mallen, et al., "Urban Heat Management in Louisville, Kentucky: A Framework for Climate Adaptation Planning," *Journal of Planning Education and Research* 43, no. 2 (October 2019): 346–58.

31 Boeckmann and Zeeb, "Justice and Equity Implications," 65; Sari Kovats and Kristie Ebi, "Heatwaves and Public Health in Europe," *European Journal of Public Health: Oxford Academic* 16, no. 6 (2006): 592–99.

32 Monika Nitschke, Antoinette Krackowizer, Alana L. Hansen, et al., "Heat Health Messages: A Randomized Controlled Trial of a Preventative Messages Tool in the Older Population of South Australia," *International Journal of Environmental Research and Public Health* 14, no. 9 (2017): 992–1002.

33 Kelsey N. Ellis., Jon M. Hathaway, Lisa Reyes Mason, et al., "Summer Temperature Variability across Four Urban Neighborhoods in Knoxville, Tennessee, USA," *Theoretical and Applied Climatology* 127 (2017): 701–10.

34 Gertrud Hatvani-Kovacs, Martin Belusko, Natalie Skinner, et al., "Heat Stress Risk and Resilience in the Urban Environment," *Sustainable Cities and Society* 26 (October 2016): 278–88.

35 Andrew M. Fraser, Mikhail V. Chester, and David Eisenman, "Strategic Locating of Refuges for Extreme Heat Events (or Heat Waves)," *Urban Climate* 25 (September 2018): 109–19.

36 Fraser, Chester, and Eisenman, "Strategic Locating of Refuges."

37 Sari Kovats and Shakoor Hajat, "Heat Stress and Public Health: A Critical Review," *Annual Rev Public Health* 29 (March 2008): 41–55.

38 Ebi, "Effective Heat Action Plans."

39 Radley Horton, Gary Yohe, William Easterling, Robert Kates, et al., "Northeast," in *Climate Change Impacts in the United States: The Third National Climate Assessment*, US Global Change Research Program, 2014, 371–95.

40 City of Boston Environment Department, *Climate Ready Boston*, accessed August 2021,https://www.boston.gov/sites/default/files/20161207_climate_ready_boston _digital2.pdf.

41 City of Boston, *Climate Ready Boston*.

42 The UHI index used in this report includes land cover type, percentage of greenery to paved areas, building height, and population density. Higher population density introduces more heat waste back into the environment due

to transportation, industrial facilities, and heating and cooling of buildings. Additionally, tall buildings provide multiple surfaces that reflect and absorb sunlight and influence how air moves through the city, which plays a prominent role in the trapping or dissipation of heat; Lei Zhao, Xuhui Lee, Ronald Smith, and Keith Oleson, "Strong Contributions of Local Background Climate to Urban Heat Islands, " *Nature* 511 (2014): 216–19.

43 City of Boston Environment Department, *Heat Resilience Solutions for Boston,* accessed June 2021, https://www.boston.gov/environment-and-energy/heat -resilience-solutions-boston.

44 Michael Adler, Samantha Harris, Megan Krey, et al., *Preparing for Cool Way to Attack Global Warming Heat Waves in Boston* (Boston, MA: City of Boston, 2010).

45 City of Boston, *Climate Ready Boston.*

46 City of Boston Environment Department, *Guidelines for High-Performance Buildings and Sustainable Development,* accessed May 10, 2021, https://www.city ofboston.gov/environment/pdfs/hpb_guidelines.pdf.

47 City of Boston Environment Department, *Guidelines,* 1.

48 Boston Planning Development Agency, *Boston Zoning Code, Article 37,* "Green Buildings," 2007, accessed August 15, 2021, http://www.bostonplans.org/getattach ment/a77140ba-cdd0-48fb-9711-84540bf31f35.

49 Boston Planning Development Agency, *Climate Resiliency Review Policy,* Boston Planning and Development Agency, 2017, accessed August 15, 2021, http:// www.bostonplans.org/getattachment/404b7556-5274-42b1-b9d5-b9493fa7c4f1.

50 "City of Boston Alerts and Notifications," May 21, 2021, https://www.boston.gov /departments/emergency-management/city-boston-alerts-and-notifications.

51 "City of Boston Emergency Management," October 10, 2023, https://www.boston .gov/departments/emergency-management.

52 US Census Bureau, *Quick Fact: State of Massachusetts,* 2020, accessed May, 2021, https://www.census.gov/quickfacts/fact/table/MA,US/PST045221.

53 US Census Bureau, *Quick Fact.*

54 Boston Planning and Development Agency, *Boston in Context: Neighborhoods: 2015–2019 American Community Survey,* BPDA research team, https://www .bostonplans.org/getattachment/e2eb8432-ac72-4a7e-8909-57aafdfbecd9.

55 Boston Planning and Development Agency, *2015–2019 American Community Survey.*

56 City of Boston, *Climate Ready Boston.*

57 City of Boston, *Heat Resilience Solutions for Boston.*

58 City of Boston, *Open Space and Recreation Plan, 2015–2021,* 2015 plan, accessed May 10, 2021, https://www.boston.gov/environment-and-energy/open-space-and -recreation-plan-2015-2021.

59 City of Boston, *Tree Canopy Assessment Boston*, Boston: Boston Parks and Recreation Department, 2020, https://www.boston.gov/sites/default/files/file/2020/09/Change-assessment_w_MJW-letter.pdf.

60 Sajani Kandel's environmental sciences dissertation focuses on this topic and deeply explores some of the questions addressed in this chapter.

61 US Energy Information Administration, *Household Energy Use in Massachusetts: A Closer Look at Residential Energy Consumption: Residential Energy Consumption Survey (RECS)*, US Energy Information Administration, 2015.

62 A Key Informant Interview (KII #1) was conducted with a resident to understand their lived experience and coping capacity of increasing extreme heat in Boston. The interview was conducted in Dorchester, Boston, in August 2021.

63 For a breakdown of median energy burdens for other groups nationally, regionally, and in twenty-five select metro areas, see Ariel Drehobl, Lauren Ross, and Roxana Ayala, *How High Are America's Residential Energy Burdens?*, September 2020, American Council Energy-Efficient Economy, https://www.aceee.org/sites/default/files/pdfs/u2006.pdf.

64 KII #2, July 2021.

65 KII #9, August 2021.

66 KII #10, August 2021.

67 KII #3, July 2021.

68 Fraser, Chester, and Eisenman, "Strategic Locating of Refuges."

69 KII #5, July 2021.

70 KII #5, July 2021.

71 City of Boston, *Climate Ready Boston*; City of Boston, *Imagine Boston* (Boston: City of Boston, https://www.boston.gov/sites/default/files/; City of Boston, *Open Space and Recreation Plan*.

72 City of Boston, *Tree Canopy Assessment Boston*.

73 City of Boston, *Open Space and Recreation Plan*.

74 KII #6, July 2021.

75 "Green infrastructure" is an interconnected network of green space that conserves natural environment values and functions (Mark Benedict and Edward McMahon, "Green Infrastructure: Smart Conservation for the 21st Century," *Renewable Resources Journal* 20, no. 3 [2002]: 12–17) and provide cobenefits to human health and well-being (Margarita Triguero-Mas, Payam Dadvand, Marta Cirach, and David Martinez, et al., "Natural Outdoor Environments and Mental and Physical Health: Relationships and Mechanisms," *Environment International* 77 [2015]: 35–41) while making cities more livable (Robert Young, Julie Zandersb, Katherine Lieberknecht, and Elizabeth Fassman-Beck, "A Comprehensive Typology for Mainstreaming Urban Green Infrastructure," *Journal of Hydrology* 519 [2014]: 2571–258); City of Boston, *Climate Ready*

Boston, 2016; City of Boston Environment Department, *Coastal Resilience Plan for Dorchester,* accessed May 10, 2021, https://drive.google.com/file/d/1sjfDuh VsKgUxnHlT_eGf4BSCuZ1Im8uw/view; City of Boston Environment Department, *Heat Resilience Solutions for Boston.*

76 KII #7, July 2021.

77 Isabelle Anguelovski, James Connolly, Hamil Pearsall, et al., "Opinion: Why Green 'Climate Gentrification' Threatens Poor and Vulnerable Populations," *Proceedings of the National Academy of Sciences* 116 (2019): 26139–43.

78 Galia Shokry, Isabelle Anguelovski, James J. T. Connolly, et al., "'They Didn't See It Coming': Green Resilience Planning and Vulnerability to Future Climate Gentrification," *Housing Policy Debate* 32, no. 1 (2022): 211–45.

79 Linda Shi, Eric Chu, Isabelle Anguelovski, et al., "Roadmap towards Justice in Urban Climate Adaptation Research," *Nature Climate Change* 6, no. 2 (2016): 131–37.

80 Annie Bolitho and Fiona Miller, "Heat as Emergency, Heat as Chronic Stress: Policy and Institutional Responses to Vulnerability to Extreme Heat," *Local Environment* 22, no. 6 (2016); Stone et al., "Urban Heat Management."

81 Stone et al., "Urban Heat Management."

82 Stone et al., "Urban Heat Management."

83 Bev Wilson, "Urban Heat Management and the Legacy of Redlining," *Journal of the American Planning Association* 86. no. 4 (2020): 443–57.

84 R. Zehra Zaidi and Mark Pelling, "Institutionally Configured Risk: Assessing Urban Resilience and Disaster Risk Reduction to Heat Wave Risk in London," *Urban Studies* 52, no. 7 (2013): 1218–33.

85 Philip Berke, "Rising to the Challenge: Planning for Adaptation in the Age of Climate Change," in *Adapting to Climate Change: Lessons from Natural Hazards Planning,* ed. B. C. Glavovic and G. P. Smith (Dordrecht, Germany: Springer, 2014), 171–90.

86 Nicole Mahlkow and Julie Donner, "From Planning to Implementation? The Role of Climate Change Adaptation Plans to Tackle Heat Stress: A Case Study of Berlin, Germany," *Journal of Planning Education and Research* 37, no. 4 (2016): 385–96.

87 Since the initial submission of this chapter, our department has implemented the first iteration of the summer program. Through this approach to resilience planning, concrete adaptation measures have been recommended for Roxbury. These include the redesign of playgrounds as open-air "cooling centers," the redesign of bus stops with heat-repellent materials, city-sponsored grants for businesses to install technological improvements for heat reduction, and the introduction of cooling rooms in publicly owned housing. The complete report from the summer school program is available at University of Massachusetts

Boston, "Cool Roxbury: Lower Roxbury's Extreme Heat Challenges and Solutions, Department of Urban Planning and Community Development," summer Immersion Program in Community Resiliency Planning, 2016, https://pauldavidoff.com/aditional-resources/#activism.

CHAPTER 8

Climate Change and Housing Instability

KIM FLIKE, SHOSHANA V. ARONOWITZ, AND TERI ARONOWITZ

Homelessness is a serious and growing global issue. It is estimated that 1.6 billion people globally live in housing that is inadequate for human habitation.[1] The absence of a universally agreed-on definition of homelessness contributes to the challenges in tackling this issue. The United Nations (UN) provides a working definition of homelessness that includes people who lack livable space, which includes those living in homeless shelters or experiencing housing insecurity, in addition to those sleeping unsheltered outside.[2] Even among the thirty-six highest-income countries in the Organization for Economic Cooperation and Development, definitions of homelessness vary despite efforts to standardize the criteria.[3] In 2020, homeless populations in these thirty-six countries was estimated to include more than 2.1 million people.[4]

In the United States definitions of homelessness vary between various federal and state organizations. The most restrictive, yet most commonly used, definition in the United States is from the Homeless Emergency Assistance and Rapid Transition to Housing Act of 2009. This law defines persons experiencing homelessness as individuals or families without a regular, stable nighttime living accommodation, or an accommodation that is either a place not meant for human habitation or is an emergency shelter or transitional housing. In addition, a person experiencing homelessness is someone who was previously homeless before incarceration and is exiting an institution such as prison or jail; a person in immediate danger (within fourteen days) of losing their housing with no other place to go; or a survivor fleeing domestic violence. This definition is used by the Department of Housing and Urban Development (HUD) in the homeless point-in-time (PIT) counts to prioritize recipients for available HUD housing.

The 2020 US PIT count, the nationwide annual count to estimate homelessness, reported an increase for the fourth consecutive year, estimated to be over 580,000 people.[5] We will use the term "people experiencing homelessness" (PEH) throughout this chapter. In the United States, 38.5 percent of PEH are female, 60.7 percent are male, 0.5 percent are transgender, and 0.3 percent are gender-nonconforming. The racial makeup of PEH is 48.3 percent White, 39.4 percent Black or African American, 1.3 percent Asian, 3.3 percent Native American, 1.5 percent Pacific Islander, and 6.1 percent multiracial. Data from the 2021 PIT count reported an 8 percent decrease in the total number of sheltered PEH between 2020 and 2021. However, due to the constraints posed by the COVID-19 pandemic, many municipalities did not conduct a count of unsheltered PEH. Additionally, the impact of reduced beds due to social distancing regulations and the number of PEH avoiding congregate shelters due to fear of COVID-19 transmission is unknown. Therefore, we will refer to the 2020 PIT data when describing the US homeless population because the 2021 national PIT count of unsheltered PEH is incomplete.

The right to adequate housing was included in the Declaration of Human Rights in 1948.[6] However, it took until 2020 for the UN to pass their first resolution on homelessness recognizing the issue as an intrusion on human dignity that affects people of all ages globally.[7] The purpose of this resolution was to recognize the diverse drivers of homelessness, often caused by systemic social and economic inequities. One such inequity singled out in this resolution was climate change, where the resolution noted with great concern the effect that climate change has and will continue to have on those experiencing and at risk for homelessness around the globe.[8]

Climate Change

Average US temperatures have increased between 1.3 and 1.9 degrees Fahrenheit, and most of this increase has occurred since 1970.[9] Climate change leads to increased morbidity and mortality secondary to exposure to extreme temperatures, respiratory illness from poor air quality, and an increase in waterborne diseases.[10] Additionally, infrastructure, including housing, is damaged by rising sea levels, heavy rains, and extreme heat.[11] The degradation of infrastructure is also damaging to water quality due to increased contamination after a heavy downpour or flooding.[12] It is predicted climate change will have devastating effects on food supplies disproportionally affecting vulnerable populations globally.[13] Climate events such as extreme temperatures, floods, and wildfires are already negatively affecting infrastructure and individuals living in extreme poverty, including PEH.[14]

Theoretical Framework

The pressure-and-release (PAR) model has been widely used by disaster researchers to understand the factors linked to social vulnerability.[15] This model outlines how PEH are especially vulnerable to harm caused by adverse weather events and natural disasters associated with climate change. The PAR model incorporates a "progression" of vulnerabilities, including systemic inequalities that deny access to power and resources. The model also emphasizes the need to examine social structures and power relations that increase risk, especially when coupled with a hazardous event such as a natural disaster.

The PAR model proposes that human vulnerability and natural hazards coincide to create disaster, wherein risk level for experiencing a disaster is a product of a natural hazard and a person or community's vulnerability. In this model, three components contribute to vulnerability: root causes, dynamic pressure, and unsafe conditions. Root causes include economic, political, and demographic processes that affect the allocation of resources and their distribution. Dynamic pressure involves economic and political circumstances locally. Unsafe conditions are identified as vulnerability expressed in time and space, such as those induced by the physical environment, the local economy, and social relations.[16] The purpose of this chapter is to present three weather events as case studies to illuminate how natural disasters caused by climate change affect PEH. We will focus on extreme temperatures, flooding, and wildfires.

CASE STUDIES

The three weather events that we will be focusing on are extreme weather temperatures (both heat and cold), flooding, and wildfires. We will discuss each case with consideration of the root causes that affect the distribution of resources as well as the potential economic and political circumstances that increase unsafe conditions for PEH.

Extreme Temperatures

Increasing global temperatures caused by climate change are resulting in both extreme heat and cold weather events.[17] Extreme heat events are defined as a series of days with higher than average temperatures for a particular location and time of year.[18] Globally, the ten-year period from 2009 to 2019 was the warmest decade on record,[19] and the summer of 2021 was the warmest in history for the contiguous United States.[20] Paradoxically, the overall increasing global temperatures have also led to an increase in extreme cold weather through weakening of the polar vortex that pushes cold air into lower latitudes. The weakening of the polar vortex causes exceptionally cold air outbreaks and increases in heavy snowfall, particularly in the northern hemisphere.[21]

Extreme temperatures are particularly distressing for people experiencing homelessness. Temperature fluctuations outside of normal ranges for the season, both heat and cold, can be deadly to those without access to shelter. Heat-related deaths among the homeless population more than doubled between 2015 and 2016 in Maricopa County, Arizona,[22] while people experiencing homelessness in Poland are thirteen times more likely to die of hypothermia than the general population in that country.[23] These increased deaths due to extreme temperatures have been linked to preexisting chronic conditions common among people experiencing homelessness, including heart disease, mental health conditions, and substance use disorders.[24]

To address these deaths, many local governments developed policies that require sheltering of PEH in extreme weather emergencies. These policies are often labeled "code red" for extreme heat and "code blue" for extreme cold. Shelters are required to take in as many people as allowed by building codes to ensure shelter and safety of PEH.[25] Enforcing these policies was complicated by the COVID pandemic as shelters had to balance providing shelter with social distancing restrictions.[26]

Extreme Heat

In the summer of 2021, Portland, Oregon, experienced a heat wave with temperatures over 103 degrees Fahrenheit (39 degrees Celsius). It was hotter than Phoenix, where the desert city hit a below-normal one hundred degrees Fahrenheit (38 degrees Celsius).[27] During the heat wave over seventy-one people died of heat-related causes in Multnomah County, where Portland is located. To prevent hyperthermia, the US Centers for Disease Control and Prevention recommend increasing fluid intake, avoiding sugary drinks, staying indoors with a fan or air conditioner, and wearing lightweight loose-fitting clothing.[28] While these are reasonable recommendations for people with adequate housing, they are often impossible for unsheltered PEH. Access to clean water can be limited, and many unsheltered PEH avoid being overhydrated due to lack of restroom access.[29] People experiencing homelessness may be reluctant to enter cooling shelters as their possessions may be confiscated, and some experience ill-treatment or violence at shelters.[30] They may not have a place to store clothing for different seasons or money to buy weather-appropriate clothing.[31] This can leave them facing the choice to leave their warmer clothing behind and risk not having anything to wear once colder weather arrives or braving heat waves with more layers than recommended.

Extreme Cold

For two weeks in February 2021, Texas experienced a winter storm followed by historic freezing temperatures,[32] which resulted in rolling electricity blackouts across the state for several days.[33] The freezing temperatures contributed to the deaths of at least six PEH,[34] and the power outages caused plumbing damage at many shelters in the state.[35] Some PEH who were able to stay in a hotel or warming center through the frigid temperatures reported their belongings being stolen or damaged on return to their encampments.[36]

Despite dangerous heat waves, icy roads, and freezing temperatures, advocates continued their outreach to PEH during weather emergencies.[37] Housed community members often step up to provide assistance to PEH, to help with a lack of coordinated, sufficient local government response. Volunteers assisted by driving unsheltered people to warming and cooling centers, staffed churches doubling as temporary shelters, and paid for hotel rooms.[38] Advocates also provided other types of services, such as delivering extra warm clothing and hot meals to warming centers during code blue events and water, fans, and information about cooling centers to PEH living in isolated encampments during code red events.[39]

Flooding

Approximately fourteen million people become homeless each year secondary to severe storms.[40] Over two decades ago it was predicted that the effect of sea level rise and more frequent occurrences of severe storms would lead to increased flooding, particularly in coastal communities and along river valleys.[41] These predictions proved true. In 2005, Hurricane Katrina hit the gulf coast and remains the costliest storm in US history. Four years after Katrina twelve thousand people were homeless in New Orleans, double the number of PEH compared to before Katrina, and sixty-five thousand buildings remained abandoned.[42] In 2011, Hurricane Irene ravaged communities on the East Coast, washing away homes under the floodwaters of swollen rivers. Atlantic seashore communities and towns all over Vermont were damaged by the storm's deluge, and hundreds of homes were damaged or destroyed, leaving millions without power.[43] A year later, in 2012, Hurricane Sandy damaged or completely destroyed 650,000 homes, and many families became homeless.[44] Over two hundred individuals were killed by the storm and there was an estimated economic loss of seventy-five billion dollars.

People experiencing homelessness are particularly vulnerable to flood events. An example is the 2013 flooding in Boulder County, Colorado.[45] This disaster, one of Colorado's most destructive, displaced approximately eighteen thousand people and caused over three billion dollars' worth of damage to state infrastructure.[46] Unfortunately, the county was not prepared to provide resources to all the people who required emergency shelter and prioritized sheltering individuals who had not been homeless prior to the flood. For example, one public shelter turned away PEH because they could not provide a home address.[47]

In the years prior to the flooding, the growth of Colorado's tech sector led to gentrification and rising housing prices, causing the displacement of many people. A camping ban in the city of Boulder effectively criminalized homelessness by increasing the risk of interactions and arrest by law enforcement, leading to subsequent imprisonment and criminal records and making future housing opportunities even harder to obtain.[48] Camping bans also further marginalize PEH by forcing them to find shelter in remote and unstable areas such as along creeks, which are particularly dangerous during flooding and other climate events.[49]

Flooding can increase risk of infection, especially when populations are displaced and water sources contaminated. Contaminated water can cause outbreaks of infectious diarrhea and was associated with an outbreak of shigella among PEH in Oregon in 2015.[50] Flooding increases vector-borne diseases such as those caused by increased mosquito populations, which can then lead to infections such as dengue in Africa and Zika virus in the United States.[51] Vector-borne disease rates increase with changes in climate globally, and PEH are vulnerable to these infections due to their lack of protection from insects, rodents, and contaminated water.[52]

Flooding from storms is costly, both in terms of infrastructure and human life. The destruction of homes by climate events such as storm flooding leads to increased home prices, displacing people with scarce economic resources. Storms have been increasing in number and severity,[53] and PEH will continue to be disproportionately affected.

Wildfires

Wildfires are a growing concern in the United States and across the globe. The 2020 Australian bushfires made international news for their unprecedented size and level of destruction, and that same year the number of wildfires on the West Coast of the United States reached a record highs.[54] This increase in the frequency and severity of wildfires has been linked to climate change. Specifically, climate change has created seasons of more concentrated rainfall, leaving summer and fall months particularly dry, has led to stronger winds caused by increasing differential temperatures between land and sea, and has increased lightning strikes, all of which increase the likelihood of devastating wildfires.[55] These conditions are expected to worsen as global temperatures increase.[56]

Wildfires have vast economic, social, and environmental impacts. The destruction of homes, contamination of water and soil with ash, and poor air quality can affect people for decades. In the 2018 California wildfires alone, the economic burden was estimated to be $148.5 billion dollars, including direct costs of damaged property, indirect costs of reduced tourism and real estate values, and healthcare costs associated with pollution caused by the fires.[57] These consequences of wildfires are felt acutely among economically vulnerable populations, particularly those experiencing homelessness.

The structural damage caused by wildfires has a twofold impact on PEH: it increases the number of PEH by damaging or destroying housing, and it reduces the available housing supply. The 2018 California Camp Fire destroyed fourteen thousand homes in an area that was already experiencing a housing shortage with a prefire housing vacancy of just 1 to 2 percent.[58] This collision of new and existing homeless populations left local officials and shelter systems with the challenging assignment of who to prioritize for limited bed space. Additional challenges arise when housed and homeless populations shelter together during evacuation events. People experiencing homelessness report feeling marginalized by housed evacuees or may be denied access to evacuation facilities.[59]

The negative health effects of wildfires are well documented. Wildfire particulate matter is especially devastating for vulnerable populations, such as those over sixty-five and those with preexisting respiratory and cardiac conditions.[60] People experiencing homelessness are disproportionally affected by cardiovascular and respiratory conditions, and the homeless population in the United States is increasing in age.[61] These factors, along with the amount of time unsheltered PEH spend outside, make them vulnerable to poor air quality caused by wildfires. In addition, there was no widespread distribution of personal protective equipment for homeless communities, leaving this task up to local advocates with limited resources.[62] There are over 150,000 PEH in California, and up to two-thirds of them are unsheltered. California accounts for over 50 percent of the unsheltered homeless in the United States.[63] The lack of shelters, coupled with the increase in the amount and severity of the wildfires, is concerning for the health and well-being of PEH.

Air pollution from wildfires is not the only concern for PEH. Ash from the wildfires creates a physical nuisance, weighing down tents and increasing the temperature in coolers holding food supplies.[64] Another concern for the California homeless population, particularly the two-thirds who sleep unsheltered in encampments, is ash contamination of the water supply. The settlement of ash on top of a water supply has immediate impact on the potability of the water,[65] and elevated levels of heavy metals have been detected in local water supplies after wildfires.[66] Homeless encampments are often found near sources of water such as rivers and streams.[67] The reduced quality of these water sources due to wildfires presents a unique challenge for PEH, particularly if they are accessing the water before it enters water treatment facilities. The ill effects of ash and heavy metal contamination of water sources are therefore felt even more acutely by the vulnerable unsheltered populations.

Increased wildfire risk along the western coast of the United States has led to new regulations prohibiting encampments in forested areas, including in the areas surrounding major cities with large unsheltered homeless populations, such as Portland, Oregon, and Los Angeles, California.[68] These bans are justified by local officials due to the high risk of wildfires and the alarming increase of fires requiring emergency response in homeless encampments; fires linked to homelessness have doubled since 2018 and are estimated to have caused over $185 million in damages in Los Angeles alone.[69] While the concerns of local officials and housed community members are valid, camping bans do little to address the root causes of wildfires or homelessness, leaving those displaced in an increasingly precarious position. For example, banning homeless encampments in the hills of unincorporated Los Angeles County pushes those staying there into areas such as "skid row" and Venice Beach, which are already overcrowded with unsheltered homeless persons.[70] Efforts to move encampments out of the wildfire high-risk areas are enforced with fines and possible jail time for those considered "noncompliant."[71] The criminalization of homelessness in the name of wildfire safety further affects PEH by burdening them with fines they can't pay and with criminal records that make it even more difficult to obtain employment and housing.[72]

The impacts of wildfires on PEH are vast. The immediate effects on respiratory and cardiovascular health are felt acutely among the unsheltered in California, and the longer-ranging impacts of reduced housing supply makes homelessness even harder to escape. The criminalization of homelessness without appropriate alternatives in the name of fire safety does nothing to improve the conditions for PEH or those living in areas at high risk for wildfires. With the increasing number and severity of wildfires around the world, the plight of PEH needs to be considered in mitigation and evacuation planning.

DISCUSSION

The rapidly increasing impacts of climate change will affect the health and safety of all of us, although individuals who lack adequate shelter are the most immediately vulnerable. Secure housing is necessary for protection from climate events such as extreme temperatures and storms. In addition, climate disasters such as wildfires and flooding destroy housing, and coupled with increasing housing prices, lead to a swelling population of unstably housed individuals who are then at risk during the next hazardous weather event. Misguided and discriminatory attempts to address homelessness, including encampment sweeps and the criminalization of people sleeping outside, ignore the root causes of homelessness and often put PEH in more danger. Unfortunately, local government responses to housing insecurity are often inadequate, and efforts by community advocates and nongovernmental organizations to support PEH are chronically underfunded. In addition, many homeless shelter policies discourage PEH from residing at them, and unsafe conditions at homeless shelters are well-documented.[73]

Guided by the PAR model, we see how PEH are particularly vulnerable during and after climate events due in part to preexisting policies that force them into precarious spaces and because of a lack of a comprehensive safety net to protect them once events happen. Many municipalities' responses to homelessness, often spurred by complaints of housed community members, worsen outcomes for PEH rather than address the issues at the root of homelessness. The criminalization of homelessness pushes PEH into areas that are potentially more dangerous in the event of a climate disaster (next to rivers prone to flooding, for example), and may contribute to future disasters such as wildfires when individuals are forced to camp in fire-prone areas. As public space dwindles and cities use criminalization and design practices to discourage PEH from spending time in well-trafficked areas,[74] PEH are pushed figuratively and literally further into the margins, into spaces where they are in more danger of injury from climate events and violence, and less likely to be noticed and assisted if something happens to them.

Addressing Housing Insecurity in the United States

California has more PEH than any other state.[75] In January 2020, 72 percent of homeless Californians were unsheltered compared to only 5 percent of PEH in New York.[76] Although several factors contribute to this disparity, including warmer weather on the West Coast, New York has had a "right to shelter" law since the 1970s. In 1979, Robert Hayes, a lawyer who cofounded the Coalition for the Homeless, brought a class action lawsuit against the city and the state (*Callahan v. Carey*) to the New York State Supreme Court, arguing that a constitutional right to shelter existed in New York. This right was not recognized in California until more recently. In 2006 six homeless individuals who were unable to find shelter in Los Angeles were arrested. They filed an Eighth Amendment challenge to the enforcement of a City of Los Angeles ordinance criminalizing sitting, lying, or sleeping on public streets and sidewalks. Although California has since recognized the right to shelter, very little has been done to help PEH find shelter.[77] Following the Centers for Disease Control and Prevention's guidelines discouraging the clearing of homeless encampments during the pandemic, many cities on the West Coast have allowed encampments. In March 2021, the American Rescue Plan was passed ,which provides billions of dollars of funding to help the lowest-income renters and PEH, and California politicians are finally working on solutions to help PEH.[78]

In response to the increasing number of natural and man-made disasters, the US Department of Veterans Affairs (VA) requires that organizations serving veterans in grant per diem (GPD) programs, who provide transitional housing to veterans experiencing homelessness, must have a disaster management plan coordinated with local emergency management agencies.[79] Research on the extent of disaster preparedness of GPD programs showed that lack of guidance and funding have hindered the planning efforts of the community organizations responsible for those programs.[80] Responsive to these findings, the VA and HUD collaborated to develop a disaster preparedness toolkit to provide strategies aimed at the integration of homeless service providers into the emergency response system during disasters.[81] The toolkit provides multilevel strategies that promote collaboration between governmental agencies and homeless service providers at the state and local level. They recommend homeless service providers be collaborators on disaster management planning teams because they are best equipped to serve PEH. Strong collaborations before an emergency occurs lead to more coordinated responses during emergencies. In addition, a focus on the needs of at-risk populations in advance of emergencies helps reduce the time and resources required during a disaster response.[82]

The toolkit identified that PEH have specific needs in a disaster. These include attending to the physical and psychological needs of PEH, and recognition that many PEH have experienced trauma that may be exacerbated by a disaster situation. After assuring physical and psychological safety, PEH should be connected with other services that can provide basic needs, such as food, clothing, and health care. Reconnection with the trusted community-based organizations PEH used prior to a disaster are important in ensuring that continuing needs are met.

Just as healthcare providers are increasingly encouraged to provide universal screening for substance use disorders, they should also regularly screen all patients for housing insecurity.[83] In addition to screening, hospitals and clinics should compile community resources for patients who screen positive and create referral pathways to easily connect patients in need to housing services. Healthcare providers may also consider partnering with shelters to provide healthcare services on-site, which can address barriers to healthcare access for PEH. "Street" medicine and nursing programs can meet individuals where they are and provide mobile healthcare services to people living unsheltered and connect individuals to higher levels of care if needed.[84]

Health care professionals addressing housing instability and specific health needs of PEH in the United States has been gaining traction with the addition of International Classification of Disease codes that address the social determinants of health.[85] However, researchers have reported that only 24 percent of hospitals and 16 percent of medical practices screen for social determinants of health, including housing and food insecurity.[86] Additionally, although four out of five health care professionals believe that unmet social determinants contribute to poor health, only one in five providers feels comfortable addressing these needs.[87]

One way to address the screening and treatment of homelessness and its associated morbidities is to increase awareness in health professions' education curriculum. Currently, homelessness is typically addressed during lectures of the social determinants of health or electives/extracurricular experiences in medical schools, and students have indicated an interest in learning more about the health needs of PEH.[88] However, some health care professional programs, including social work programs,[89] schools of nursing,[90] and occupational therapy programs,[91] are beginning to include increasing didactic and experiential learning about housing insecurity in their curriculum. Continuing to increase exposure to the health and social needs of PEH in health care professional education is imperative in increasing the level of comfort health care providers feel to meet these needs and reduce the stigma associated with homelessness.[92] As climate change continues to affect safe and affordable housing, these initiatives will become increasingly important. The dearth of affordable housing contributes to the problems PEH face when trying to exit homelessness. One innovative solution to the housing crisis is the construction of tiny home communities for PEH. Tiny home communities contain individual units ranging from eighty to seven hundred square feet of living space and provide a flexible housing option that can be adapted to the needs of the local community and adjusted to meet local regulations.[93] Researchers examined six effective tiny home villages and determined there were four common threads that contributed to their success.[94] The first was a shared governance, where the residents of the tiny homes participated in determining how the program was run. The feeling of ownership over their community fostered independence and self-efficacy among the residents. The second theme was public support; although many of the tiny home villages initially encountered resistance from residents, after successful implementation of the program, local communities often rallied behind the tiny home villages. Third, the availability of affordable permanent housing was found to be a contributing factor to long-term success. Although many tiny home communities were initially designed as transitional housing, some became permanent housing in locations with a lack of affordable housing options. Finally, adequate funding was essential for success. Because the goal of the tiny home communities was to create a safe, affordable place to live, they relied on funding sources that were often pieced together through private donations, public grant funding, and the rent collected from residents.[95]

CONCLUSION

The PAR theoretical framework helps to explain the experiences of PEH during hazardous climate events.[96] Root causes such as stigma led to lack of access to services, as well as a dearth of funding allocated to resources serving marginalized PEH. Dynamic pressures such as the lack of affordable housing and living-wage jobs lead to unsafe living conditions in remote and potentially dangerous geographical locations. The accelerating impacts of climate change have increased the number and intensity of hazardous climate events as described in the case studies, which often lead to disastrous conditions for PEH.

As hazardous climate events occur more frequently, it is vital for healthcare and public policy professionals to focus attention on those most vulnerable to their dangers, including PEH. In addition, the population of homeless individuals will continue to grow as disasters associated with climate change destroy housing and make large swaths of areas inhospitable to humans. The United States is finally beginning to acknowledge our growing lack of stable and secure housing, but until we take urgent action to address both affordable housing and climate change, this crisis will continue to worsen.

NOTES

Authors have no conflict of interest.
No funding was received for this work.

1 Habitat for Humanity, "Statement by Habitat for Humanity about the Future of the United Nations Settlement Program, UN-Habitat," 2017, https://www.habitat.org/newsroom/2017/statement-habitat-humanity-about-future-united-nations-human-settlement-programme-un.

2 United Nations, "Affordable Housing and Social Protection Systems for All to Address Homelessness," 2020, https://www.un.org/development/desa/dspd/wp-content/uploads/sites/22/2020/06/chair-summary-csocd58-priority-theme-panel.pdf.

3 Organization for Economic Cooperation and Development (OECD), "HC3.1. Homeless Population," 2020, https://www.oecd.org/els/family/HC3-1-Homeless-population.pdf.

4 OECD, "HC3.1. Homeless Population."

5 Meghan Henry, Tanya de Sousa, Caroline Roddey, et al., "The 2020 Annual Homeless Assessment Report to Congress," Department of Housing and Urban Development, 2021, https://www.huduser.gov/portal/sites/default/files/pdf/2020-AHAR-Part-1.pdf.

6 United Nations, Universal Declaration of Human Rights (UN General Assembly, 1948).

7 United Nations 2020.

8 United Nations 2020.

9 Russell S. Vose, Derek Arndt, Viva F. Banzon, et al., "NOAA's Merged Land-Ocean Surface Temperature Analysis," *Bulletin of the American Meteorological Society* 93, no. 11 (2012): 1677–85, https://doi.org/10.1175/BAMS-D-11-00241.1.

10 Martine Dennekamp and Marion Grace Carey, "Air Quality and Chronic Disease: Why Action on Climate Change Is Also Good for Health," *New South Wales Public Health Bulletin* 21, no. 5–6 (2010): 115–21.

11 Jerry M. Melillo, Terese Richmond, and Gary W. Yohe, "Climate Change Impacts in the United States: The Third National Climate Assessment," US Global Change Research Program, 2014, https://www.globalchange.gov/browse/reports/climate -change-impacts-united-states-third-national-climate-assessment-0.

12 Frank C. Curriero, Jonathan A. Patz, et al., "The Association between Extreme Precipitation and Waterborne Disease Outbreaks in the United States, 1948–1994," *American Journal of Public Health* 91, no. 8 (2001): 1194–99, https://doi.org /10.2105/ajph.91.8.1194.

13 Roni A. Neff, Cindy L. Parker, Frederick L. Kirschenmann, et al., "Peak Oil, Food Systems, and Public Health," *American Journal of Public Health* 101, no. 9 (2011): 1587–97, https://doi.org/10.2105/AJPH.2011.300123.

14 Sean A. Kidd, Susan Greco, and Kwame McKenzie, "Global Climate Implications for Homelessness: A Scoping Review," *Journal of Urban Health* 98, no. 3 (2021): 385–93, https://doi.org/10.1007/s11524020-00483-1; Carole Ziegler, Vincent Morelli, and Omotayo Fawibe, "Climate Change and Underserved Communities," *Physician Assistant Clinics* 4 (2019): 203–6, https://doi.org/10.1016/j.cpha .2018.08.008.

15 On the PAR model, see Piers Blaikie, Terry Cannon, Ian Davies, and Ben Wisner, *At Risk: Natural Hazards, People's Vulnerability and Disaster* (London: Routledge, 1994). On factors linked to social vulnerability, see Christopher Burton, Samuel Rufat, and Eric Tate, "Social Vulnerability," in *Vulnerability and Resilience to Natural Hazards*, ed. Sven Fuchs and Thomas Thaler (Cambridge: Cambridge University Press, 2018), 53–81.

16 Burton, Rufat, and Tate, "Social Vulnerability."

17 Garth Heutel, David Molitor, and Nolan Miller, "Climate Change and Weather Extremes: Both Heat and Cold Can Kill," *Conversation*, January 3, 2018, https:// www.theconversation.com/climate-change-and-weather-extremes-both-heat -and-cold-can-kill-77449.

18 US Environmental Protection Agency, "Climate Change and Extreme Heat: What You Can Do to Prepare," 2016, https://www.epa.gov/sites/default/files/2016 -10/documents/extreme-heat-guidebook.pdf.

19 National Aeronautics and Space Administration, "NASA, NOAA Analyses Reveal 2019 Second Warmest Year on Record," 2020, https://www.nasa.gov /feature/nasa-noaa-analyses-reveal-2019-second-warmest-year-on-record.

20 Chris Dolce, "Summer 2021 Was Hottest on Record in the Contiguous U.S., NOAA Says," *Climate and Weather*, September 9, 2021, weather.com/news /climate/news/2021-09-09-summer-hottest-on-record-united-states-noaa.

21 Judah Cohen, Laurie Agel, Mathew Barlow Mathew, et al., "Linking Arctic Variability and Change with Extreme Winter Weather in the United States," *Science* 373, no. 6559 (2021): 1116–21, https://doi.org/10.1126/science.abi9167.

22 Paul Chakalian, "Extreme Heat Is Killing People Who Lack Shelter," *High Country News*, June 20, 2018, https://www.hcn.org/articles/climate-desk-extreme-heat-is-killing-people-who-lack-shelter.

23 Jerzy Romaszko, Iwona Cymes, Ewa Dragańska, et al., "Mortality among the Homeless: Causes and Meteorological Relationships," *PloS one* 12, no. 12 (2017): e0189938, https://doi.org/10.1371/journal.pone.0189938.

24 Joseph Rampulla, "Hyperthermia and Heat Stroke: Heat Related Conditions," in *The Health Care of Homeless Persons: A Manual of Communicable Diseases and Common Problems in Shelters and On the Streets,* Boston Health Care for the Homeless Program with the National Health Care for the Homeless Council, 2004, https://www.bhchp.org/publications/the-health-care-of-homeless-persons-a-manual-of-communicable-diseases-and-common-problems-in-shelters-and-on-the-streets/.

25 National Coalition for the Homeless (NCH) Staff, "Response to Homelessness: Hot or Cold?," *National Coalition for the Homeless*, 2010, https://nationalhomeless.org/tag/hyperthermia/.

26 Melissa Perri, Naheed Dosani, and Stephen W. Hwang, "COVID-19 and People Experiencing Homelessness: Challenges and Mitigation Strategies," *Canadian Medical Association Journal* 192, no. 26 (2010): E716, https://doi.org/10.1503/cmaj.200834.

27 Associated Press, "Number of Damaging Fires in Los Angeles Homeless Camp Grows," *Seattle Times*, May 13, 2021, https://www.seattletimes.com/seattle-news/health/number-of-damaging-fires-in-los-angeles-homeless-camps-grows/.

28 NCH Staff, "Response to Homelessness."

29 Alex Brown, "The Pandemic Has Closed Public Restrooms and Many Have Nowhere to Go," *PEW Charitable Trusts*, 2020, https://www.pewtrusts.org/en/research-and-analysis/blogs/stateline/2020/07/23/the-pandemic-has-closed-public-restrooms-and-many-have-nowhere-to-go.

30 Adam Mahoney, "COVID Left Portland's Homeless Population in Crisis Mode. Then the Heat Wave Hit," *Grist*, July 12, 2021, https://grist.org/cities/portland-heat-wave-homeless-support/.

31 NCH Staff, "Response to Homelessness."

32 "Historic 2021 Cold Outbreak: February 6–18th," National Weather Service, 2021, https://www.weather.gov/ict/historicCold.

33 Michael Giberson, "Texas Power Failures: What Happened in February 2021 and What Can Be Done?," *Reason Foundation*, 2021, https://reason.org/wp-content/uploads/texas-power-failures-what-happened-what-can-be-done.pdf#page=4.

34 Juan Pablo Garnham, "At Least Six People Experiencing Homelessness Died during the Winter Storm. That Number Could Rise," *Texas Tribune*,

February 22, 2021, https://www.texastribune.org/2021/02/22/texas-winter-storm
-homeless-deaths/.

35 Garnham, "Six People Experiencing Homelessness.

36 Pauline Pineda, "Austin Aid Groups, Volunteers Stepped up to Shelter Homeless
during February Storms," *Austin American Statesman*, March 10, 2021, https://
www.statesman.com/story/news/2021/03/10/austin-aid-groups-stepped-up
-shelter-homeless-during-texas-freeze/6886608002/.

37 Chloe Atkins, "Texas Homeless Population Seeks Refuge amid Bone-Chilling
Temperatures," *NBC News*, February 18, 2021, https://www.nbcnews.com/news
/us-news/texas-homeless-population-seeks-refuge-amid-bone-chilling-tem
peratures-n1258273.

38 Pineda, "Austin Aid Groups."

39 On code blue events, see Penny Kmitt, "Code Blue Warming Station Needs Supplies Ahead of Longest Activation Period," *MSN.com*, February 10, 2021, https://
www.newschannel10.com/2021/02/10/code-blue-warming-station-needs
-supplies-ahead-longest-activation-period/. On code red events, see the Associated Press, "Volunteers Are Rushing to Help Homeless People in the North-
west Cope with the Heat," *NPR*, August 13, 2021, https://www.npr.org/2021/08/13
/1027364299/northwest-heat-wave-homeless-oregon-volunteers-relief-shelters.

40 Adela Suliman, "Disasters Make 14 Million People Homeless Each Year—UN,"
Reuters, October 12, 2017, https://www.reuters.com/article/un-disaster-displace
ment-idUKL8N1MM2N7.

41 Pablo Suarez, William Anderson, Vijay Mahal, and T. R. Lakshmanan,
"Impacts of Flooding and Climate Change on Urban Transportation: A Sys-
temwide Performance Assessment of the Boston Metro Area," *Transportation
Research Part D: Transport and Environment* 10, no. 3 (May 2005): 231–44,
https://doi.org/10.1016/j.trd.2005.04.007.

42 Wendy Grace Evans, "Reaching Out, Four Years after Hurricane Katrina," *Canadian Observatory on Homelessness*, 2009, https://www.homelesshub.ca/resource
/reaching-out-four-years-after-hurricane-katrina.

43 Frances L. Edwards, "Homes Wiped Away by Natural Disasters," *Public Manager*
40, no. 4 (Winter 2011): https://link.gale.com/apps/doc/A275414858/EAIM?u
=mlin_c_umassmed&sid=bookmark-EAIM&xid=4df4037b.

44 Alyssa L. Basile, "Disaster Relief Shelter Experience during Hurricane Sandy: A
Preliminary Phenomenological Inquiry," *International Journal of Disaster Risk
Reduction* 45 (May 2020): 101466, https://doi.org/10.1016/j.ijdrr.2019.101466.

45 Jamie Vickery, "Homelessness and Inequality in the US: Challenges for Community Disaster Resilience," in *Emerging Voices in Natural Hazards Research*,
ed. Fernando Rivera (Oxford: Butterworth-Heinemann, 2019), 145–72.

46 John Aguilar and Joey Bunch, "Two Years Later, 2013 Colorado Floods
Remain a Nightmare for Sone," *Denver Post*, September 12, 2015, https://www

.denverpost.com/2015/09/12/two-years-later-2013-colorado-floods-remain-a
-nightmare-for-some/.

47 Cecelia Gilboy, "Boulder Homeless Turned Away at Red Cross Shelter," *Boulder Weekly*, September 19, 2013, https://www.boulderweekly.com/news/boulder -homeless-turned-away-at-red-cross-shelter/.

48 Nantiya Ruan, "Too High a Price: What Criminalizing Homelessness Costs Colorado," SSRN, April 27, 2018, https://papers.ssrn.com/sol3/papers.cfm?abstract _id=3169929.

49 Mark Kammerbauer and Christine Wamsler, "Social Inequality and Marginalization in Post-Disaster Recovery: Challenging the Consensus?," *International Journal of Disaster Risk Reduction* 24 (September 2017): 411–18. https://doi.org /10.1016/j.ijdrr.2017.06.019.

50 Jonas Z. Hines, Meredith A. Jagger, Thomas L. Jeanne, et al., "Heavy Precipitation as a Risk Factor for Shigellosis among Homeless Persons during an Outbreak— Oregon, 2015–2016," *Journal of Infection* 76, no. 3 (2018): 280–85, https://doi.org /10.1016/j.jinf.2017.11.010.

51 On dengue in Africa, see M. Vyawahare, "As Climate Change Brings More Floods, Mosquito Numbers Could Swell: Study," *Mongabay*, 2021, https://news.mongabay.com/2021/03/as-climate-changebringsmore-floods-mosquito-numbers-could-swell-study/amp/?print. On Zika virus in the United States, see Jennifer Kay and Kelli Kennedy, "'Zika Is Now Here': Mosquitos Now Spreading Virus in the US," *Associated Press*, July 29, 2016, https://apnews.com/63778f53d2e34bbb8c199c859c0d797f.

52 Andy Haines, R. Sari Kovats, Diarmid Campbell-Lendrum, and Carlos Corvalán, "Climate Change and Human Health: Impacts, Vulnerability, and Mitigation," *Lancet* 367, no. 9528 (2006): 2101–9; Marc R. Settembrino, "Hurricane Sandy's Impact on the Predisaster Homeless and Homeless Shelter Services in New Jersey," *Journal of Emergency Management* 14, no. 1 (2016): 7–16.

53 NOAA Office for Coastal Management, "Hurricane Costs," 2021, https://coast.noaa.gov/states/fast-facts/hurricane-costs.html.

54 United Nations, "Affordable Housing and Social Protection Systems."

55 Rongbin Xu, Pei Yu, Michael J. Abramson, et al., "Wildfires, Global Climate Change, and Human Health," *New England Journal of Medicine* 383, no. 22 (2020): 2173–81, https://doi.org/10.1056/NEJMsr2028985.

56 Marco Turco, Juan José Rosa-Cánovas, Joaquín Bedia, et al., "Exacerbated Fires in Mediterranean Europe Due to Anthropogenic Warming Projected with Non-Stationary Climate-Fire Models," *Nature Communications* 9, no.1 (2018): 3821, https://doi.org/10.1038/s41467-018-06358-z.

57 Daoping Wang, Dabo Guan, Shupeng Zhu, "Economic Footprint of California Wildfires in 2018," *Nature Sustainability* 4, no. 3 (2021): 252–60, https://doi.org/10.1038/s41893-020-00646-7.

58 Alexandra S. Levine, "After a California Wildfire, New and Old Homeless Populations Collide," *New York Times*, December 3, 2018, https://www.nytimes.com/2018/12/03/us/california-fire-homeless.html.

59 On marginalization, see June L. Gin, Claudia Der-Martirosian, Christine Stanik, and Aram Dobalian, "Roadblocks to Housing after Disaster: Homeless Veterans' Experiences after Hurricane Sandy," *Natural Hazards Review* 20, no. 3 (2019): 04019005. On denied access, see J. Dearen and K. Kennedy, "Yellow Wristbands, Segregation for Florida Homeless in Irma," *Chicago Tribune*, September 29, 2017, https://apnews.com/article/us-news-ap-top-news-staugustine-hurricane-matthew-hurricanes-813b55bfe9b2441fb6f7c0928fdf1684.

60 Xu et al., "Wildfires."

61 On cardiovascular and respiratory conditions, see Travis P. Baggett, Samantha S. Liauw, and Stephen W. Hwang, "Cardiovascular Disease and Homelessness," *Special Focus Issue: Cardiovascular Health Promotion* 71, no. 22 (2018): 2585–97, https://doi.org/10.1016/j.jacc.2018.02.077.

62 Margot Kushel, "Homelessness among Older Adults: An Emerging Crisis," *Generations* 44, no. 2 (2020): 1–7.

63 Vivian Ho, "Homeless Californians Face New Crisis: Living Outside in Smoke Filled Air," *Guardian*, September 7, 2020, https://www.theguardian.com/us-news/2020/sep/07/homeless-wildfire-smoke-breathing-california.

64 Henry et al., "2020 Annual Homeless Assessment Report."

65 Ho, "Homeless Californians Face New Crisis."

66 Environmental Protection Agency, "Wildfires: How Do They Affect Our Water Supplies?," 2019, https://www.epa.gov/sciencematters/wildfires-how-do-they-affect-our-water-supplies.

67 Hugh G. Smith, Gary J. Sheridan, Patrick N. J. Lane, et al., "Wildfire Effects on Water Quality in Forest Catchments: A Review with Implications for Water Supply," *Journal of Hydrology* 396, no. 1 (2011): 170–92, https://doi.org/10.1016/j.jhydrol.2010.10.043.

68 Jeremiah Lehman, "Homelessness and Urban Water Quality," *Stormwater Solutions*, October 29, 2019, https://www.stormwater.com/stormwater-management/blog/33053345/homelessness-and-urban-water-quality; Zoe Strozewski, "Portland Bans Homeless Encampments in Forested Areas over Concerns of Wildfire," *Newsweek*, July 29, 2021, https://www.newsweek.com/portland-bans-homeless-encampments-forested-areas-over-concerns-wildfire-1614445.

69 Piper French, "Banning the Homeless Won't Keep California from Burning," *New Republic*, September 22, 2021, https://newrepublic.com/article/163739/los-angeles-homeless-encampments-wildfire.

70 Strozewski, "Portland Bans Homeless Encampments."

71 Associated Press, "Damaging Fires."

72 French, "Banning the Homeless."

73 NBC News, "LA Approves a Plan Making It Easier to Clear Homeless Encampments in Fire Zones," September 4, 2019, https://www.nbclosangeles.com/news/local/fires-los-angeles-brush-fire-homeless-encampments-wildfires/1964831/.

74 Amanda Aykanian and Sondra J. Fogel, "The Criminalization of Homelessness," in *Homelessness Prevention and Intervention in Social Work* (New York: Springer, 2019), 185–205.

75 Vickery, "Homelessness and Inequality."

76 Robert Rosenberger, *Callous Objects: Designs Against the Homeless* (Minneapolis: University of Minnesota Press, 2017).

77 Greg Rosalsky, "How California Homelessness Became a Crisis," *National Public Radio*, June 8, 2021, https://www.npr.org/sections/money/2021/06/08/1003982733/squalor-behind-the-golden-gate-confronting-californias-homelessness-crisis.

78 Wendy Carillo and David Chiu, "California's Homelessness Challenges in Context," Legislative Analyst's Office, 2021, https://lao.ca.gov/handouts/localgov/2021/Homelessness-Challenges-in-Context-012121.pdf.

79 US Department of Veterans Affairs, "VA Homeless Providers Grant and per Diem Program," 2013, https://www.va.gov/homeless/gpd.asp.

80 J. L. Gin, C. Der-Martirosian, C. Stanik, and A. Dobalian, "Roadblocks to Housing after Disaster: Homeless Veterans' Experiences after Hurricane Sandy" *Natural Hazards Review* 20, no. 3 (2019), https://doi.org/10.1061/(ASCE)NH.1527-6996.0000330.

81 US Department of Veterans Affairs and Department of Housing and Urban Development, "Disaster Preparedness to Promote Community Resilience," 2017, retrieved from https://www.va.gov/HOMELESS/nchav/docs/VEMEC-Disaster-Preparedness-508.pdf.

82 US Department of Veterans Affairs and Department of Housing and Urban Development, "Disaster Preparedness."

83 S. J. Gordon, K. Grimmer, A. Bradley, "Health Assessments and Screening Tools for Adults Experiencing Homelessness: A Systematic Review," *BMC Public Health* 19, no. 1 (2019): 994, https://doi.org/10.1186/s12889-019-7234-y.

84 Noemi C. Doohan and Ranit Mishori, "Street Medicine: Creating a 'Classroom without Walls' for Teaching Population Health," *Medical Science Educator* 30, no. 1 (2020): 513–21, https://doi.org/10.1007/s40670-019-00849-4; Jim Withers, "Street Medicine: An Example of Reality-Based Health Care," *Journal of Health Care for the Poor and Underserved* 22 (February 2011): 1–4, https://doi.org/10.1353/hpu.2011.0025.

85 Robert Wood Johnson Foundation, "Health Care's Blind Side: Unmet Social Needs Leading to Worse Health," 2011, https://www.rwjf.org/en/library/articles-and-news/2011/12/health-cares-blind-side-unmet-social-needs-leading-to-worse-heal.html.

86 Taressa K. Fraze, Amanda L. Brewster, Valerie A. Lewis, et al., "Prevalence of Screening for Food Insecurity, Housing Instability, Utility Needs, Transportation Needs, and Interpersonal Violence by US Physician Practices and Hospitals," *JAMA Network Open* 2, no. 9 (2019): e1911514, https://doi.org/10.1001/jamanetworkopen.2019.11514.

87 Robert Wood Johnson Foundation, "Health Care's Blind Side."

88 Syeda Shanza Hashmi, Ammar Saad, Caroline Leps, "A Student-Led Curriculum Framework for Homeless and Vulnerably Housed Populations," *BMC Medical Education* 20, no. 1 (2020): 232, https://doi.org/10.1186/s12909-020-02143-z.

89 Donna M. Aguiniga and Pam H. Bowers, "Teaching Note—Partnering Macro Social Work Students and Agencies Addressing Youth Homelessness: A Model for Service Learning," *Journal of Social Work Education* 54, no. 2 (2018): 379–83, https://doi.org/10.1080/10437797.2017.1336138.

90 Diana M. Cavazos, "Street Nursing: Teaching and Improving Community Health," *Hispanic Health Care International*, May 2022, https://doi.org/10.1177/15404153221098958; Andra S. Opalinski, Danielle B. Groton, Donna Linette, et al., "Immersion Experiences bout Homelessness and Psychological Processes of Nursing Students: A Pilot Study," *Journal of Nursing Education* 60, no. 4 (2021): 216–19.

91 Tracy Van Oss, Sydney Barnes, Christina Carmona, et al., "Homelessness: Understanding Unmet Needs and Identifying Resources," *Work* 65, no. 2 (2020): 257–63.

92 Sophie Nadia Gaber, Andreas Karlsson Rosenblad, Elisabet Mattsson, and Anna Klarare, "The Relationship between Attitudes to Homelessness and Perceptions of Caring Behaviours: A Cross-Sectional Study among Women Experiencing Homelessness, Nurses and Nursing Students," *BMC Women's Health* 22, no. 1 (2022): 159, https://doi.org/10.1186/s12905-022-01744-8.

93 Anson Wong, Jerry Chen, Renée Dicipulo, et al., "Combatting Homelessness in Canada: Applying Lessons Learned from Six Tiny Villages to the Edmonton Bridge Healing Program," *International Journal of Environmental Research and Public Health* 17, no. 17 (2020): 6279, https://doi.org/10.3390/ijerph17176279.

94 Wong et al., "Combatting Homelessness in Canada."

95 Wong et al., "Combatting Homelessness in Canada."

96 Blaikie et al., *At Risk.*

Part 3
Valuing Ecosystem Thinking

CHAPTER 9

Advancing Climate Justice to Achieve Birth Equity

An Intersectionality-Based Policy Approach

LISA HEELAN-FANCHER AND LAURIE NSIAH-JEFFERSON

Climate change is the greatest threat to human health of the twenty-first century.[1] The global climate crisis is anticipated to have far-reaching effects—from rising sea levels to hotter temperatures to more frequent extreme weather events—all of which are expected to negatively affect human health. Systematic reviews of the literature have highlighted how climate change, including excessive heat, which is the focus of this chapter, will negatively affect infant health outcomes that include increased risk of preterm birth (PTB), low birth weight, stillbirths, and infant mortality.[2] Maternal health is also at risk with heat exposure, with studies identifying increased incidence of maternal hypertensive disease,[3] placental abruption,[4] and pregnancy-related complications such as gestational diabetes[5] and poor maternal mental health.[6] As such, climate change is negatively affecting the health of pregnant women and developing fetuses and poses long-term problems for individuals, families, and society.

Racial disparities are evident in birth outcomes, with Black, American Indian, Alaska Native, and Pacific Islander infants having the highest rates of mortality, PTB, and low birth weight compared to infants born to White women.[7] Birth equity is defined as all mothers and infants having the necessary conditions present for optimal births.[8] Low-resourced households or communities and populations that have historically been marginalized or excluded, need more resources to achieve birth equity compared to groups that have not been marginalized.[9] As we consider how to address climate change and birth equity, through policy and related initiatives, we must consider the limitations of traditional public policy, and highlight the rationale for using an intersectional approach.

INTERSECTIONALITY AND ITS RELATIONSHIP TO TRADITIONAL POLICY DEVELOPMENT

Policy analysis is concerned with the outcomes of a given policy and the impact a particular policy will have on people.[10] Historically, health policy analysis has prioritized efficiency, effectiveness, and the political feasibility of advancing a particular policy. Although this method may imply that public policies are neutral and benefit all citizens equally, they are imbued with power and privilege, and thus affect populations differently.[11]

Race- and gender-based policy analysis is grounded in social justice; however, each offers a policy analytical perspective that is focused on either race or gender. Kimberle Crenshaw viewed this as a limitation to understanding the complexity and multidimensionality of marginalized women's experiences, as each perspective focuses on only one facet of her social identify, such as race or gender, and not race *and* gender.[12] To more fully understand the experiences of women of color, Crenshaw proposed the term "intersectionality."[13] An intersectional approach recognizes the interconnection between race and gender and the many social identities pregnant persons have, including the environment in which they live, work, and play, and the social processes and structures that are shaped by power dynamics, such as their experiences of systemic racism, sexism, and homophobia at the micro and macro level.

An intersectionality-based policy analysis (IBPA) framework stems from the above theory and, unlike gender- or race-specific analytic frameworks and social determinants of health frameworks, was designed to guide health policy from a social justice and equity perspective while recognizing how identities are mutually constituted.[14] Furthermore, intersectionality has been identified as a theoretical, research, practice, and policy paradigm for fundamentally altering the ways in which social problems are identified, experienced, understood, and addressed to reflect the multiplicity of a person's lived experience.[15] Using the IBPA framework, therefore, represents a novel approach to inform public policy by understanding and addressing the impact of climate change on pregnant persons, their infants, and communities.

AIMS

The aim of this chapter is first to use the IBPA framework as a method to describe the authors' knowledge, values, and experiences that were brought to this policy analysis; review the literature on the physiologic impact of climate change, with a focus on excessive heat on birth outcomes and the role of social determinants of health, and social injustices, as a contributing factor in poor birth outcomes; analyze existing US federal government legislative policies and initiatives as they relate to, and affect climate change, climate justice, and birth equity, and review who the key stakeholders are in advancing legislation and initiatives; examine the populations targeted for policy interventions; and identify the inequities that currently exist. The second aim is to address and critique the adequacy of policy responses and solutions and suggest alternatives or revisions to advance climate justice to achieve birth equity utilizing an intersectional lens.

IBPA FRAMEWORK AND METHODS

The IBPA framework includes nine principles and twelve questions to help shape health policy analysis.[16] Among the nine principles that guide the IBPA framework are power, social justice and equity, intersecting categories, multilevel analysis, reflexivity, diverse knowledge, and alternative time and space of marginalized groups, all principles that are emphasized in this chapter. The twelve questions are divided into two categories: descriptive questions that provide background knowledge of the problem from an intersectional perspective, and transformative questions that will focus on and critique current policies to facilitate consideration of alternative policy responses and strategies to advance climate justice to achieve birth equity.[17]

Our IBPA analysis will be augmented with a reproductive justice framework. The reproductive justice framework asserts that maintaining personal bodily autonomy, a choice of having children or not having children, and parenting children in *safe and sustainable communities*, is a human right and is directly influenced by the conditions of communities and how women are personally affected by reproductive and systematic oppression.[18] By using a reproductive justice lens for policymaking and practice concerning the relationship between climate change and birth equity, the roots and intertwined nature of inequities can be addressed, and health equity can be achieved.

Data Collection

Legislative bills and initiatives for this analysis were chosen following a review of federal public policies, executive orders, and other initiatives focused on protecting people from excessive heat exposure (a specific aspect and outcome of climate change), including mitigation efforts and training and research relating to climate change and birth equity. These data primarily came from congressional member and caucus websites, bill tracking (Congress.gov, GovTrack, Billmap), White House websites, and gray literature from advocacy organizations, professional organizations, and governmental agency websites. Data were also obtained from academic journals, budget analyses, testimonies, and blogs.

Descriptive Questions

Descriptive questions are intended to generate essential background information about the policy problem, including, the values and experiences of the health policy analysts (reflexivity), how specific policy actors identify, frame and seek to address the problem, how different group and subgroups are privileged or harmed by the framing of the problem and the intended policy, and what are the current policy responses to the problem. According to Olena Hankivsky and colleagues,[19] the purpose of these questions is to "reveal assumptions that underpin existing government priorities, including the populations targeted for the intervention."

Reflexivity

The IBPA principle of reflexivity invites researchers and policymakers to identify how their knowledge, experience, values, standpoints, and social positions inform their analysis of birth equity and climate change. Consequently, our work is grounded in our: (1) intersecting relationships to Black and Brown communities, (2) institutional and personal networks, (3) pregnancy and birth experience and that of those whom we know, (4) personal experiences of racism and/or sexism, and (5) professional training and work.

The two authors work at an urban minority-serving research institution (situated in a diverse community) that is also renowned for its community engagement. We hold diverse social positions. One is a cisgender, able-bodied married White middle-class woman who grew up in a large northeastern city and was educated and served as a nurse practitioner for many years. She is currently a researcher and professor who focuses on maternity-related issues primarily in low- and middle-income communities. The other is a cisgender, able-bodied Black senior-aged individual who has experienced motherhood and pregnancy. She delivered her full-term infant via a cesarean section and had a mediocre birth experience while observing poor treatment of Black women during her hospital stay. However, she herself was born prematurely and of low birth weight. Finally, she has experienced interpersonal, institutional, and structural racism, and has devoted much of her professional career to addressing issues related to women of color and reproductive and maternal health, with a focus on policy. She has studied the role of racism and discrimination on health outcomes, including pregnancy.

Representation of the Problem
The practice of IBPA is to provide the historic, political, policy, and social context of the problem and highlight the ways the problem is framed. In the United States, the issue of climate change has been politicized, which has led to a flip-flop in policy direction. One example is the Paris Climate Agreement, which President Barack Obama (Democrat) signed on behalf of the US people in 2016, then President Donald Trump (Republican) withdrew from in 2020, and then President Joe Biden (Democrat) rejoined on his first day of office in 2021.[20] Nevertheless, there has been increasing attention to climate justice and birth equity by government, labor, religious, professional, and advocacy organizations, and this has fueled congressional action to address climate change and/or birth equity.

How Are Groups Differentially Affected by This Representation of the "Problem"?
Policy language is filled with metaphors and symbols.[21] The implicit biases that we all have, the subtle meaning behind the words that we do use, and our word choices affect how policymakers craft policy, distribute resources, and title bills.[22] To assess how the representation of the problem of birth equity is addressed, we examined three bills introduced in the 117th Con-

gress (2021–22), Protecting Moms and Babies Against Climate Change Act (HR 957, S 423),[23] the Asunción Valdivia Heat Illness and Fatality Prevention Act of 2021 (HR 2193, S 1068),[24] and The Pregnant Workers Fairness Act (HR 2417, S 1486).[25]

In the Protecting Moms and Babies Against Climate Change Act, low-income individuals, women of color, and those who live in high-risk environmental communities will benefit most from the bill in that the framing of the challenge of birth equity and climate change in the bill highlight the risk to pregnant persons who are from these groups and communities. The second bill, the Asunción Valdivia Heat Illness and Fatality Prevention Act of 2021, focused on mitigating extreme heat conditions for workers who work either indoors or outside. Although this bill recognized that Black and Brown persons were most affected by working in extreme heat conditions, it did not identify pregnant persons as needing special consideration while working in extreme heat conditions nor did it consider that their needs are probably greater than those of a person who is not pregnant. This suggests that pregnant persons as a group were generally neglected or invisible in the policy language of this bill. And the third bill, the Pregnant Workers Fairness Act, which was signed into law in 2022, prohibits employment practices that discriminate against making reasonable accommodations for qualified employees affected by pregnancy, childbirth, or related medical conditions. One difference between the Pregnant Workers Fairness Act and the Americans with Disabilities Act is that a pregnant employee meets the criteria "qualified" if she is only temporarily unable to perform the essential functions of her job and if what she needs in the way of accommodations are reasonable.[26] For pregnant persons who are exposed to excessive heat, a reasonable accommodation could be frequent water and bathroom breaks.

All women are harmed by the framing of needing protection because it perpetuates the view of women as being overly dependent, which sets them up for stigma and social control. Notably, low-income individuals and women of color are considered the most vulnerable.

What Are the Current Policy Responses to the Problem?
The Biden-Harris administration is very aware that climate change (including excessive heat) threatens to exacerbate inequities for people of color and low-income households and communities. Since being sworn into office,

President Biden has signed several executive orders that establish climate change as a national security priority.[27] He also embraced the Black Maternal Health Momnibus Act of 2021 (HR 959, S 346),[28] which was created by the Black Maternal Health Caucus led by representatives Lauren Underwood (D-IL) and Alma Adams (D-NC), two Black women addressing Black maternal health and climate change.[29] The Momnibus Act was incorporated into President Biden's proposed Build Back Better Act (HR 5376, S 484), which included legislation to address the link between climate change and its impact on pregnancy and environmental justice.[30]

Although the Build Back Better Act narrowly passed the House in November 2021, it was not brought to a vote in the Senate, most likely because it would not have passed. In its place, the Inflation Reduction Act of 2022, a reconciliation bill, was negotiated between senators Charles Schumer (D-NY) and Joseph Manchin (D-WV) and passed in both chambers of Congress with bipartisan support before being signed into law by President Biden in August 2022.[31] A strength of this bill is that it addresses greenhouse gas emissions and invests in environmental justice communities across the United States by providing grants of over sixty billion dollars to historically marginalized and disadvantaged communities most affected by environmental pollution and climate change.[32] However, a limitation of this bill is that it invests more in the physical infrastructure of clean energy and less in the human-social infrastructure to advance birth equity, such as diversifying the perinatal health workforce or addressing the social determinants of health like housing and nutrition that were originally proposed in the Build Back Better Act.[33] Hence, there is still a need for additional legislation to address the inequities of maternal and infant health outcomes, especially among persons of color.

Although most of the twelve bills of the Momnibus Act (with some modifications) were included in the Build Back Better Act, many of them were not taken up in the Inflation Reduction Act of 2022 that was signed into law. What is not known is if congressional lawmakers were aware that if they passed the entire Momnibus Act, their actions would restore justice to traditionally marginalized communities. The inability to pass a comprehensive bill such as the Momnibus Act suggests that we as a society are only chipping away at injustice as foundational discrimination and structural racism persist. Deborah Stone suggests that incremental change occurs because policymakers view the world in a certain way, such as Black and White or sameness and difference, when a kaleidoscope of color (or viewpoints) is what is present.[34] Yet inherent delays in enacting legislation occur even when there is broad social support for an issue. For example, even though there was a steep increase in women entering the workforce in the 1970s, it took until the early 1990s for childcare legislation to be enacted at the federal level.[35] Even today, childcare remains an issue on the government agenda highlighting that even with consensus, change takes time.

Transformative Questions

The second set of IBPA questions identify alternative policy responses and develop solutions to reduce inequities. The transformative questions in this section report on the inequities that exist in relation to the problem; identify interventions to improve the problem; describe the feasibility of short-, medium-, and long-term solutions; propose policies that will reduce inequities; and evaluate if inequities have been reduced as a result of the policy intervention.

What Inequities Actually Exist in Relation to the Problem?

In 2018, 10.02 percent of all infants in the United States were born preterm,[36] and Black women were about 50 percent more likely to deliver their infants preterm compared to White women.[37] Yet it should come as no surprise that poor birth outcomes are exacerbated by climate-related events. For example, Black and Hispanic women with heat exposure had double or more the risk of PTB or stillbirth compared to White women,[38] and this birth inequity was also found among infants born to college-educated Black women, who experienced a higher relative risk of PTB following a heat wave compared to White women with college degrees.[39] Another study found that for Black women an additional day of exposure to extreme heat during the third trimester doubled the odds of hospitalization compared to White women (5 percent versus 2.5 percent).[40] Other studies have found that in cities with higher rates of residential segregation and lower socioeconomic means, temperatures in the summer are much higher in redlined areas compared to nonredlined areas.[41]

Additionally, we know that people of color have fewer ways to mitigate exposure to extreme heat, and when they do have air conditioning, a greater proportion of their income is spent on paying for it.[42] Residing in an inner city also contributes to higher exposure to heat compared to other parts of the city, and this is generally due to black asphalt and the lack of tree cover. These environmental exposures exacerbate and worsen birth outcomes for pregnant women of color.[43] Finally, it is important to note that pregnant farmworkers labor in fields where temperatures can reach 120 degrees.[44] These workers usually lack information about the symptoms of heat-related issues, and many of these workers are Latinx.

Harm to health as a result of the climate crisis in the United States should be understood in the wider social and political context of structural racism, including the unequal distribution of wealth and resources.[45] This is most evident when Black women are three to four times more likely to die from pregnancy-related complications compared to White women.[46] Although most of the literature on birth inequity is focused on White-Black disparity, which is real, we need to consider that there are other women and birthing persons in the United States who have poor birth outcomes that may be exacerbated by climate change, including women with disabilities,[47] Native American women,[48] women who are undocumented,[49] women who are homeless,[50] women who live in rural America instead of urban America,[51] women who are poor and with fewer years of formal education,[52] LBGTQ women,[53] and adolescents,[54] to name just a few groups of women often not mentioned publicly yet who are often not identified or mentioned when birth outcomes are discussed.

Where and How Can Interventions Be Made to Improve the Problem?
Several interventions can be implemented to address the problem of climate change (excessive heat) and birth equity, including lowering the exposure and susceptibility of individuals to extreme heat by adding trees and vegetation, green roofs, cooling pavements, public education and information, and employing mitigation efforts once individuals have been exposed, such as financial assistance for energy costs, distribution of air conditioning units, air filtration units and other equipment to support indoor cooling, water stations, energy bill assistance and restrictions so that electricity cannot be turned off during warmer months of the year, cooling stations, and easy access to clean water.[55]

These interventions can be achieved by utilizing policy levers, including taxing and incentives, penalties, laws, regulations, employer standards, community ordinances, education, information and persuasion of key stakeholders, data and research, and government programs and services and can be implemented at the federal, state, and local levels.[56] In addition, communities can now apply for monies to support some of their community initiatives through the Environmental and Climate Justice Block grants funded through the Inflation Reduction Act of 2022.[57] Furthermore, consumers can be made aware of incentives to lower their energy costs, also funded by the Inflation Reduction Act.[58]

Although the Inflation Reduction Act does address climate change, it falls short of the policy and funding requests contained within the Protecting Moms and Babies Against Climate Change bill that was not passed by the 117th Congress, including educating health professionals about climate change, establishing a Consortium on Birth and Climate Change to coordinate research on climate change, and having the US Centers for Disease Control and Prevention develop a strategy to identify areas where there is a high risk of adverse maternal or infant health outcomes due to climate change.[59] We would also recommend that education on climate change be required for students in schools of social work, public policy, and business and that this education should include the broader social risks and causes of inequities. For example, social worker students could learn about the resources available to mitigate heat-related injuries and how to treat the mental health and social effects of climate change. Furthermore, those in policy programs could learn more about conducting intersectional policy analysis on proposed and current laws and the potential political minefields to create better and more effective laws and regulations. Additionally, there needs to be more investment in research and data with an intersectional lens on the social, economic, and political dimensions of climate change and birth equity. This work can be implemented using community-university engaged partnerships.

Pregnancy health has largely been omitted from heat awareness efforts, legislation, agency regulations and evaluation, and equity protocols, suggesting, therefore, that efforts need to be greatly increased and improved. For example, the Occupational Safety and Health Administration (OSHA) recently released a fact sheet announcing the need to protect workers from the effects of heat to prevent heat illness and death; however, there is no mention of the impact of heat on pregnant persons, nor are specific guidelines offered to protect this population.[60] It is our recommendation that OSHA and other regulating bodies consider including information on pregnant persons when inviting input for formulating standards and regulations. Other regulations could include banning utility companies from disconnecting electricity to any household during periods of high heat, particularly if there are pregnant and postpartum persons in the household, developing a national heat vulnerability index that recognizes the susceptibility of pregnant women and their fetuses to extreme heat, and threading climate change awareness throughout all aspects of government.[61]

What Are Feasible Short-, Medium-, and Long-Term Solutions?
When considering climate change and birth equity we can parse out solutions and strategies that are short-, medium-, and long-term. In this section, we will include strategic as well as practical measures. For brevity we will focus only on legislative and executive efforts and will not address what individuals or organizations are doing at the local, state, or national level.

Short-term strategies include drafting legislation that can address climate change and birth equity, directly and indirectly, and identifying key, well-positioned, and influential legislators to support and endorse legislation that has been proposed. Likewise, it would be prudent to reintroduce relevant legislation that did not make it successfully through the legislative process of the 117th Congress and garner ways to improve the chances of some of the bills passing. This may include an analysis of barriers to success, including political climate, timing, cost, caution due to upcoming elections, inadequate or ineffective stakeholder outreach, and bill content.[62] Furthermore, it may be plausible to use a multistrand analysis approach on relevant bills to highlight how they are of benefit to different subpopulations of women.[63] Other short-term strategies to advance climate justice and achieve birth equity can include local public health departments disseminating much more information on heat risks to pregnant women and their infants. It is a sad commentary that there are more heat warnings for pets than there are for pregnant women.[64]

Medium-term strategies would include evaluating the outcomes, impacts, and processes of climate change laws/regulations and other initiatives that have been implemented and revising those efforts, if necessary. In addition, it may be advisable to create a new White House office to focus on sexual and reproductive health and well-being while also ensuring that climate equity is infused in all relevant federal, state, and local offices.[65]

Long-term strategies include initiatives that have in most instances five-year timelines or longer. These include addressing issues that are at the root cause of inequities in exposure to excessive heat for pregnant women and others due to political and structural inequities that are deemed the underlying drivers of climate susceptibility.[66]

We assert that change takes time, and historically, legislative action occurs incrementally. Although the Momnibus Act was comprehensive in that many of the bills addressed climate change and/or birth equity issues, policymakers did not use an intersectionality-based policy approach to guide their policy formation. Nevertheless, we feel that the policymakers when writing the Momnibus Act demonstrated that they were heading in the direction of using an intersectional-based policy approach, which when used in policy development can advance the operationalization of equity in public policy. However, to reach consensus and get a bill passed, much of the social-infrastructure guardrails of the Momnibus Act that addressed birth equity and Black maternal health were gutted and not included in the Inflation Reduction Act. As such, this is an area where more work is needed to advance legislative efforts so that birth equity can be realized.

How Will Proposed Policy Responses Reduce Inequities?
The Inflation Reduction Act of 2022 is the largest investment to date that addresses the negative impact of climate change, especially for persons living in environmental justice communities. Transitioning from fossil fuels to cleaner energy is anticipated to reduce carbon emissions by 40 percent in less than ten years, and reducing air pollution will inevitably improve the health of pregnant persons and their infants.[67] Additionally, ten billion dollars has been allocated for grants that will be invested in low-resourced communities to address the public health harms related to pollution and climate change, and these monies can be used to support clean transportation infrastructure, invest in clean heavy-duty vehicles such as school and transit buses and garbage trucks, and purchase and install zero-emission equipment and technology at seaports.[68] This is a step in the right direction to reduce inequities for persons living in low-resourced communities and has the potential to improve Black maternal health and infant outcomes.

However, the Inflation Reduction Act of 2022 did little to nothing to directly address the inequities of Black maternal health. Legislative bills were introduced during the 117th Congress to address the social determinants of maternal health such as housing and nutrition, award community-based grants to mitigate heat effects related to maternal and infant health risks, increase health research at minority-serving institutions, expand access to maternal health equity digital tools, and protect workers from the harm of climate change, to name just a few initiatives.[69] Some of the above initiatives, such as awarding community-based grants to mitigate public health harms related to pollution and climate change in environmental justice communities, were included in the Inflation Reduction Act, while many of the other proposed initiatives, such as legislation that centered on directly improving Black maternal health and protecting workers from harm of climate change, were not included.

How Will You Know if Inequities Have Been Reduced?

When measuring progress toward addressing climate change and birth equity we need to evaluate the process, outcomes, and impacts of this work. Evaluating climate change efforts includes asking if the policy achieved its stated goals and objectives, and whether the program had unanticipated consequences, particularly negative ones.[70] It is also important to look at the goals of the implementation process for the policy. This includes the interpretation and implementation of the law at the agency and street level,[71] and determining whether key stakeholders are being held accountable for the law or regulation. Implementation will also be looked at in terms of budget constraints (underfunded mandates) and the political stance of agency heads and the executive branch. We also need to examine the uptake, attitude toward, and adherence to laws by implementing agencies and the impact of public health education campaigns. Additionally, we need to evaluate the likelihood of the success of a proposed legislative initiative and the influence of key stakeholders on this success.

When evaluating outcomes of implemented policy on birth outcomes, we need to review national and local-level data using time series data on preterm birth, low birth weight, stillbirths, and other relevant birth outcomes related to excessive heat exposure, such as maternal heat-related death, illness, and hospitalization. Data on birth and maternal outcomes should be collected and analyzed by race and ethnicity, within and between racial groups by geographic area, socioeconomic status, age, immigrant status, and other characteristics. Interdisciplinary teams and community members should then examine and analyze data at these relevant intersections using multilevel modeling when appropriate. Mixed-methods research would also include rich qualitative data about the how and why of a process or outcome.[72] An intersectional evaluation should also determine which subpopulations are benefiting from a specific policy and which are not and why. This might include, for example, looking closely at methods of outreach/education or messaging as well as social or political feasibility. The above are just a sample of some of the questions that can be asked; there are many others.

How Has the Process of Engaging in an Intersectionality-Based Policy Analysis Transformed Your Thinking?

The Inflation Reduction Act of 2022 is a huge investment in reducing emissions and putting the United States on the path toward cleaner energy; however, it is unknown how this legislation will directly benefit women of color and their infants, and pregnant persons who live in environmental justice communities. Unlike the Black Maternal Health Momnibus of 2021, the Inflation Reduction Act of 2022 doesn't address the social determinants of health initiatives that were specifically designed to address birth inequities in the United States. As such, there will be a need to evaluate the impact of the Inflation Reduction Act on birth outcomes in the context of climate justice and reintroduce legislation that specifically addresses birth inequities.

The process of engaging in this legislative review using an IBPA framework has made us much more aware of how important words are in describing and identifying populations when writing bills and framing policy. The legislative policies included in this analysis addressed birth and climate justice inequities from a gender, gender and race, or ethnicity perspective and considering its intersection with pregnancy. However, pregnant persons may use identities other than race to describe themselves, such as living with a disability or living in a high-risk community, or they may hold gender identities other than female, to name just a few ways pregnant persons may identify themselves and the communities in which they live or feel most comfortable.

Although some legislative bills introduced during the 117th Congress included pregnant persons living in high-risk or environmental justice communities, the titles of the bills showed that they did not incorporate most of the nine principles that guide the IBPA framework when shaping their health policy initiatives, nor did they consider power dynamics. Using an IBPA framework can show policymakers and interested parties how the same policy affects individuals and communities differently, which informs us about where the resources are most needed, and where the funding will have the greatest impact in advancing climate justice and birth equity.

NOTES

1 World Health Organization, "Climate Change and Health," last modified October 30, 2021, https://www.who.int/newsroom/fact-sheets/detail/climate-change -and-health.

2 Bruce Bekkar, Susan Pacheco, Rupa Basu, and Nathaniel DeNicola, "Association of Air Pollution and Heat Exposure with Preterm Birth, Low Birth Weight, and Stillbirth in the US," *JAMA* 3, no. 6 (June 2020): e208–43, https:// doi.org/10.1001/jamanetworkopen.2020.8243. Matthew F. Chersich, Minh D. Pham, Ashtyn Area, et al., "Associations between High Temperatures in Pregnancy and Risk of Preterm Birth, Low Birth Weight, and Stillbirths: Systematic Review and Meta Analysis," *BMJ* 371 (November 2020): 1–13, http://dx.doi.org /10.1136/bmj.m3811.

3 Tao Xiong, Peiran Chen, Yi Mu, Xiaohong Li, et al., "Association between Ambient Temperature and Hypertensive Disorders in Pregnancy in China," *Nature Communication* 11, no. 1 (June 2020): 1–7, https://doi.org/10.1038/s41467 -020-16775-8.

4 Siyi He, Tom Kosatsky, Audrey Smargiassi, et al., "Heat- and Pregnancy-Related Emergencies: Risk of Placental Abruption during Hot Weather," *Environmental International* 111 (February 2018): 295–300, https://doi.org/10.1016/j.envint .2017.11.004.

5 Emma V. Preston, Claudia Eberle, Florence M. Brown, and Tamarra James-Todd, "Climate Factors and Gestational Diabetes Mellitus Risk: A Systematic Review," *Environmental Health* 19, no. 1 (November 2020): 1–19, https://doi.org/10.1186 /s12940-020-00668-w.

6 Linda C. Giudice, Erlidia F. Llamas-Clark, Nathaniel DeNicola, Santosh Pandipati, Marya G. Zlatnik, Ditas Cristina D. Decena, Tracey J. Woodruff, Jeanne A. Conry, and FIGO Committee on Climate Change and Toxic Environmental Exposures, "Climate Change, Women's Health, and the Role of Obstetricians and Gynecologists in Leadership," *International Journal of Gynecology and Obstetrics* 155, no. 3 (December 2021): 345–56, https://doi.org/ 10.1002/ijgo.13958.

7 Emily E. Petersen, Nicole L. Davis, David Goodman, et al., "Racial/Ethic Disparities in Pregnancy-Related Deaths: United States, 2007–2016," *Morbidity Mortality Weekly Report* 68, no. 35 (May 2019): 762–65, https://www.10.15585/mmwr .mm6835a3.

8   March of Dimes, "Birth Equity for Moms and Babies Consensus Statement: Advancing Social Determinants Pathways for Research, Policy, and Practice," last modified October 18, 2018, https://www.marchofdimes.org/professionals /Birth-Equity-for-Moms-and-Babies-Consensus-Statement.aspx. National Birth Equity Collaborative, "Birth Equity for All Black Birthing People," 2022, https://birthequity.org.

9   March of Dimes, "Birth Equity."

10 Tea Collins, "Health Policy Analysis: A Simple Tool for Policy Makers," *Public Health* 199, no. 3 (March 2005): 192–96, https://www.10.1016/j.puhe.2004.03.006.

11 Thomas A. Birkland, *An Introduction to the Policy Process: Theories, Concepts, and Models of Public Policy Making* (New York: Routledge, Taylor and Francis Group, 2020). Olena Hankivsky, Daniel Grace, Gemma Hunting, et al., "An Intersectionality-Based Policy Analysis Framework: Critical Reflections on a Methodology for Advancing Equity," *International Journal for Equity in Health* 13, no. 119 (December 2014): 1–16, http://www.equityhealthj.com/content/13/ 1/119.

12 Kimberle Crenshaw, "Demarginalizing the Intersection of Race and Sex: A Black Feminist Critique of Antidiscrimination Doctrine, Feminist Theory, and Antiracist Politics," *University of Chicago Legal Forum* 1, no. 8 (1989): 139–67, http://chicagounbound.uchicago.edu/uclf/vol1989/iss1/8.

13 Crenshaw, "Demarginalizing the Intersection."

14 Hankivsky et al., "Policy Analysis Framework."

15 Bonnie T. Dill and Ruth E. Zambrana, *Emerging Intersections: Race, Class, and Gender in Theory, Policy, and Practice* (New Brunswick, NJ: Rutgers University Press, 2009).

16 Hankivsky et al., "Policy Analysis Framework."

17 Hankivsky et al., "Policy Analysis Framework."

18 SisterSong Women of Color Reproductive Justice Collective, "Birth Justice," accessed June 20, 2022, https://www.sistersong.net/justice-statement.

19 Hankivsky et al., "Policy Analysis Framework," 3.

20 Antony J. Blinken, "The United States Officially Rejoins the Paris Agreement," US Department of State, February 19, 2021, https://www.state.gov/the-united -states-officially-rejoins-the-paris-agreement/.

21 Deborah Stone, *Policy Paradox: The Art of Political Decision Making* (New York: Norton, 2011).

22 Stone, *Policy Paradox.*

23 HR 957, Protecting Moms and Babies Against Climate Change Act, 117th Congress (2021–22), accessed July 12, 2022, https://www.congress.gov/bill/117th-congress/house-bill/957. S 423, Protecting Moms and Babies Against Climate Change Act, 117th Congress (2021–22), accessed July 12, 2022, https://www.congress.gov /bill/117th-congress/senate-bill/423/.

24 HR 2193, Asunción Valdivia Heat Illness and Fatality Prevention Act of 2021, 117th Congress (2021–22), accessed July 12, 2022, https://www.congress.gov/bill /115th-congress/house-bill/2193. S 1068 Asunción Valdivia Heat Illness and Fatality Prevention Act of 2021, 117th Congress (2021–22), accessed July 12, 2022, https://www.congress.gov/bill/117th-congress/senate-bill/1068.

25 HR 2417, Pregnant Workers Fairness Act, 117th Congress (2021–22), accessed July 12, 2022, https://www.congress.gov/bill/115th-congress/house-bill/2417. S 1486, Pregnant Workers Fairness Act, 117th Congress (2021–22), accessed July 12, 2022, https://www.congress.gov/bill/117th-congress/senate-bill/1486.

26 Kara E. Stockdale, "President Biden Signs into Law the Pregnant Workers Fairness Act and the Providing Urgent Maternal Protections for Nursing Mothers Act," January 9, 2023, https://www.bairdholm.com/blog/president-biden-signs-into-law-the-pregnant-workers-fairness-act-and-the-providing-urgent-mater-nal-protections-for-nursing-mothers-act/#:~:text=on%20Monday%2C%20 9%20January%202023%20in%20Labor%20%26,Act"%29%20and%20the%20 Pregnant%20Workers%20Fairness%20Act%20%28"PWFA"%29.

27 US White House, "Fact Sheet: President Biden Takes Executive Actions to Tackle the Climate Change Crisis at Home and Abroad, Create Jobs, and Restore Scientific Integrity across Federal Government," January 27, 2021, https:// www.whitehouse.gov/briefing-room/statements-releases/2021/01/27/fact-sheetpresident-biden-takes-executive-actions-to-tackle-the-climate-crisis -at-home-and-abroad-create-jobs-and-restore-scientific-integrity-across-fed-eral-government/.

28 HR 959, Black Maternal Health Momnibus Act of 2021, 117th Congress (2021–22), accessed July 12, 2022, https://www.congress.gov/bill/117th-congress/house-bill/959. S 346, Black Maternal Health Momnibus Act of 2021,117th Congress (2021–22), accessed July 12, 2022, https://www.congress.gov/bill/117th-congress /senate-bill/346.

29 US House of Representatives Black Maternal Health Caucus, "Black Maternal Health Caucus Celebrates Historic Passage of "Momnibus" in Build Back Better Act," November 19, 2021, https://blackmaternalhealthcaucus-underwood.house .gov/media/press-releases/black-maternal-health-caucus-celebrates-historic-passage-momnibus-build-back.

30 US White House, "Fact Sheet: How the Build Back Better Plan Will Create a Bet-ter Future for Young Americans," July 22, 2021, https://www.whitehouse.gov/ briefing-room/statements-releases/2021/07/22/fact-sheet-how-the-build-back -better-plan-will-create-a-better-future-for-young-americans/.

31 US Senate Democrats, "Joint Statement from Leader Schumer and Senator Man-chin Announcing Agreement to add the Inflation Reduction Act of 2022 to the FY 2022 Budget Reconciliation Bill and Vote in Senate Next Week," July 27, 2022, https://www.democrats.senate.gov/newsroom/press-releases/senate-majority

-leader-chuck-schumer-d-ny-and-sen-joe-manchin-d-wv-on-wednesday-announced-that-they-have-struck-a-long-awaited-deal-on-legislation-that-aims-to-reform-the-tax-code-fight-climate-change-and-cut-health-care-costs.

32 US Senate Democrats, "Summary of the Energy Security and Climate Change Investments in the Inflation Reduction Act of 2022," n.d., https://www.democrats.senate.gov/imo/media/doc/.

33 US White House, "Fact Sheet: The Inflation Reduction Act Supports Workers and Families," August 19, 2022, https://www.whitehouse.gov/briefing-room/statements-releases/2022/08/19/fact-sheet-the-inflation-reduction-act-supports-workers-and-families.

34 Sally Cohen, *Championing Child Care* (New York: Columbia University Press, 2001).

35 Stone, *Policy Paradox*.

36 Joyce A. Martin, Brady E. Hamilton, Michelle J. K. Osterman, and Anne K. Driscoll, "Births: Final Data for 2018," *National Vital Statistics Reports* 68, [no. 13] (April 2021): 1–51. https://pubmed.ncbi.nlm.nih.gov/32501202/.

37 US Centers for Disease Control and Prevention, "Preterm Birth," last reviewed November 1, 2021, https://www.cdc.gov/reproductivehealth/maternalinfanthealth/pretermbirth.htm.

38 M. Luke Smith and Rachel R. Hardeman, "Association of Summer Heat Waves and the Probability of Preterm Birth in Minnesota: An Exploration of the Intersection of Race and Education," *International Journal of Environmental Research and Public Health* 17 (September 2020): 1–12, https://doi.org/10.3390/ijerph17176391.

39 Smith and Hardeman, "Summer Heat Waves."

40 Jiyoon Kim, A. Jin Lee, and Maya Rossin-Slater, "What to Expect When It Gets Hotter: The Impact of Prenatal Exposure to Extreme Heat on Maternal and Infant Health," *NBER Working Paper Series, no. A26384* (October 2019). http://ssrn.com/abstract=3472819.

41 Jeremy S. Hoffman, Virek Shandas, and Nicholas Pendleton, "The Effects of Historical Housing Policies on Resident Exposure to Intra-Urban Heat: A Study of 108 US Urban Areas," *Climate* 8 (January 2020): 1–15, https://doi.org/10.3390/cli8010001.

42 A Better Balance, Black Women's Health Imperative, Human Rights Watch, et al., "Fact Sheet: Increasing Temperatures because of the Climate Crisis Is a Reproductive Justice Issue in the United States," October 2020, https://www.hrw.org/sites/default/files/media_2020/10/climatecrisis-reproductivejustice-US_1020_web.pdf.

43 Bekkar et al., "Air Pollution and Heat Exposure," 1–13.

44 Joan Flocks, Valerie Vi Thien Mac, Jennifer Runkle, et al., "Female Farmworkers' Perceptions of Heat-Related Illness and Pregnancy Health," *Journal of Agromed-*

icine 18, no. 4 (January 2013): 350–58, https://doi.org/10.1080/1059924X.2013
.826607.

45 National Academies of Sciences, Engineering, and Medicine, "Communities in Action: Pathways to Health Equity," Washington, DC: The National Academies Press, 2016, https://doi.10.17226/24624.

46 US Centers for Disease Control and Prevention, "Preterm Birth." National Academies of Sciences, Engineering, and Medicine. 2016. "Communities in Action: Pathways to Health Equity." Washington, DC: The National Academies Press. doi: 10.17226/24624, https://www.ncbi.nlm.nih.gov/books/NBK425845/. National Academies of Sciences, Engineering, and Medicine, "Communities in Action: Pathways to Health Equity" (Washington, DC: The National Academies Press, 2016), doi: 10.17226/24624https://www.ncbi.nlm.nih.gov/books/NBK425845/.

47 Monika Mitra, Karen M. Clements, Jianying Zhang, et al., "Maternal Characteristics, Pregnancy Complications and Adverse Birth Outcomes among Women with Disabilities," *Medical Care* 53, no. 12 (December 2015): 1027–32, https://doi.org/10.1097/MLR.0000000000000427.

48 Jennifer L. Heck, Emily J. Jones, Diane Bohn, et al., "Maternal Mortality among American Indian/Alaska Native Women: A Scoping Review," *Journal of Women's Health* 30, no. 2 (February 2021): 220–29, https://10.1080/jwh.2020.8890.

49 Noor C. Gieles, Julia B. Tankink, Myrtha van Midde, et al., "Maternal and Perinatal Outcomes of Asylum Seekers and Undocumented Migrants in Europe: A Systematic Review," *European Journal of Public Health* 29, no. 4 (August 2019): 714–23, https://doi.org/10.1093/eurpub/ckz042.

50 Judith A. Stein, Michael C. Lu, and Lillian Gelberg, "Severity of Homelessness and Adverse Birth Outcomes," *Health Psychology* 19, no. 6 (November 2000): 524–34, https://doi.org/10.1037/0278-6133.19.6.524.

51 National Advisory Committee on Rural Health and Human Services, *Maternal and Obstetric Care Challenges in Rural America: Policy Brief and Recommendations to the Secretary*, May 2020, https://www.hrsa.gov/sites/default/files/hrsa/advisory-committees/rural/publications/2020-maternal-obstetric-care-challenges.pdf.

52 Paula Braverman, Katherine Heck, Susan Egerter, et al., "Worry about Racial Discrimination: A Missing Piece of the Puzzle of Black-White Disparities in Preterm Birth?," *PLOS-One* 12, no. 10 (October 2017): 1–17, https://doi.10.1371/journal.pone.0186151.

53 Bethany G. Everett, Michelle A. Kominiarek, Stephanie Mollborn, et al., "Sexual Orientation Disparities in Pregnancy and Infant Outcomes," *Maternal and Child Health Journal* 23 (January 2019): 72–81, https://doi.org/10.1007/s10995-018-2595-x.

54 Togoobaatar Ganchimeg, Rintaro Mori, Erika Ota, et al., "Maternal and

Perinatal Outcomes among Nulliparous Adolescents in Low- and Middle-Income Countries: A Multi-Country Study," *BJOG: An International Journal of Obstetrics and Gynaecology* 120, no. 13 (December 2013): 1622–30, https://doi.org/10.1111/1471-0528.12391.

55 National Birth Equity Collaborative, "Birth Equity."

56 Michael E. Kraft and Scott R. Furlong, *Public Policy: Politics, Analysis, and Alternatives*, 7th ed. (Thousand Oaks, CA: CQ Press, 2021).

57 US Senate Democrats, "Summary."

58 US Senate Democrats, "Summary."

59 HR 957 (2021–22). S 4231 (2021–22).

60 Occupational Safety and Health Administration (OSHA), "Heat Injury and Illness Prevention in Outdoor and Indoor Work Settings," *US Department of Labor*, October 2021, https://www.federalregister.gov/documents/2021/10/27/2021-23250/heat-injury-and-illnessprevention-in-outdoor-and-indoor-work-settings.

61 Blosemeli Leon-DePass and Carol Sakala, "Higher Temperatures Hurt Moms. Moms and Babies, Series 01," *National Partnership for Women and Families and the National Birth Equity Collaborative*, May 2021, https://www.nationalpartnership.org/our-work/health/moms-and-babies/highertemperatures-hurt-moms.html. National Birth Equity Collaborative, "Birth Equity."

62 John Kingdon, *Agendas, Alternatives, and Public Policies: Update Edition, with an Epilogue on Health Care* (London: Pearson Education UK, 2013).

63 Alison Parken, "A Multi-Strand Approach to Promoting Equalities and Human Rights in Policy Making," *Policy and Politics* 38, no. 1 (January 2010): 79–100, https://doi.org/10.1332/030557309X445690.

64 Cara Korte, "How Climate Change Threatens Pregnant Women and Their Fetuses," *CBS News*, November 9, 2021, https://www.cbsnews.com/news/pregnancy-women-fetuses-risks-climate-change/.

65 On sexual and reproductive health and well-being, see National Birth Equity Collaborative, "National Birth Equity Collaborative Pens Open Letter to Vice President Kamala Harris," December 2021, https://birthequity.org/news//maternal-health-day-of-action/. On climate equity at all offices, see US White House, "The Path to Achieving Justice40," July 7, 2021, https://www.whitehouse.gov/omb/briefing-room/2021/07/20/the-path-to-achieving-justice40/.

66 Sumi Cho, Kimberle Crenshaw, and Leslie McCall, "Toward a Field of Intersectionality Studies Theory, Applications, and Praxis," *Signs: Journal of Women in Culture and Society* 38, no. 40 (May 2013): 785–810, https://doi.org/10.1086/669608. Hoffman, Shandas, and Pendleton, "Effects of Historical Housing Policies."

67 US Senate Democrats, "Summary."

68 US Senate Democrats, "Summary."

69 US House of Representatives Black Maternal Health Caucus, "Black Maternal Health Caucus."

70 Kraft and Furlong, *Public Policy*.

71 Marcia K. Meyers and Vibeka Nielsen, "Street-Level Bureaucrats and the Implementation of Public Policy," *The SAGE Handbook of Public Administration* (January 2012): 305–18, https://dx.doi.org/10.4135/9780857020970.n13.

72 Kraft and Furlong, *Public Policy*.

CHAPTER 10

At Risk but Overlooked and Underserved

Older Adults and Vulnerabilities to Climate Change Impacts

CAITLIN CONNELLY, KATHRIN BOERNER, NATASHA BRYANT, AND
ROBYN STONE

Due to climate change, natural disasters are increasing in frequency and severity.[1] Natural disasters bring with them increases in mortality rates, displacement, and many other negative physical and psychological health effects.[2] These negative outcomes of disasters are often worse for adults age sixty-five and older, relative to adults at younger ages. Older adults who have higher rates of chronic conditions, functional limitations, and social isolation compared to the general population are more vulnerable during and immediately following natural disasters.[3]

This chapter provides an overview of the health impacts of climate change–related extreme weather and natural disasters on older adults, the vulnerability factors that make older adults more susceptible, and discussion of relevant policy and service provision. The first section describes the population of older adults in the United States. The following section recounts the adverse impacts of heat waves, hurricanes, flooding, and wildfires that older adults have experienced over the past twenty years. The last section describes the current role of service providers and policy in helping older adults prepare for and deal with climate change–related weather events as well as recommendations for improvements in how service providers and policies address these concerns.

OLDER ADULT POPULATION

Adults age sixty-five and older currently make up 16 percent of the population in the United States. This percentage is expected to increase over the next forty years to 23 percent of older adults by 2060, or approximately 95 million older adults.[4] The demographic composition of older adults is also shifting, with an increasing proportion of Black, Indigenous, and other people of color. By 2050, the proportion of non-Hispanic Black older adults is predicted to increase to 12 percent (from 9 percent in 2014) and the percentage of Hispanic older adults is predicted to increase to 18 percent (from 7 percent in 2014).[5]

Physical and cognitive health are significant determinants of the degree of vulnerability to climate change–related weather impacts. As of 2019, 85 percent of older adults had at least one chronic condition.[6] These rates are even higher for older adults of color relative to White older adults.[7] Black and Hispanic adults are predicted to have the greatest increases in dementia prevalence over the next forty years, relative to other racial/ethnic groups.[8] This contributes to a cumulative disadvantage in the aging experience for older adults of color in the United States. In the case of disasters, those with health challenges are more susceptible to harm, as they may require assistance and may be reliant on medical devices and electricity and possibly regular health care and medicine. Disasters can result in disruptions to critical infrastructure systems (electricity, other power sources, water) and can adversely affect these individuals. Physical impairments limit older adults' ability to quickly and efficiently evacuate their homes before, during, and after a disaster. Cognitive impairment can interfere with the information processing and decision making needed to prepare for and navigate through a disaster event.[9]

The living situation for older adults also plays a role during extreme weather. More than one-quarter of community-dwelling older adults live alone, and rates of living alone are highest among those age eighty-five and older.[10] There is a growing preference for aging in place and utilizing home- and community-based services (HCBS) when the need arises. In 2018, nearly eight million older adults, 90 percent of which lived in a community, needed long-term services and supports (LTSS).[11] A nationally representative study found that in 2011, approximately 6 percent of adults age sixty and older used HCBS.[12] Long-term care facilities still have an important function; approximately 37 percent of older adults will require nursing home care at some point in their lives.[13] Older adults who live alone may require assistance obtaining food and traveling to a safer location in a disaster event but may not have the social resources to do so.[14]

Older adults in residential care facilities are at heightened risk of harm during disasters because of their physical, mental, and functional impairments and dependence on care staff to meet their basic functional needs.[15]

Many older adults have limited financial resources. In 2017, the median income of older adult males was $32,545, and females sixty-five and older had a lower median income at $19,180.[16] Almost 10 percent of older people lived below the poverty level (4.7 million people) and another 4.9 percent were classified as "near poor," which means their incomes were between the poverty level and 125 percent of the poverty level.[17] Older adults who are economically disadvantaged may lack the financial resources to prepare for, respond to, and recover from disasters and climate-related events. They may be at greater risk during heat waves and high heat periods because they do not have air conditioning or are reluctant to use it due to cost.[18] The issues that come with a lower socioeconomic status may burden more older adults of color than White older adults as White households have overall substantially more wealth than Black and Hispanic households.[19] Additionally, older adults living in rural areas are more likely to be poorer than those in urban areas and to live in housing considered to be at higher risk for adverse disaster impacts.[20] Having insurance and owning a home, which occurs at higher rates for White adults, are also protective factors against physical health declines and economic difficulties following disasters.[21]

HEALTH EFFECTS OF EXTREME WEATHER EVENTS FOR OLDER ADULTS

Older adults are at greater risk of harm in cases of extreme heat, wildfires and air pollution, and extreme weather disasters such as hurricanes and flooding.[22] Worsening health, rates of hospitalization, and mortality are greater for adults age sixty-five and older during natural disasters and prolonged weather events compared to adults at younger ages.[23] Older adult disaster survivors may have an increased risk of negative mental health outcomes, both short- and long-term.[24]

Over the past thirty years, there has been an increase in awareness of the detrimental effects of natural disasters on older adults. In 1995, the Chicago heat wave brought attention to more vulnerable adults when the majority of heat-related deaths were of people who were older, Black, and poor.[25] "This was a disaster . . . that targeted the most vulnerable people in the city," said Eric Klinenberg, who wrote the 2002 book Heat Wave: A Social Autopsy of Disaster in Chicago analyzing the heat wave and its aftermath. In 2003, a heat wave devastated Paris, and again older adults suffered disproportionately more than younger adults. Approximately 80 percent of the fifteen thousand excess deaths in the Paris heat wave were of people age seventy-five years and older.[26] The vulnerability of older adults in natural disasters was brought to the forefront again when Hurricane Katrina hit New Orleans in 2005. In addition to older adults having the highest mortality rate of any age group, Katrina also highlighted the lack of preparedness in long-term care facilities for this type of event.[27] Nearly every natural disaster since has continued to demonstrate that older adults are disproportionately susceptible to disaster-related harm. This section describes the physical and mental health effects of extreme weather for older adults.

Hurricanes and Flooding

One early and devastating example of the high mortality rate of older adults during and after hurricanes was Hurricane Katrina in 2005, in which 71 percent of the deaths were of adults age eighty-five and older.[28] During Hurricane Maria in 2017, older adults had the greatest mortality risk relative to adults younger than age sixty-five.[29] Older adult mortality in New Jersey was 13 percent higher in the four months following Hurricane Sandy compared with nonhurricane periods.[30] Older adults residing in nursing homes also have higher mortality risk in the context of hurricanes. This increased mortality risk may continue well after the hurricane, as was found ninety days after Hurricane Katrina, when nursing home residents still had a 2.6 percent higher mortality risk relative to nondisaster periods.[31]

Older adults also have an increased risk of declining health and hospitalization associated with hurricanes. In the three weeks following Hurricane Sandy, emergency department visits of older adults increased significantly, with rates highest among those age eighty-five and older.[32] Following Katrina, the thirty-day and ninety-day hospitalization rates increased by 2.5 percent and 2.2 percent, respectively, relative to nonhurricane periods.[33] After precipitation-related flooding, older adults are more susceptible to negative physical health effects than younger adults because of medication interruption, existing illness, lack of education, and low socioeconomic status.[34]

It is important to note that health impacts of hurricanes are not equally distributed among different racial and ethnic groups of older adults. Hospitalization rates due to cardiovascular disease tend to be higher for Black older adults as compared to White older adults, but during and following Hurricane Katrina, this disparity grew, indicating that cardiovascular health aggravation was worse for Black older adults.[35] In fact, an examination of storms from 1999 to 2017 showed that the mortality rate of Black older adults was more than twice that of White older adults.[36]

Evacuation and relocation of older adults during hurricanes may also be detrimental to their health. Due to the flooding during Hurricane Katrina there were over two hundred thousand displaced or isolated individuals with chronic illnesses that were unable to access medical care and take medications.[37] Maintaining continuity of care throughout a disaster event can be quite challenging, particularly for older adults who rely on durable medical equipment, ventilators, and dialysis to manage their chronic medical conditions. People of low socioeconomic status stay in temporary shelters longer than homeowners and wealthier individuals.[38] This contributes to negative health outcomes, as temporary, crowded shelters are not conducive to good mental health and often serve food with low nutritional value or food that is not appropriate for individuals with diabetes.[39]

Evidence derived from nursing home residents that were evacuated in hurricanes has demonstrated that evacuation—while often necessary—is harmful for older adults in facilities, and in some cases may be more harmful than sheltering in place. Nursing home residents who were evacuated during Hurricane Katrina developed more pressure ulcers and had a higher mortality rate after the storm than a matched sample of residents that were not evacuated.[40] After Hurricane Gustav in 2008, nursing home residents with dementia who were evacuated had elevated mortality rates relative to nondisaster periods thirty and ninety days after the storm.[41]

In addition to physical health impacts, older adults are susceptible to mental health impacts following disasters.[42] Hurricanes, earthquakes, and precipitation-related flooding have all been found to be associated with an increase in post-traumatic stress disorder symptoms for older adults.[43] Earthquakes and precipitation-related flooding events have also been found to be associated with higher psychological distress and anxiety for older adults.[44] Studies have shown a significant association between hurricanes and subsequent increases in depressive symptoms.[45]

The impact of hurricanes on mental health may continue years after the weather event. Three years after Hurricane Katrina, older adults who survived the hurricane had higher stress levels than younger adult survivors.[46] Six years after Hurricane Sandy in the United States, older adults that experienced traumatic stress during the hurricane had significantly higher rates of functional limitations than those who did not experience stress during the hurricane.[47]

Heat Waves

Older adults are at an increased risk of hospitalization during heat waves and sustained high temperatures relative to periods without high heat. Hospitalization during heat waves is primarily due to heat stroke and sunstroke but also fluid and electrolyte disorders, renal failure, acute kidney failure, urinary tract infections, and septicemia.[48] Older adults taking psychotropic medication to treat mental illnesses are at even greater risk of hospitalization and death because side effects of psychotropic medication include reduced ability to regulate body temperature.[49]

Mortality risk can also increase during heat waves and high temperatures. Higher and variable temperatures in the summer season in the southeast United States were associated with an increased mortality rate for older adults.[50] Older adults had the highest mortality rates relative to other age groups following the 1995 Chicago heat wave, the 2003 Paris heat wave, and the 2011 Texas heat wave.[51] Moreover, risk of mortality increases with age among older adults, with the risk increasing for adults age seventy-five to eighty-four, and again for adults over age eighty-four.[52] The increased risk of mortality for older adults was found to continue up to ten days following heat waves in Texas.[53]

Electricity-dependent populations and older adults of lower socioeconomic status may be more at risk during high heat. During Hurricane Irma in 2017, an assisted living facility lost power, and fourteen residents subsequently suffered heat-related deaths.[54] Among people of all ages, 16 percent of heat-related deaths during Hurricane Irma were due to power outages—primarily because of a loss of air conditioning and also due to interruption of electricity-dependent medical treatment.[55] Power outages and failure to use sufficient air conditioning are risk factors for hospitalization and mortality during high heat.

Wildfires

Older adults and persons with preexisting cardiovascular or respiratory diseases are at increased risk of negative health effects due to air pollution, including air pollution from wildfire smoke.[56] During the 2003 California wildfires, respiratory hospital admissions were increased and highest for older adults.[57] Similarly, cardiovascular, cerebrovascular, and respiratory emergency department visits were highest among older adults during the 2015 California wildfires.[58] The vast majority of excess deaths in California wildfire smoke–affected areas in 2003 were of older adults, relative to other ages.[59]

Future Research Directions

There are notable gaps in research examining the impact of climate change on older adults and the challenges and needs of older adults using HCBS. Future studies should consider how the impact of climate change on the health of older adults using HCBS might differ from the general population of older adults or their peers living in nursing homes. Additionally, most studies of health impacts on older adults have focused on acute weather events and failed to provide information on other common occurrences, such as worsening air quality, wildfires, and longer-term high heat periods.

The cumulative disadvantage of increasing age and structural and institutional racism experienced over the life course—presenting in lower socioeconomic status, living in more vulnerable housing, and having poorer health than White older adults—leaves older adults of color more likely to experience harm from disasters.[60] In the few studies that examined preparedness by race and ethnicity, Hispanic and Black older adults were found to be less prepared than White older adults in two studies,[61] but another study did not find statistically significant differences in preparedness levels by race.[62] Regarding disaster effects, Rachel Adams and colleagues found that between 1999 and 2017, among all older adults, American Indian/Alaska Native males had the highest mortality rate in extreme cold events, and Black males had the highest mortality rates from hurricanes.[63]

Older adults in rural housing and older adults of color are more likely to be at greater risk of harm from disasters, due in part to lower socioeconomic status and living in more at-risk housing, and yet there is a paucity of research on disaster impacts for these populations.[64] During disasters, means of communication and transportation become vital, and the pathways to secure these needs are likely different for rural older adults compared to those in urban settings. To fully understand the risks, more research in gerontology is needed on the disparate health effects of natural disasters for older adults of color, rural populations, and socioeconomically disadvantaged older adults.

SERVICE PROVISION AND POLICY

Public health departments have the primary responsibility for ensuring public health and safety during natural disasters. However, they do not necessarily focus on older adults in their planning and programs, even in designated age-friendly communities.[65] While not for older adults specifically, some public health departments have programs for people of all ages with functional limitations, chronic conditions, and greater fall risk.[66]

The Older Americans Act requires State Units on Aging and local Area Agencies on Aging (AAAs) to have emergency preparedness plans for their community. These plans are required to have information on how AAAs will coordinate activities and develop long-term emergency plans with state and local governments and emergency response agencies, relief organizations, and other entities responsible for emergency preparedness and response.[67]

The US Department of Health and Human Services of the Assistant Secretary for Preparedness and Response (ASPR) worked with Centers for Medicare and Medicaid Services (CMS) to develop the Emergency Preparedness Rule in 2016.[68] This rule requires disaster planning for all seventeen Medicare/Medicaid providers, which covers CMS-regulated HCBS, long-term care facilities, home health agencies, hospitals, hospices, other medical centers, Programs of All-Inclusive Care for the Elderly, and community mental health services. This rule requires providers to (1) create a disaster plan, (2) regularly train employees on disaster preparedness, and (3) conduct testing exercises of the plan. Compliance with this rule was required by November 2017 and is assessed by state-conducted surveys.

Under the Republican Trump administration, revisions were made to the 2016 version of the CMS Emergency Preparedness Rule to lessen restrictions. One revision was to rescind the requirement to document collaboration efforts between providers and state and local disaster management agencies. Additionally, the frequency of testing and reviewing plans was decreased from annually to biennially, with the exception of long-term care facilities, which are still required to do so annually.[69]

In August 2021, during the Democratic Biden administration, the US Department of Health and Human Services opened the Office of Climate Change and Health Equity as part of President Biden's executive order on tackling climate change. This office will focus on equitable solutions to reduce racial and ethnic disparities as well as age disparities in climate change–related health impacts.[70]

Recommendations for Policy and Practice

Research on disaster communication has highlighted that older adults are getting their disaster-related information primarily through television, as well as from friends and family.[71] Current studies of social media use for disaster communication rarely examine use by age, and thus the applicability of this method for older adults is still unknown.[72] Storm warnings and recovery resources are primarily in English and thus do not reach populations who do not speak English.[73] Older adults report that their perceived trustworthiness of the source of information is of great import to them, and it affects their decision of whether or not to evacuate.[74] Thus, trusted community leaders, such as faith leaders and local council members, could disseminate information to their older community members.

Social workers, case managers, and medical care professionals can also serve as trusted sources of information for older adults.[75] These frontline professionals can provide informed support tailored to the older adults they serve, using their knowledge of individual risk factors such as social isolation, frailty, sensory impairment, reduced mobility, and dementia.[76] They also can provide help aiding in recovery following disasters. Health risks may continue months or years after a disaster. Physicians and social workers should be aware of the physical and mental health risks predisaster and postdisaster to better educate their older adult clients about the risks and to provide care after the event.[77]

Studies of employees in public health departments found that while the departments engaged in disaster preparedness and resilience activities, their efforts were not tailored to older adults and programs focused on individuals with functional limitation did not meet the needs of older adults.[78] In qualitative interviews, public health department officials reported that they believed their overall goal was disaster preparedness for all age groups and that it was not their responsibility to prepare the sixty-five-and-older population.[79] Only about half of the public health department staff members in one study worked with LTSS facilities or other residential organizations for older adults to help them plan for emergencies.[80]

Studies among HCBS agency administrators find that although administrators are responsible for developing emergency preparedness plans for their clients, they feel unprepared to create the plans.[81] Within HCBS, home care aides are prime for being first responders to help prepare, assist during disaster, and follow up with clients. To do this, they need training and support, which should be built into state and federal training requirements and competency standards. Long-term care ombudsmen also have reported that they needed comprehensive, performance-based emergency preparedness curriculum and ongoing training to effectively perform their appropriate emergency service roles.[82]

There are many online resources to support and assist providers with meeting the CMS Emergency Preparedness Rule standards. ASPR Technical Resources, Assistance Center, and Information Exchange and CMS have compiled resources online, including templates and checklists, that providers can use to help create, revise, and maintain their disaster plans.[83] Other federal resources for provider planning include the Administration for Community Living and the US Centers for Disease Control and Prevention.[84] Further resources for providers that are both covered and not covered by CMS are available through nonprofits and advocacy groups, including LeadingAge, the American Health Care Association, and the National Center for Assisted Living.[85] However, it is still unclear how and if the disaster planning resources are being used.[86]

All facilities and providers, including those that do not have to comply with CMS regulations, would benefit from following the guidelines set by the Emergency Preparedness Rule and using the resources available for meeting it. Assisted living facilities do not fall under national jurisdiction but instead are regulated at the state level, and vary from state to state.[87] To ensure a similar level of preparedness across all facility types, the American Academy of Nursing recommends that assisted living facilities be required to submit an emergency preparedness plan to the state.[88]

While the majority of nursing homes provide mental health services during normal operations, this is often interrupted and not prioritized during disasters, especially when residents are evacuated.[89] Providers also should train staff to provide mental health services during and immediately following a disaster.[90]

Local and state planners need to think about the impact of evacuation and relocation on the health and well-being of older adults during disasters. As noted above, some studies have suggested that during natural disasters evacuation can sometimes be more harmful than sheltering in place. Local governments should consider these unintended consequences before ordering mandatory evacuations.[91] If evacuation occurs, policies and procedures to address the special needs of at-risk older adults need to be put in place.

Shelters are often incompatible with the complex needs of nursing home residents and community-dwelling older adults with multiple functional and/or cognitive impairments. Shelters are frequently not staffed by medical professionals who have experience working with older adults.[92] Some states mandate that shelters provide space for populations with special needs, such as vulnerable older adults. Special needs shelters are operated separately from general shelters and can accommodate and care for people with severe medical needs during a disaster.[93] In Florida, the State Emergency Management Act requires the Division of Emergency Management to submit a state plan that identifies the location and size of special needs shelters.[94] Nursing homes residents who were relocated to nonclinical locations, such as school gymnasiums or churches, have faced additional challenges that would not have arisen in a medical facility, such as inappropriate sleeping conditions, numerous risks to resident safety, and lack of equipment to support residents with limited mobility.[95] To provide more appropriate shelters, facilities that are not evacuated during an emergency should be prepared to accommodate residents from other relocated facilities as well as those in need from the community.

Inadequate transportation and financial feasibility may act as barriers to evacuation for older adults living in the community. During Hurricane Katrina, older adults were less likely to have access to a vehicle for evacuation. This is especially true for more vulnerable older adults with functional impairments and those receiving LTSS.[96] A study of older adults with low socioeconomic status in Georgia found that 46 percent of respondents would need assistance with transportation in order to evacuate.[97]

Many of the current weaknesses in disaster planning and response could be addressed by increasing cross-sector collaboration. Local, state, and federal agencies' disaster plans too often leave out the unique needs of some older adults and are not well-coordinated.[98] For LTSS that directly serve older adults, providers often do not have enough resources in terms of time, disaster-related knowledge, and finances to adequately prepare for, respond to, and address the aftermath of disasters. Long-term services and supports providers, as well as older adults themselves, could inform emergency and public health agencies of the pressing need for plans and resources aimed at supporting more vulnerable older adults. In return, the resources that emergency management agencies have could be used to strengthen disaster preparedness and response for LTSS.

Cross-sector collaboration could be encouraged by requiring partnerships between older adult-serving agencies and government agencies, where both parties reach out to one another. As noted previously, the 2018 revision to the CMS Emergency Preparedness Rule took away the requirement to document collaboration efforts between CMS providers and local emergency management personnel.[99] However, increased collaboration is a critical protective action to support organized local- and state-wide response to disasters. Another means of encouraging collaboration could be via extant regional emergency preparedness health care coalitions, as suggested by Debra Dobbs and colleagues.[100] Including a diverse group of LTSS providers in health care coalitions can provide a platform for providers to speak to the unique needs of older adults living in the community receiving HCBS and living in long-term care facilities.

Florida, a state that experiences many hurricanes each year, practices coordination across agencies that can serve as an example for other states. The Florida Department of Health developed a data collection tool and the state provides each county with a description of the population to help counties identify the specific needs of older adults in an emergency. The Florida Department of Elder Affairs works with the state's emergency operations centers and AAAs to prepare for and respond to disasters. In addition, Florida's LTSS organizations are required to file an emergency preparedness plan with the county emergency operations center for approval.

CONCLUSION

Natural disasters were first conceptualized in academic literature as external forces, outside of our control, that equally distributed damage and loss to humankind. The literature has since evolved to recognize that, while natural disasters are uncontrollable forces, their destruction does not occur equally across individuals.[101] In the United States, older adults experience a disproportionate weight of disasters' burdens, especially those with functional impairment, cognitive impairment, and multiple chronic conditions, and those of advanced age.[102] Older adults have had the highest mortality rate of any age group throughout hurricanes, floods, and heat waves. They are also more likely than other age groups to be hospitalized and experience health declines during extreme weather.

There is a need to examine the relative risks and potential negative consequences of disasters for subgroups of older adults. The high prevalence of functional and cognitive impairments as well as chronic health conditions necessitates continuity of care that may be interrupted during a natural disaster, which further exacerbates health challenges. HCBS recipients, who are more likely to be in poorer health than other community-dwelling older adults, are an understudied subgroup of older adults in the current disaster literature. More research is also needed to examine the differences in risks and impacts among various racial and ethnic groups. A better understanding of the subgroups of older adults and climate change–related health disparities will assist policymakers, public health and other emergency climate change–related response systems, service providers, and others to better predict, prepare for, and adequately respond to the short- and long-term needs of older adults and better target resources for groups most in need, especially given the anticipated increase in frequency and intensity of climate-related events.

Public health departments, policymakers, AAAs, LTSS providers, frontline professionals, family caregivers, and community members can all play an important role in supporting older adults through natural disasters and extreme weather exposure. These organizations and providers need to be better prepared to fulfill the role of improving outcomes for older adults during disasters and dangerous weather. Policymakers should be informed of the specialized needs of subpopulations of older adults and target resources to improve outcomes for those in greatest need. Increased collaboration across these organizations will improve disaster response. Caution needs to be remembered when making the decision of whether or not older adults should evacuate or shelter in place. Evacuation sites should be adapted to be more appropriate for more vulnerable populations. Implementing more equitable and age-appropriate disaster relief policies can ensure that everyone has an equal opportunity and access to the support they need in the short and long term.

NOTES

1 John Balbus et al., "Climate Change and Human Health," in *The Impacts of Climate Change on Human Health in the United States: A Scientific Assessment* (Washington, DC: US Global Change Research Program, 2016), 25–42.

2 Balbus et al., "Climate Change and Human Health"; Janet L. Gamble et al., "Climate Change and Older Americans: State of the Science," *Environmental Health Perspectives* 121, no. 1 (2013): 15–22, https://doi.org/10.1289/ehp.1205223.

3 Gamble et al., "Climate Change and Older Americans."

4 Mark Mather, Paola Scommegna, and Lillian Kilduff, "Fact Sheet: Aging in the United States," Population Reference Bureau, July 2020, https://www.prb.org/resources/fact-sheet-aging-in-the-united-states/.

5 Jennifer M. Ortman, Victoria A. Velkoff, and Howard Hogan, *An Aging Nation: The Older Population in the United States*, US Census Bureau, Economics and Statistics Administration, US Department of Commerce, 2014.

6 "About Chronic Diseases," US Centers for Disease Control and Prevention, accessed 2019, https://www.cdc.gov/chronicdisease/about/index.htm.

7 Kenneth F. Ferraro, Blakelee R. Kemp, and Monica M. Williams, "Diverse Aging and Health Inequality by Race and Ethnicity," *Innovation in Aging* 1, no. 1 (2017), https://doi.org/10.1093/geroni/igx002.

8 Kevin A. Matthews et al., "Racial and Ethnic Estimates of Alzheimer's Disease and Related Dementias in the United States (2015–2060) in Adults Aged ≥ 65 Years," *Alzheimer's & Dementia* 15, no. 1 (2019): 17–24, https://doi.org/10.1016/j.jalz.2018.06.3063.

9 Laura Banks, "Caring for Elderly Adults during Disasters: Improving Health Outcomes and Recovery," *Southern Medical Journal* 106 (2013): 94–98, https://doi.org/10.1097/smj.0b013e31827c5157.

10 National Academies of Sciences, Engineering, and Medicine, *Social Isolation and Loneliness in Older Adults: Opportunities for the Health Care System* (Washington, DC: The National Academies Press, 2020).

11 Edem Hado and Harriet Komisar, *Fact Sheet Long-Term Services and Supports*, AARP Public Policy Institute, August 2019, https://www.aarp.org/content/dam/aarp/ppi/2019/08/long-term-services-and-supports.doi.10.26419-2Fppi.00079.001.pdf.

12 Amanda Sonnega, Kristen Robinson, and Helen Levy, "Home and Community-Based Service and Other Senior Service Use: Prevalence and Characteristics in a National Sample," *Home Health Care Services Quarterly* 36, no. 1 (2017): 16–28, https://doi.org/10.1080/01621424.2016.1268552.

13 Richard W. Johnson et al., *Most Older Adults Are Likely to Need and Use Long-Term Services and Supports*, ASPE, January 31, 2021, https://aspe.hhs.gov/reports/most-older-adults-are-likely-need-use-long-term-services-supports-issue-brief-0.

14 Banks, "Caring for Elderly Adults."

15 H. Wayne Nelson et al., "State Long-Term Care Ombudsmen's Perceptions of Their Program's Disaster Preparedness Roles and Readiness," *Journal of Applied Gerontology* 32, no. 8 (2013): 952–74, https://doi.org/10.1177/073347481244865.

16 Administration for Community Living and Administration on Aging, *2018 Profile of Older Americans*, 2018, https://acl.gov/sites/default/files/Aging%20and%20Disability%20in%20America/2018OlderAmericansProfile.pdf.

17 Administration for Community Living and Administration on Aging, *2018 Profile of Older Americans*.

18 Gamble et al., "Climate Change and Older Americans."

19 Benjamin W. Veghte, E. Schreur, and Mikki Waid, "Social Security and the Racial Gap in Retirement Wealth," *National Academy of Social Insurance* 48 (2016), https://www.nasi.org/research/2016/social-security-racial-wealth-gap.

20 Robyn Stone, "Housing," in *Handbook of Rural Aging*, ed. Lenard W. Kaye (London: Routledge, 2021), 40.

21 Jee Young Lee and Shannon Van Zandt, "Housing Tenure and Social Vulnerability to Disasters: A Review of the Evidence," *Journal of Planning Literature* 34, no. 2 (2019): 156–70, https://doi.org/10.1177/0885412218812080.

22 Gamble et al., "Climate Change and Older Americans"; Erwin William A. Levya, Adam Beaman, and Patricia M. Davidson, "Health Impact of Climate Change in Older People: An Integrative Review and Implications for Nursing," *Journal of Nursing Scholarship* 49, no. 6 (2017): 670–78, https://doi.org/10.1111/jnu.12346.

23 Executive Office of the President, Assistant to the President for Homeland Security, and Counterterrorism, *The Federal Response to Hurricane Katrina: Lessons Learned* (Washington, DC: Government Printing Office, 2006); Zachary S. Wettstein et al., "Cardiovascular and Cerebrovascular Emergency Department Visits Associated with Wildfire Smoke Exposure in California in 2015," *Journal of the American Heart Association* 7, no. 8 (2018): e007492, https://doi.org/10.1161/JAHA.117.007492.

24 Yoshinori Kamo, Tammy L. Henderson, and Karen A. Roberto, "Displaced Older Adults' Reactions to and Coping with the Aftermath of Hurricane Katrina," *Journal of Family Issues* 32, no. 10 (2011): 1346–70, https://doi.org/10.1177/0192513 X11412495; Levya, Beaman, and Davidson, "Health Impact of Climate Change."

25 Klinenberg, *Heat Wave.*

26 Anne Fouillet et al., "Excess Mortality Related to the August 2003 Heat Wave in France," *International Archives of Occupational and Environmental Health* 80 (2006): 16–24, https://doi.org/10.1007/s00420-006-0089-4.

27 David Dosa et al., "Effects of Hurricane Katrina on Nursing Facility Resident Mortality, Hospitalization, and Functional Decline," *Disaster Medicine and Public Health Preparedness* 4, no. S1 (2010): S28–S32. https://doi.org/10.1001/dmp .2010.11; Executive Office of the President, *Federal Response to Hurricane Katrina.*

28 Executive Office of the President, *Federal Response to Hurricane Katrina.*

29 Raul Cruz-Cano and Erin L. Mead, "Causes of Excess Deaths in Puerto Rico after Hurricane Maria: A Time-Series Estimation," *American Journal of Public Health* 109, no. 7 (2019): 1050–52, https://doi.org/10.2105/ajph.2019.305015.

30 Soyeon Kim et al., "Hurricane Sandy (New Jersey): Mortality Rates in the Following Month and Quarter," *American Journal of Public Health* 107, no. 8 (2017): 1304–7, https://doi.org/10.2105/AJPH.2017.303826.

31 Dosa et al., "Effects of Hurricane Katrina."

32 Sidrah Malik et al., "Vulnerability of Older Adults in Disasters: Emergency Department Utilization by Geriatric Patients after Hurricane Sandy," *Disaster Medicine and Public Health Preparedness* 12, no. 2 (2018): 184–93, https://doi.org /10.1017/dmp.2017.44.

33 Dosa et al., "Effects of Hurricane Katrina."

34 Dianne Lowe, Kristie L. Ebi, and Bertil Forsberg, "Factors Increasing Vulnerability to Health Effects before, during and after Floods," *International Journal of Environmental Research and Public Health* 10, no. 12 (2013): 7015–67. https:// doi.org/10.3390/ijerph10127015.

35 Ninon A. Becquart et al., "Cardiovascular Disease Hospitalizations in Louisiana Parishes' Elderly before, during and after Hurricane Katrina," *International Journal of Environmental Research and Public Health* 16, no. 1 (2019): 74, https:// doi.org/10.3390/ijerph16010074.

36 Rachel M. Adams et al., "Mortality from Forces of Nature among Older Adults by Race/Ethnicity and Gender," *Journal of Applied Gerontology* 40, no. 11 (2021): 1517–26, https://doi.org/10.1177/0733464820954676.

37 Executive Office of the President, *Federal Response to Hurricane Katrina.*

38 Richard M. Zoraster, "Vulnerable Populations: Hurricane Katrina as a Case Study," *Prehospital and Disaster Medicine* 25, no. 1 (2010): 74–78, https://doi. org/10.1017/S1049023X00007718.

39 Baylor College of Medicine and the American Medical Association, *Recommendations for Best Practices in the Management of Elderly Disaster Victims*, 2006, https://www.bcm.edu/pdf/bestpractices.pdf; Zoraster, "Vulnerable Populations."

40 Nicholas G. Castle and John B. Engberg, "The Health Consequences of Relocation for Nursing Home Residents following Hurricane Katrina," *Research on Aging* 33, no. 6 (2011): 661–87, https://doi.org/10.1177/0164027511412197.

41 Lisa M. Brown et al., "The Effects of Evacuation on Nursing Home Residents with Dementia," *American Journal of Alzheimer's Disease & Other Dementias* 27, no. 6 (2012): 406–12, https://doi.org/10.1177/1533317512454709.

42 Lisa M. Brown et al., "Use of Mental Health Services by Nursing Home Residents after Hurricanes," *Psychiatric Services* 61, no. 1 (2010): 74–77, https://doi.org/10.1176/ps.2010.61.1.74; Daniel R. Levinson and Inspector General, "Gaps Continue to Exist in Nursing Home Emergency Preparedness and Response during Disasters: 2007–2010," *Washington, DC: US Department of Health and Human Services* (2012).

43 Bei Bei et al., "A Prospective Study of the Impact of Floods on the Mental and Physical Health of Older Adults," *Aging & Mental Health* 17, no. 8 (2013): 992–1002, https://doi.org/10.1080/13607863.2013.799119; Kamo, Henderson, and Roberto, "Displaced Older Adults"; Levya, Beamen, and Davidson, "Health Impact of Climate Change."

44 Bei et al., "Prospective Study"; Levya, Beaman, and Davidson, "Health Impact of Climate Change."

45 Allison R. Heid et al., "Exposure to Hurricane Sandy, Neighborhood Collective Efficacy, and Post-Traumatic Stress Symptoms in Older Adults," *Aging & Mental Health* 21, no. 7 (2017): 742–50, https://doi.org/10.1080/13607863.2016.1154016; Kamo, Henderson, and Roberto, "Displaced Older Adults."

46 Francis O. Adeola and J. Steven Picou, "Social Capital and the Mental Health Impacts of Hurricane Katrina: Assessing Long-Term Patterns of Psychosocial Distress," *International Journal of Mass Emergencies & Disasters* 32, no. 1 (2014): 121–56.

47 Rachel Pruchno et al., "Type of Disaster Exposure Affects Functional Limitations of Older People 6 Years Later," *Journals of Gerontology: Series A* 75, no. 11 (2020): 2139–46, https://doi.org/10.1093/gerona/glz258.

48 Stephanie Hopp, Francesca Dominici, and Jennifer F. Bobb, "Medical Diagnoses of Heat Wave-Related Hospital Admissions in Older Adults," *Preventive Medicine* 110 (2018): 81–85, https://doi.org/10.1016/j.ypmed.2018.02.001; Rebecca K. McTavish et al., "Association between High Environmental Heat and Risk of Acute Kidney Injury among Older Adults in a Northern Climate: A Matched Case-Control Study," *American Journal of Kidney Diseases* 71, no. 2 (2018): 200–208, https://doi.org/10.1053/j.ajkd.2017.07.011.

49 Abderrezak Bouchama et al., "Prognostic Factors in Heat Wave–Related Deaths: A Meta-Analysis," *Archives of Internal Medicine* 167, no. 20 (2007): 2170–76, https://doi.org/10.1001/archinte.167.20.ira70009.

50 Regina A. Shih et al., *Improving Disaster Resilience among Older Adults: Insights from Public Health Departments and Aging-in-Place Efforts* (Santa Monica, CA: RAND Corporation, 2018), https://www.rand.org/pubs/research_reports/RR2313.html.

51 Fouillet et al., "Excess Mortality"; Steven Whitman et al., "Mortality in Chicago Attributed to the July 1995 Heat Wave," *American Journal of Public Health* 87, no. 9 (1997): 1515–18, https://doi.org/10.2105/AJPH.87.9.1515; Kai Zhang, Tsun-Hsuan Chen, and Charles E. Begley, "Impact of the 2011 Heat Wave on Mortality and Emergency Department Visits in Houston, Texas," *Environmental Health* 14, no. 1 (2015): 1–7, https://doi.org/10.1186/1476-069X-14-11.

52 Jeffrey Berko et al., "Deaths Attributed to Heat, Cold, and Other Weather Events in the United States, 2006–2010," *National Health Stat Report* 76 (2014): 1–15.

53 Lung-Chang Chien, Yuming Guo, and Kai Zhang, "Spatiotemporal Analysis of Heat and Heat Wave Effects on Elderly Mortality in Texas, 2006–2011," *Science of the Total Environment* 562 (2016): 845–51, https://doi.org/10.1016/j.scitotenv.2016.04.042.

54 Anindita Issa et al., "Deaths Related to Hurricane Irma—Florida, Georgia, and North Carolina, September 4–October 10, 2017," *Morbidity and Mortality Weekly Report* 67, no. 30 (August 2018): 829, https://.doi.org/10.15585%2Fmmwr.mm6730a5.

55 Issa et al., "Deaths Related to Hurricane Irma."

56 Anjali Haikerwal et al., "Impact of Fine Particulate Matter (PM2.5) Exposure during Wildfires on Cardiovascular Health Outcomes," *Journal of the American Heart Association* 4, no. 7 (2015): e001653, https://doi.org/10.1161/JAHA.114.001653; Jia C. Liu et al., "A Systematic Review of the Physical Health Impacts from Non-occupational Exposure to Wildfire Smoke," *Environmental Research* 136 (2015): 120–32, https://doi.org/10.1016/j.envres.2014.10.015.

57 Ralph J. Delfino et al., "The Relationship of Respiratory and Cardiovascular Hospital Admissions to the Southern California Wildfires of 2003," *Occupational and Environmental Medicine* 66, no. 3 (2009): 189–97, https://doi.org/10.1136/oem.2008.041376.

58 Wettstein et al., "Emergency Department Visits."

59 Ikuho Kochi et al., "Valuing Mortality Impacts of Smoke Exposure from Major Southern California Wildfires," *Journal of Forest Economics* 18, no. 1 (2012): 61–75, https://doi.org/10.1016/j.jfe.2011.10.002.

60 Bob Bolin and Liza C. Kurtz, "Race, Class, Ethnicity, and Disaster Vulnerability," *Handbook of Disaster Research* (2018): 181–203, https://doi.org/10.1007/978-3

-319-63254-4_10; Jeremy S. Hoffman, Vivek Shandas, and Nicholas Pendleton, "The Effects of Historical Housing Policies on Resident Exposure to Intra-Urban Heat: A Study of 108 US Urban Areas," *Climate* 8, no. 1 (2020): 12, https://doi.org/10.3390/cli8010012; David R. Williams, Jourdyn A. Lawrence, and Brigette A. Davis, "Racism and Health: Evidence and Needed Research," *Annual Review of Public Health* 40 (2020): 105–25, https://doi.org/10.1146/annurev-publhealth-040218-043750.

61 Katherine Cox and BoRin Kim, "Race and Income Disparities in Disaster Preparedness in Old Age," *Journal of Gerontological Social Work* 61, no. 7 (2018): 719–34, https://doi.org/10.1080/01634372.2018.1489929; HaeJung Kim and Michael Zakour, "Disaster Preparedness among Older Adults: Social Support, Community Participation, and Demographic Characteristics," *Journal of Social Service Research* 43, no. 4 (2017): 498–509, https://doi.org/10.1080/01488376.2017.1321081.

62 Tala M. Al-Rousan, Linda M. Rubenstein, and Robert B. Wallace, "Preparedness for Natural Disasters among Older Us Adults: A Nationwide Survey," *American Journal of Public Health* 105, no. S4 (2015): S621–S626, https://doi.org/10.2105/AJPH.2013.301559.

63 Adams et al., "Mortality from Forces of Nature."

64 Adams, "Mortality from Forces of Nature"; Stone, "Housing."

65 Shih et al., *Improving Disaster Resilience.*

66 Shih et al., *Improving Disaster Resilience*; Amit Suneja et al., "Chronic Disease after Natural Disasters: Public Health, Policy, and Provider Perspectives," National Center for Disaster Preparedness, Columbia University Earth Institute, New York, 2018, https://doi.org/10.7916/D8ZP5Q23.

67 Diane Elmore and Lisa Brown, "Emergency Preparedness and Response: Health and Social Policy Implications for Older Adults," *Generations* 31, no. 4 (2008): 66–79.

68 Assistant Secretary for Preparedness and Response Technical Resources, Assistance Center, and Information Exchange (ASPR TRACIE), "Emergency Preparedness Requirements for Medicare and Medicaid Participating Providers and Suppliers (CMS EP Rule)," accessed September, 2021, https://asprtracie.hhs.gov/cmsrule.

69 Centers for Medicare and Medicaid Services, "Quality, Safety and Oversight Group—Emergency Preparedness," 2018, http://www.cms.hhs.gov/SurveyCert EmergPrep/.

70 US Department of Health and Human Services, "HHS Establishes Office of Climate Change and Health Equity," August 30, 2021, https://www.hhs.gov/about/news/2021/08/30/hhs-establishes-office-climate-change-and-health-equity.html.

71 Sue Anne Bell et al., "Supporting the Health of Older Adults before, during and after Disasters," *Health Affairs Blog*, 2019, https://doi.org/10.1377/hblog20191126

.373930; Crystal Kwan and Christine A. Walsh, "Seniors' Disaster Resilience: A Scoping Review of the Literature," *International Journal of Disaster Risk Reduction* 25 (2017): 259–73.

72 Stine Eckert et al., "Health-Related Disaster Communication and Social Media: Mixed-Method Systematic Review," *Health Communication* 33, no. 12 (2018): 1389–400, https://doi.org/10.1080/10410236.2017.1351278.

73 Zoraster, "Vulnerable Populations"; Ariane Prohaska, "Still Struggling: Intersectionality, Vulnerability, and Long-Term Recovery after the Tuscaloosa, Alabama USA Tornado," *Critical Policy Studies* 14, no. 4 (2020): 466–87, https://doi.org/10.1080/19460171.2020.1724549.

74 Kwan and Walsh, "Seniors' Disaster Resilience."

75 Haley B. Gallo et al., "Social Workers Can Help Older Adults Prepare for and Respond to Natural and Man-Made Emergencies," *Journal of Gerontological Social Work* 61, no. 7 (2018): 697–700, https://doi.org/10.1080/01634372.2018.14 32737; Nancy Kusmaul, Allison Gibson, and Skye N. Leedahl, "Gerontological Social Work Roles in Disaster Preparedness and Response," *Journal of Gerontological Social Work* 61, no. 7 (2018): 692–96, https://doi.org/10.1080/01634372.2 018.1510455.

76 World Health Organization, "Older Persons in Emergencies: An Active Ageing Perspective," 2008, https://www.who.int/ageing/publications/Emergencies English13August.pdf.

77 Elizabeth Fussell and Sarah R. Lowe, "The Impact of Housing Displacement on the Mental Health of Low-Income Parents after Hurricane Katrina," *Social Science & Medicine* 113 (2014): 137–44, https://doi.org/10.1016/j.socscimed.201405.025.

78 Shih et al., *Improving Disaster Resilience*; Suneja et al., "Chronic Disease."

79 Shih et al., *Improving Disaster Resilience*; Suneja et al., "Chronic Disease."

80 Shih et al., *Improving Disaster Resilience*.

81 Maria L. Claver, Tamar Wyte-Lake, and Aram Dobalian, "Disaster Preparedness in Home-Based Primary Care: Policy and Training," *Prehospital and Disaster Medicine* 30, no. 4 (2015): 337–43, https://doi.org/10.1017/S1049023X1500 4847; Jill D. Daugherty et al., "Disaster Preparedness in Home Health and Personal-Care Agencies: Are They Ready?," *Gerontology* 58, no. 4 (2012): 322–30, https://doi.org/10.1159/000336032.

82 Wayne Nelson et al., "Community Long Term Care Ombudsman Program Disaster Assistance: Ready, Willing or Able?," *Journal of Homeland Security and Emergency Management* 17, no. 2 (2020): 20190015, https://doi.org/10.1515/jhsem-2019-0015.

83 ASPR TRACIE, "Emergency Preparedness Requirements,"; Centers for Medicare and Medicaid Services, "Emergency Preparedness Rule," 2016, https://www.cms.gov/Medicare/Provider-Enrollment-and-Certification/SurveyCert EmergPrep/Emergency-Prep-Rule.

84 Administration for Community Living, "Emergency Preparedness," accessed August 2021, https://acl.gov/emergencypreparedness; US Centers for Disease Control and Prevention, "Planning Resources by Setting: Long-Term, Acute, and Chronic-Care," accessed October 2020, https://www.cdc.gov/cpr/readiness/healthcare/longterm.htm.

85 American Health Care Association and National Center for Assisted Living, "Emergency Preparedness," accessed June 5, 2020, https://www.ahcancal.org/Survey-Regulatory-Legal/Emergency-Preparedness/Pages/default.aspx; "Emergency Preparedness Toolkit," LeadingAge, accessed June 5, 2020, https://www.leadingage.org/emergency-preparedness-toolkit.

86 Hilary Eiring, Sarah C. Blake, and David H. Howard, "Nursing Homes' Preparedness Plans and Capabilities," *American Journal of Disaster Medicine* 7, no. 2 (2012): 127–35, https://doi.org/10.5055/ajdm.2012.0088; Suneja et al., "Chronic Disease."

87 Lisa M. Brown et al., "Experiences of Assisted Living Facility Staff in Evacuating and Sheltering Residents during Hurricanes," *Current Psychology* 34 (2015): 506–14, https://doi.org/10.1007/s12144-015-9361-7.

88 Wanda R. Spurlock et al., "American Academy of Nursing on Policy Position Statement: Disaster Preparedness for Older Adults," *Nursing Outlook* 67, no. 1 (2019): 118–21, https://doi.org/10.1016/j.outlook.2018.12.002.

89 Brown et al., "Use of Mental Health Services,"; Levinson, "Gaps Continue to Exist."

90 Brown et al., "Use of Mental Health Services."

91 J. Rush Pierce et al., "Improving Long-Term Care Facility Disaster Preparedness and Response: A Literature Review," *Disaster Medicine and Public Health Preparedness* 11, no. 1 (2017): 140–49, https://doi.org/10.1017/dmp.2016.59; Elmore and Brown, "Emergency Preparedness and Response," 66–79.

92 William F. Benson and Nancy Aldrich, "CDC's Disaster Planning Goal: Protect Vulnerable Older Adults," *CDC Health Aging Program*, 2007, https://www.cdc.gov/aging/pdf/disaster_planning_goal.pdf.

93 Benson and Aldrich, "CDC's Disaster Planning Goal."

94 Florida State Emergency Management Act, Florida State Title XVII, Statues 252.311 and 252.355(1), retrieved from https://www.flsenate.gov/Laws/Statutes/2016/Chapter252/All.

95 Janelle J. Christensen, Lisa M. Brown, and Kathryn Hyer, "A Haven of Last Resort: The Consequences of Evacuating Florida Nursing Home Residents to Nonclinical Buildings," *Geriatric Nursing* 33, no. 5 (2012): 375–83, https://doi.org/10.1016/j.gerinurse.2012.03.014.

96 Zoraster, "Vulnerable Populations."

97 Marlene Rosenkoetter et al., "Perceptions of Older Adults Regarding Evacuation in the Event of a Natural Disaster," *Public Health Nursing* 24, no. 2 (2007): 160–68, https://doi.org/10.1111/j.1525-1446.2007.00620.x.

98 Shih et al., *Improving Disaster Resilience*; Suneja et al., "Chronic Disease."

99 Centers for Medicare and Medicaid Services, "Oversight Group."

100 Debra Dobbs et al., "Protecting Frail Older Adults: Long-Term Care Administrators' Satisfaction with Public Emergency Management Organizations during Hurricane Irma and COVID-19," *Public Policy & Aging Report* 31, no. 4 (2021): 145–50, https://doi.org/10.1093/ppar/prab019.

101 Deborah Thomas S. K. et al., *Social Vulnerability to Disasters* (Boca Raton, FL: CRC Press, 2013).

102 Gamble et al., "Climate Change and Older Americans."

CHAPTER 11

Climate Displacement and Migration after Hurricane Maria

Implications for Puerto Ricans' Mental Health

ROSALYN NEGRÓN, LORENA M. ESTRADA-MARTÍNEZ, MARISOL NEGRÓN, ORLANDO MALDONADO-MELÉNDEZ, AND DANIEL ALDRICH

Climate change predictions show that already vulnerable places, such as Puerto Rico, will be disproportionately affected by climate-driven natural disasters.[1] Human displacement is one of many dire effects of climate-related hazards like hurricanes, although the impact is distributed unequally along axes of class, racial, gendered, and sexual vulnerabilities as well as physical disabilities.[2] Immediately before or in the immediate aftermath, such displacement takes the form of evacuation. In the face of a disaster, deciding to stay or leave is critically complex. Perceptions of risk, access to information, and social influence are among the factors that shape disaster evacuation decisions,[3] which are typically made in the days and hours before an impending hazard. In some cases, such as in Puerto Rico before Hurricane Maria, most doubted the impact of the hazard and therefore few, of those who could, evacuated. Some disasters are slow-moving, so their full impacts may take not hours or days but months and even years to manifest.[4] Here, too, Puerto Rico is instructive. After Hurricane Maria, evacuation factors such as risk, information, and social influence evolved against the backdrop of shifting personal, environmental, community, and institutional contexts. Protracted disruptions to daily life and uncertainties about the months after the hurricane struck contributed to a mental health crisis in Puerto Rico highlighted by rising post-Maria suicide trends in Puerto Rico and an increase in the prevalence of chronic diseases.[5]

Hurricane Maria was a high-end category-four hurricane that struck Puerto Rico on September 20, 2017, two weeks after Hurricane Irma disrupted access to power and water. Five years later, Puerto Rico was continuing to recover from the disaster, which compounded existing structural vulnerabilities. For those who left, the loss of community and social networks, challenges of resettlement, and disaster-related traumas speak to the cumulative stress of post-Maria disruptions and displacement.[6] The emerging literature on the public health consequences of climate-related migration has emphasized the challenges of providing healthcare services to climate migrants and the physical health impacts of climate displacement.[7] Here we focus on factors affecting Puerto Ricans' mental health, both of those displaced and of those residing in Puerto Rico. We cannot divorce an analysis of climate migrants' mental health from the mental health of those who stay. Puerto Ricans who left the archipelago after Maria carried the traumas of the disaster and the heartache of saying goodbye. But many also have survivor's guilt, grounded in the knowledge that their loved ones continued to struggle. Discourses of resiliency and debates about whether it was moral or unpatriotic to leave a nation in need intensified the affective contours of the decision process faced by Puerto Ricans considering leaving.[8]

We interviewed seventy-two Puerto Ricans between March 2021 and May 2022. Of those, forty-seven had remained in Puerto Rico, and twenty-five had left for the continental United States. Some left Puerto Rico as soon as they could board a flight, while others left in the months to come. Several had already considered migrating before Maria due to an economic crisis that placed unsustainable pressures on their ability to live safe, healthy, and economically secure lives. Others who had never considered leaving found the protracted post-Maria recovery, what one interviewee called a "four-year emergency" in 2021, a sign of irrevocable decline.

The designation "climate migrants" describes persons displaced by the effects of climate change and disasters. It may refer to those who migrate after an environmental disaster such as a hurricane; due to the consequences of long-term conditions, such as coastal erosion or deforestation; or to avoid the risks posed by climate change, such as rising costs and housing shortages. It includes both temporary and permanent and planned migration, whether within a country or across national borders.[9] While the term is often used interchangeably with "climate refugee" to highlight the severity of displacement, the latter categorization does not imply legal status under international law.[10] Neither term adequately accounts for the entanglement of climate change and migration with capitalism and racial, gendered, and sexual hierarchies.[11] As Richard Black and colleagues argue, the social, moral, and political dimensions of a combination of factors make it challenging to label climate-related disaster mobilities as simply "climate migration" or migrants in such contexts as "climate migrants" or "climate refugees."[12] For Puerto Ricans, these terms also risk occluding the role of racial-colonial governance in displacement.

In the months following Maria, the imbrication of disaster and migration with racial-colonial governance became increasingly apparent. A web of colonial debt, distributed unequally across racial, gendered, class, and sexual hierarchies,[13] included an outdated electrical grid, dependence on fossil fuels, food insecurity, questionable water quality, a fragile hospital system near collapse, and crushing public debt had initiated an ongoing mass migration to the United States a decade earlier. The inadequate response to requests for federal aid failed to consider the structural context and, like president Donald Trump's insistence that Puerto Ricans "want everything done for them" and strain the US economy, was laced with "reminders of the island's debtor status and outspoken disdain for a child-like dependency of a people too accustomed to having their problems solved by the federal government."[14] A dispute between Puerto Rico's local government and the federal fiscal management board, established by the Obama administration in 2016 to restructure Puerto Rico's seventy-two-billion-dollar debt, underscored the limits of Puerto Rico's colonial autonomy. Expressing concern about misuse and mismanagement of funds, the federal oversight board took over administration of federal funding and recovery efforts.[15]

THE ACCUMULATION OF STRESS AND TRAUMA DRIVING PUERTO RICO'S POST-MARIA MENTAL HEALTH CRISIS

Hurricane Maria's devastation—and the subsequent mismanagement of the recovery process—severely affected the mental health of Puerto Ricans. Nine months to a year after the hurricane, more than 20 percent of the Kaiser Family Foundation survey participants reported needing mental health services.[16] The mental health crisis is seen most starkly in the suicide statistics. Data show a significant increase in completed suicides, suicide attempts, and suicidal ideation among Puerto Rico residents in November and December 2017 compared to the same months in 2016.[17] While calls to the crisis intervention hotline Primera Ayuda Psicosocial due to suicidal ideation initially decreased after Maria, they increased by 55 percent in November, 92 percent in December, and 38 percent in January 2018. Similarly, the number of calls due to suicide attempts was 228 percent higher in November, 386 percent higher in December, and 183 percent higher in January 2018.[18] As Susanna Hoffman suggests, in the immediate days after a catastrophe strikes, people come together to help each other.[19] In Puerto Rico, this contributed to an intense sense of community and connection, a form of "therapy," as numerous Maria survivors described it, that warded off despair.[20] However, as several women interviewees also indicated, the mounting stress over the months became unbearable. Sixty-six-year-old Laura, a retiree from the town of Morovis, explained, "[At first], well, it was like a challenge. But later, you get tired, you get anxious . . . [becoming emotional]. Cooking, doing the laundry [by hand], not combing your hair with a blower because you don't have electricity after a month. . . . It was very challenging, very challenging. I had to leave because after six months I couldn't. . . . After two or three months, I couldn't take it anymore. So I went with my children, and I took the youngest granddaughter with me."[21] Camila, a fifty-two-year-old single parent of three adult children and caretaker to her grandmother, described the guilt that compelled her to delay addressing her own needs after floods destroyed her home in the municipality of Bayamón and exposure to contaminated floodwaters led to her hospitalization:

> I felt ashamed because I was not a person of—I have always been a person that I have given, that I have never asked for anything; that gives of the little I have and shares the little I have. Because even before Maria, I always gave what little I had. I will never forget, it was July 22, 2018, when all the aid, all the help centers, closed, and . . . my [help center] boss tells me, "You know what? I've seen you all year, more than a year, giving everything you had to the whole world. But none of the times we've been out on the island have you told me 'I need.' My question is, Camila, have you already fixed your house?" And that's when I sat down to think, and I told him, "No. No, because my priority is to help others." I spent a year helping, helping, and I forgot that I was the same as many of those I helped. And that's why I say I felt ashamed because my situation is—. My situation is—. Sometimes I'm even embarrassed when I have to say, "I need x thing." And that's why I don't have my house finished.

While Camila's story suggests the therapeutic effects of participating in mutual aid, a year after Maria, she experienced severe depression and described an urge to jump out the window of the hotel where the Federal Emergency Management Agency had housed her without her family. Camila's experience highlights how unprocessed trauma and persistent insecurity challenged public mental health among Puerto Ricans after Maria.

The frustration with inadequate institutional responses, which increased as time passed and expectations for timely relief were unmet, likely contributed to the mental health crisis. Delays in the distribution of federal aid and other catastrophic flaws in the federal disaster response in Puerto Rico generated confusion and anger among residents.[22] The disillusionment with the federal government contributed to an increasing sense that the colonial relationship between the United States and Puerto Rico was untenable.[23] The extent to which these dynamics raised feelings of despair, hopelessness, and a sense of urgency to leave Puerto Rico remains unclear. Still, evidence from our interviews strongly points to a cumulative effect of stress that hit a tipping point for many, thereby leading to a decision to leave.

MENTAL HEALTH AMONG THOSE WHO LEFT PUERTO RICO AND THE INTERACTION AMONG MATERIAL, SOCIAL, AND MORAL DILEMMAS

As we have learned from Hurricane Katrina, the hardships that displaced people face in the months after the initial shocks of disaster compound the traumas of injury, death of loved ones, and property loss.[24] In New Orleans, diminished access to social support, increased exposure to discrimination, and cultural dislocation and disorientation compounded the unlivable conditions created when Hurricane Katrina decimated whole wards of the city.[25] Similarly, in Puerto Rico, the stress and long-term trauma described by interviewees exacerbated the loss or destruction of homes, lack of electrical power vital to maintaining medication and life-sustaining health equipment, and water and food scarcity that propelled the evacuation of residents for the US mainland after Maria.

As the months after Maria stretched on and some essential public services were restored in some places, for many, unlivability took a complex valence: many Puerto Ricans experienced the return of uneasy normalcy (e.g., resuming work and school routines) alongside the persistence of deep material, social, and environmental loss further exacerbated by the ongoing debt crisis. The out-migration of relatives, neighbors, classmates, and coworkers depopulated neighborhoods and schools, compounding the sense of uncertainty about the future. Several interviewees remarked that the persistent unlivability "removed the veil" that had previously prevented them from "seeing" the depth of social vulnerabilities. Marelis, a twenty-three-year-old special education teacher from San Juan, was at the time of our interview actively planning to move to the continental United States within a year. This period revealed to her "how little the [local] government cared about its people," and she describes her epiphany as follows:

> I knew that people had needs, I knew that people did not have the same opportunities, but it was something that I knew; it was not something that I *saw*. . . . I thought that [because] I worked full time and studied full time that it was difficult for me. But next to those people, I'm lucky; I'm lucky that I was able to make the decision to study. There are people who had to drop out of school because they lost their parents and had to provide for their family. There are people who clocked out from one job and then clocked into another simply to be able to pay [for necessities] because gasoline was expensive, food was expensive. There was no way to bathe. It was like, I already knew. I already knew, but the truth is that that was what made me realize that there is not much future here, and it is not the best place to be in a time of crisis.

Marelis, who attributed unnecessary suffering to the Puerto Rican government's neglect, speaks to a few ways to understand what unlivability may mean to displaced Puerto Ricans. First, there is the usual way that unlivability is understood in the context of disaster, namely material conditions that make health and well-being difficult, if not impossible. Second, she speaks to institutional failure and abandonment, contributing to material suffering, emotional distress, and moral frustration. Finally, she describes a deepening sense of hopelessness and uncertainty. As we discuss below, even those with the means to live relatively well in Puerto Rico found the uncertainty and diminished control over their future untenable.

Debates about the moral implications of leaving Puerto Rico, which first emerged in 2016 amid the largest wave of out-migration to the continental United States since the 1950s, gained renewed urgency after Maria.[26] The initiative #YoNoMeQuito (*Yo no me quito* means "I'm not giving up," "I'm not leaving") started by Carlos López-Ley, president of a transportation distribution company and one of Puerto Rico's biggest employers, invoked hard work, honesty, perseverance, generosity, unity, strength, positivity, and love of country as necessary for recovery.[27] Given its moral connotations, the hashtag generated debate, as those who left experienced the declaration as an implicit critique of their decision.[28]

When we asked interviewees about their perceptions of #YoNoMeQuito, we heard a generalized ambivalence toward these discourses. There were those who felt hurt by these discourses. A forty-eight-year-old nurse who settled in Florida described it as such: "That was very sad. *Yo no me quito.* I make effort, I give my best, as much as I can. But thinking of it another way, the government doesn't help you. It's awful. I don't know why they created the slogan or what the intention was. But the real intention of '*Yo no me quito*' doesn't help you with the government." Those who felt favorably toward #YoNoMeQuito pointed to their own resilience and patriotism. Still others—including most of our interviewees—saw the campaign's merits but insisted that no one should pass moral judgment on those who left. Other interviewees mocked #YoNoMeQuito as the empty rhetoric of business people with the means to live comfortably in Puerto Rico, as suggested by the sarcastic response of Nelson, a fifty-eight-year-old resident of Bayamón who, along with his wife, was planning to join his adult children in Texas: "Fine. Fine. It's pretty, 'Yo no me quito.' Are you going to help me so that 'yo no me quitarme [sic]' [so that I don't have to leave]? What are you going to do for me? It's good for you to say 'Yo no me quito.' But what's going to help me to 'yo no me quitarme [sic]?' [What's going to help me stay]?" Myriam (age not disclosed), who left for Springfield, Massachusetts, because she could not get medical treatment in Puerto Rico, also criticized #YoNoMeQuito: "How cute of them, because if I stayed, I would go blind. 'Yo no me quito,' but blind. . . . You have to see not everyone has the same situation. Some left out of necessity and others out of desire, but we can't generalize to all."

Myriam's distinction between those who leave out of "necessity" (e.g., health, homelessness, or children) and those motivated by "desire" highlights the implicit judgment expressed by some interviewees: that some who left did not have the strength to confront the challenges of life in Puerto Rico. The following statement by Carmen, a sixty-year-old preschool teacher, exemplifies this sentiment:

> There are people who leave because they don't know what it means to sacrifice. . . . Nobody wants to suffer, but everyone has to. . . . I always said the light will come; the water will come, there will be more food. . . . No, I am

staying here, I stay here, and if I have to help, I help. But many people left because . . . in part, my husband is right, as if out of cowardice, for not staying to fight. Others left, well, because they needed medical services, obviously, and then they ended up staying there. Others, because there were no schools for their children, and they didn't want their children to miss the school year, and they stayed there. But I think there was a lot, a lot. . . . We [as a people] don't have this . . . this sense of fighting for what you want, and that we're going to stay, and we are going to support each other.

As Carmen's statement suggests, one of the circulating discourses in Puerto Rico society frames leaving Puerto Rico as individual failure and abdication of duty to the community. Conversely, staying framed hard work and patriotism as essential Puerto Rican values while eliding structural inequalities and social hierarchies that disproportionately affected vulnerable communities. How such discourses have shaped the mental health and well-being of post-Maria migrants requires further examination.

Our interviews suggest that such moralization may manifest in feelings of shame or guilt among those who left. For still others, feelings of guilt are tied to concerns about the well-being of their loved ones in Puerto Rico. For example, Sofia, a forty-year-old lawyer who was recruited by a Florida company after Maria, articulated her feelings this way:

You feel guilty because, you know, so many people, they lost their jobs, they lost their homes, they have been struggling, still to this day are struggling because of the aftermath. Because anybody that tells you that things are back to normal is lying. Things are not great . . . they just aren't as bad as they were a week after the hurricane flattened the entire thing. You feel a little bit of . . . you feel a lot of guilt because you are okay, and they are not. I feel guilty because my mom will call, frustrated because she's sixty-five, my dad's eighty, and they can't find a handyman to move something from their house because the service is just not the same. She's been waiting two months for someone to fix the cable . . . it's just. I feel bad. I feel bad when my mom goes to the supermarket and spends three hundred dollars and has five bags. I feel bad. I feel terrible, and I understand the shame.

Like most people interviewed, Sofia understood that leaving after Maria was the right choice for her and her family. However, like other interviewees she held complicated and mixed feelings about her family's departure. Some who left were reluctant but did not want to be separated from departing family members. For others Maria made already existing challenges unbearable. Among those who left, there were those who felt a constant desire to return to Puerto Rico and report an improved quality of life and a renewed sense of security amid a sense of loss of their relationships. In this last part of the chapter, we turn to understanding the ambivalent experiences of post-Maria migration, which complicates easy designations of Puerto Ricans' mobilities as belonging to discrete categories of evacuation, displacement, economic migration, or climate migration.

A NEW PHASE IN PUERTO RICO'S LONG HISTORY OF MIGRATION

Given its colonial relationship to the United States, Puerto Rico has a long history of migration, including circular migration, to the continental United States. Today, Puerto Rico is one among few nations with more people in the diaspora than in the homeland. The census estimates that over five million Puerto Ricans live in the mainland United States (versus 3.3 million in Puerto Rico), and approximately one-third of those were born in Puerto Rico.[29] Even as the diaspora's significant posthurricane mobilization boosted Puerto Rico's recovery,[30] strong ties to diasporic Puerto Ricans facilitated evacuation and resettlement for some.

Where Post-Maria Migrants Have Settled

Historically, the largest Puerto Rico diasporic community has been in New York State, particularly New York City. However, after Maria, Florida became the top destination for Puerto Rican evacuees.[31] Indeed, 53 percent of our interviewees in the continental United States lived in Florida, primarily in and around Orlando, where a Puerto Rican community has existed since the 1960s but experienced a population boom in the 1990s.[32] Puerto Ricans in the area organized to receive evacuees, and the Federal Emergency Management Agency established congregate shelters and hotel rooms for the displaced, where many of the most vulnerable evacuees lived for months.[33] An estimate from the Multi-Agency Resources Center at Orlando International Airport indicates that fifty-three thousand Puerto Ricans and US Virgin Islanders, who also experienced displacement due to Maria, moved to Florida in the first year after Maria.[34] Within two years, the Puerto Rican population in Florida grew by over 120,000 people (more than 11.5 percent).[35]

The Ambivalence of Climate Displacement and Migration

While a quite diverse set of circumstances informed decisions to leave Puerto Rico, all interviewees shared a sense of dissatisfaction with the political, economic, labor, and housing conditions aggravated by Maria. A prominent theme was concern about the quality of medical and educational services, particularly for children with special needs. For example, Eduardo, a thirty-seven-year-old father of an autistic child, left for Massachusetts specifically because his son needed better occupational and therapeutic services.

Having lived in Puerto Rico all their lives, with regular routines, embeddedness in family networks, amassed possessions, and established work lives, interviewees described their decision to leave as a decision to leave everything behind and "start from zero" (*empezar desde cero*). As a sixty-two-year-old preschool teacher in Florida put it, "What did I leave? I left family, I left my house, I left everything." Nayla, a forty-six-year-old resident of Kissimmee, Florida, explains, "I left practically everything and started over at my age. . . . I had so many plans to reach fifty-five [and retire]. In the job I was in, I could have done it."

In the early 2000s, Nayla lived in the continental United States, first in Florida and then in Washington State. She returned to Puerto Rico in 2007 and describes an economically comfortable life with her husband. But after the hurricane, having experienced property damage and loss of material possessions, her stateside family encouraged her to leave. While her eldest daughter left because she needed access to insulin, Nayla decided to leave in 2020 after a two-month earthquake swarm in southern Puerto Rico. She left with one of her daughters, although her husband remained in Puerto Rico for some time, and settled into a community in Florida close to family.

For parents like Eduardo and Nyla, migration decisions were guided not just by material and economic considerations but also by responsibility for the well-being of their children. Nayla articulated it in this way: "I think it was more this maternal instinct, or what it means to have my people safe."

For some displaced persons, migration and resettlement instantiated unexpected disruptions in access to social support, such as the dissolution of intimate relationships. Pedro, a sixty-six-year-old retiree who relocated to Florida, felt forced to migrate because his wife wanted to join her parents and siblings in Florida. But migration did not lead to an improved quality of life for him. As he put it, "Right now, I tell you, I have found myself here in an extremely cold country. You barely know your neighbor; there is no small business on the corner where you go and play your pool, even if you don't drink, your little domino games, saying hello to your friends, going to the bakery, the beaches in all four corners of the island." He and his wife eventually separated, and he found himself without friends and distanced from his family: "Here it has been a life of suffering, a lot of work, a lot of sacrifice. They even gave me a cornea transplant, and my body rejected it. I assume because of the problems in my marriage. Now they have to do the other one for me, supposedly. All the money I've earned, I don't have five dollars saved. I live in a mobile home, modestly, as I never thought I would live. Loneliness consumes me. I have to do everything, wash my clothes, iron my clothes, cook."

Pedro's post-Maria migration experiences add a caveat to a common theme among interviewees who spoke about achieving a better quality of life and acquiring amenities in the United States, even if they needed to work more and deal with a higher cost of living. Sofia experienced "stress that there was no stress" in the early days of resettlement. In Puerto Rico she felt the subtle but constant stress of multiple complications; something was always going wrong somewhere. Once in Florida and in the calm of their new home, she worried that there was some problem she was overlooking. For Nyla, relocation led to the loss of her job and lower wages in the United States.

The adaptation process was also full of uncertainty, fear, and, for some, questions about whether they had made the right decision. One common challenge expressed by interviewees was experiencing ethnic/racial and linguistic discrimination. The xenophobia and anti-Black racism in Massachusetts described by Eduardo have disrupted his life and led him to consider returning to Puerto Rico. He explains, "Sometimes when I receive discriminatory treatment, I feel like buying a ticket and getting out. . . . I feel racism more here. . . . sometimes with the language, sometimes they ask questions or microaggressions that tell you you don't belong, or they doubt your qualities, your abilities."

Eduardo's, Pedro's, and Sofia's stories show how the possibility of return, the longing for home, the loss of social support, and concerns about loved ones in the archipelago have shaped the post-Maria migratory experience.

Mental health and well-being in this context are not a uniform or constant state of either good or bad but might oscillate over days or across months. Nevertheless, some who have visited family and friends in Puerto Rico expressed a renewed sense that they cannot return, given the generalized stressors and everyday hassles of state neglect and dysfunction.

CONCLUSION

Material and social concerns are most often examined in climate migration literature. Here, we add drivers at the level of beliefs and values, namely the moral considerations people weighed in choosing whether to leave or stay. In the wake of federal inaction in response to Maria, a public discourse that initially demanded federal accountability shifted to an emphasis on individual resiliency and *staying* in Puerto Rico.[36] Yet, as multiple interviewees contend, resettlement in the United States wasn't simple or uncomplicated, bringing mixed results and an ambivalent set of emotional experiences that Puerto Ricans continue to negotiate. Evidence suggests that poor mental health has persisted in the years since Maria. Results from a 2019 survey of Puerto Rican adults on the island also indicate that over one-third continued to experience high levels of psychological distress two years after the hurricane.[37] Calls to the intervention hotline Primera Ayuda Psicosocial, which rose 30 percent between 2017 and 2018, rose another 470 percent in 2020, reflecting anxieties created by the earthquake swarm and the COVID-19 pandemic. However, there are reasons for optimism. The suicide rate decreased from nine per one hundred thousand in 2017 to 5.6 per one hundred thousand in 2021, below pre-Maria levels.[38] Similarly, Josiemer Mattei and colleagues noted a 20 percent drop in the prevalence of depression among Puerto Rican adults from 2015 (two years before the storm) to 2019 (two years after the storm).[39] This decrease may reflect improvements in the archipelago's physical infrastructure, residents' ability to adapt to new social conditions, and the out-migration (or death) of those at the highest risk of poor mental health.

Puerto Rico's preparedness for future climate shocks calls for consideration of how colonialism shapes its disaster adaption systems, including migration. Understanding the role of connections among Puerto Ricans in the archipelago and diaspora should also inform policies. Diasporic Puerto Rican postdisaster activism in the United States provided millions of dollars in relief funds and supplies. It also aided those who evacuated to the United States. While the Puerto Rico government showed increased interest in fostering and managing the diaspora's engagement in Puerto Rican development and politics after Maria,[40] it must be responsive to the needs of the newest members of the diaspora. Beyond economic concerns, our work, along with that of others writing about Maria's aftermath,[41] shows the need for reform of local and federal governments. Most of our interviewees lamented the Puerto Rican government's ineptitude and corruption in the post-Maria recovery process alongside federal neglect.

Migration flows can also undermine a country's ability to adapt to climate change. Maria's devastation has led to significant population decline, led primarily by migration.[42] Climate change will continue to magnify Puerto Rico's economic vulnerabilities, which are the main drivers of migration,[43] affect nearly every municipality,[44] and decrease community cohesion and social capital.

Puerto Rican civil society was forced into the role of "first responders" in the face of climate crisis, but community leaders who we spoke with felt inadequately prepared to meet the mental health needs of the residents they served. Many interviewees involved in mutual aid felt they had to subsume their own traumas to support others who were traumatized. Future public mental health initiatives in Puerto Rico must invest directly in civil society groups and mutual aid networks through the provision of mental health training and supports.

One cannot fully understand the mental health and well-being of Puerto Ricans who reresettled in the continental United States without considering the cascade of disasters that have plagued Puerto Rico under colonial rule. The imperial-racial hierarchies mobilized through public policy to produce an "archipelago of racialized neglect" ties Puerto Rico to other sites of climate disaster, such as New Orleans; of state neglect, as in Detroit; and of environmental racism, as in Flint, Michigan's water crisis.[45] Climate change scientists and social scientists note that the world's socioeconomically marginalized populations are especially vulnerable to natural disasters,[46] and their ability to recover is exacerbated by reduced resources, inadequate infrastructure, and limited administrative capacity.[47] Puerto Rico provides a unique case study, given its colonial status and accelerating social and economic vulnerability to natural disasters.

This research was funded by a grant from the National Science Foundation, Cultural Anthropology Program, Award No. BCS 1951148.

NOTES

1 Donald J. Wuebbles, David W. Fahey, and Kathy A. Hibbard, "Climate Science Special Report: Fourth National Climate Assessment," *U.S. Global Change Research Program*, NCA4, vol. 1, 2017; David Eckstein, Vera Künzel, Laura Schäfer, and Maik Winges, "Global Climate Risk Index 2020" (Bonn: Germanwatch, 2019).

2 Kyle Whyte, "Indigenous Climate Change Studies: Indigenizing Futures, Decolonizing the Anthropocene," *English Language Notes* 55, no. 1 (2017): 153–62; Zachary A. Morris, R. Anna Hayward, and Yamirelis Otero, "The Political Determinants of Disaster Risk: Assessing the Unfolding Aftermath of Hurricane Maria for People with Disabilities in Puerto Rico," *Environmental Justice* 11, no. 2 (2018): 89–94; Lewis Williams, "Climate Change, Colonialism, and Women's Well-Being in Canada: What Is to Be Done?," *Canadian Journal of Public Health* 109 (2018): 268–71.

3 Daniel P. Aldrich, *Building Resilience: Social Capital in Post-disaster Recovery* (Chicago: University of Chicago Press, 2012); Jasmin K. Riad, Fran H. Norris, and R. Barry Ruback, "Predicting Evacuation in Two Major Disasters: Risk Perception, Social Influence, and Access to Resources 1," *Journal of Applied Social Psychology* 29, no. 5 (1999): 918–34.

4 Susanna M. Hoffman and Anthony Oliver-Smith, "Anthropology and the Angry Earth: An Overview," *The Angry Earth: Disaster in Anthropological Perspective* (1999): 1–16.

5 On the mental health crisis, see the Comision para la Prevencion del Suicidio, *Estadisticas preliminares de casos de suicidio* (Puerto Rico: Departmento de Salud, Gobierno de Puerto Rico, 2018). Josiemer Mattei, Martha Tamez, June O'Neill, et al., "Chronic Diseases and Associated Risk Factors among Adults in Puerto Rico after Hurricane Maria," *JAMA Network Open* 5, no. 1 (2022): e2139986–e2139986.

6 Maria T. Padilla and Nancy Rosado, *Tossed to the Wind: Stories of Hurricane Maria Survivors* (Gainesville: University Press of Florida, 2020).

7 Ibrahim Abubakar, Robert W. Aldridge, Delan Devakumar, et al., "The UCL–Lancet Commission on Migration and Health: The Health of a World on the Move," *Lancet* 392, no. 10164 (2018): 2606–54; Lori M. Hunter and Daniel H.

Simon, "Climate Change, Migration and Health," in *A Research Agenda for Migration and Health,* ed. K. Bruce Newbold and Kathi Wilson (Cheltenham, UK: Edward Elgar, 2019), 50–66; Valéry Ridde, Tarik Benmarhnia, Emmanuel Bonnet, et al., "Climate Change, Migration and Health Systems Resilience: Need for Interdisciplinary Research," *F1000 Research* 8, no. 22 (2019); Patricia Schwerdtle, Kathryn Bowen, and Celia McMichael, "The Health Impacts of Climate-Related Migration," *BMC Medicine* 16, no. 1 (2018): 1–7.

8 On morality and patriotism, see Danica Coto, "¿Irse or quedarse? El dilema moral de los puertorriqueños," *Associated Press,* November 9, 2017.

9 Khawar Abbas Khan, Khalid Zaman, Alaa Mohamd Shoukry, Abdelwahab Sharkawy, Showkat Gani, Sasmoko, Jamilah Ahmad, Aqeel Khan, Sanil S. Hishan, "Natural Disasters and Economic Losses: Controlling External Migration, Energy and Environmental Resources, Water Demand, and Financial Development for Global Prosperity," *Environmental Science and Pollution Research* 26, no. 14 (2019): 14287–99, https://doi.org/10.1007/s11356-019-04755-5.

10 Satchit Balsari, Caleb Dresser, and Jennifer Leaning, "Climate Change, Migration, and Civil Strife," *Current Environmental Health Reports* 7 (2020): 404–14; Issa Ibrahim Berchin, Isabela Blasi Valduga, Jéssica Garcia, and José Baltazar Salgueirinho Osório de Andrade, "Climate Change and Forced Migrations: An Effort towards Recognizing Climate Refugees," *Geoforum* 84 (2017): 147–50; Oli Brown, *Migration and Climate Change,* IOM Migration Research Series, United Nations, 2008, 31–29; Jane McAdam, "Displacement in the Context of Climate Change and Disasters," in *The Oxford Handbook of International Refugee Law,* ed. Cathryn Costello, Michelle Foster, and Jane McAdam (Oxford: Oxford University Press, 2021), 832–47.

11 Whyte, "Indigenous Climate Change Studies"; Williams, "Climate Change."

12 Richard Black, Dominic Kniveton, and Kerstin Schmidt-Verkerk, "Migration and Climate Change: Toward an Integrated Assessment of Sensitivity," *Disentangling Migration and Climate Change: Methodologies, Political Discourses and Human Rights* (2013): 29–53.

13 Yarimar Bonilla, "The Coloniality of Disaster: Race, Empire, and the Temporal Logics of Emergency in Puerto Rico, USA," *Political Geography* 78 (2020): 102181; Marisol LeBrón, "Policing Coraje in the Colony: Toward a Decolonial Feminist Politics of Rage in Puerto Rico," *Signs: Journal of Women in Culture and Society* 46, no. 4 (2021): 801–26; Rocio Zambrano, *Colonial Debts: The Case of Puerto Rico* (Durham, NC: Duke University Press, 2021).

14 Hilda Lloréns, "Ruin Nation: In Puerto Rico, Hurricane Maria Laid Bare the Results of a Long-term Crisis Created by Dispossession, Migration, and Economic Predation," *NACLA Report on the Americas* 50, no. 2 (2018): 154–59.

15 Mary Williams Walsh, "In Puerto Rico, Law Passed for Fiscal Crisis Hampers Storm Recovery," *New York Times,* November 16, 2017. The breadth of corruption and inefficiency of local officials that have since come to light, and

which led to the July 2019 uprising that ultimately ousted Puerto Rico's governor, highlight what Hilda Lloréns Torres describes as the "broken rule of law" ("Ruin Nation," 156) at the local level that colludes with colonial governance.

16 Bianca DiJulio, Cailey Muñana, and Mollyann Brodie, "Views and Experiences of Puerto Ricans One Year after Hurricane Maria," *Henry J. Kaiser Family Foundation*, September 2018.

17 Comisión para la Prevención del Suicidio, 2018, https://estadisticas.pr/files/Inventario/publicaciones/Febrero%202018.pdf.

18 These may be conservative estimates; only clear cases of suicide are classified as such by healthcare professionals; Comisión para la Prevención del Suicidio.

19 Hoffman and Oliver-Smith, "Anthropology."

20 A dip in suicidal outcomes in September and October may also reflect the collapse of the electrical and cellular grid, which may have impeded attempts to call.

21 All names used are synonyms to protect participants' identities. All participants provided written or oral consent prior to interviews.

22 Danny Vinik, "How Trump Favored Texas over Puerto Rico," *Politico*, March 27, 2018.

23 B. Morales Meléndez, "Puerto Rico y Estados Unidos: Una relación en quiebra, *La Diaria*, April 7, 2018; Benjamin Torres-Gotay, "'I'm Quite Comfortable': Abandonment and Resignation after Hurricane Maria," in *Aftershocks of Disaster: Puerto Rico before and after the Storm*, ed. Yarimar Bonilla and Marisol LeBrón (Chicago: Haymarket Books, 2019), 82–89.

24 Priscilla Dass-Brailsford, "After the Storm: Recognition, Recovery, and Reconstruction," *Professional Psychology: Research and Practice* 39, no. 1 (2008): 24; Alexis A. Merdjanoff, "There's No Place Like Home: Examining the Emotional Consequences of Hurricane Katrina on the Displaced Residents of New Orleans," *Social Science Research* 42, no. 5 (2013): 1222–35.

25 On cultural dislocation and disorientation, see Yonatan Carl, Rosa L. Frias, Sara Kurtevski, Tamara González Copo, et al., "The Correlation of English Language Proficiency and Indices of Stress and Anxiety in Migrants from Puerto Rico after Hurricane Maria: A Preliminary Study," *Disaster Medicine and Public Health Preparedness* 14, no. 1 (2020): 23–27.

26 Jessica P. Cámara, "La campaña publicitaria que divide a Puerto Rico," *Univision Puerto Rico*, February 23, 2016.

27 The original web pages included a photogallery and a store, along with a description of the project, at yonomequitopr.com. A link through the wayback machine can be found at http://web.archive.org/web/20160529012649/http://www.yonomequitopr.com/.

28 Cámara, "La campaña publicitaria."

29 US Census 2010, Puerto Rico Profile, https://www2.census.gov/geo/pdfs/reference/guidestloc/72_PuertoRico.pdf.

30 Julio López Varona, "El premio de la diáspora boricua," *El Nuevo Día*, May 20, 2018.

31 Ying Wang and Stefan Rayer, "Growth of the Puerto Rican Population in Florida and on the US Mainland," *Florida Bureau of Economic and Business Research*, February 9, 2018.

32 Jorge Duany and Patricia Silver, "The 'Puerto Ricanization' of Florida: Historical Background and Current Status," *Centro Journal* 22, no. 1 (2010): 4–31.

33 Padilla and Rosado, *Tossed to the Wind*.

34 Demographic Estimating Conference, "Florida Demographic Estimating Conference Tables," December 13, 2021, http://edr.state.fl.us/content/conferences/population/ConferenceResults.pdf.

35 Center for Puerto Rican Studies, "Enduring Disasters: Puerto Rico, Three Years after Hurricane María," September 2020, https://centropr.hunter.cuny.edu/reports reports/enduring-disasters-puerto-rico-three-years-after-hurricane-maria/.

36 On individual resiliency, see Bonilla, "Coloniality of Disaster."

37 Mattei et al., "Chronic Diseases."

38 María Coss-Guzmán and Nayda Román-Vázquez, "Informe Mensual de Suicidios en Puerto Rico," *Comisión para la Prevención del Suicidio* (San Juan, Puerto Rico: Departamento de Salud, June 2021).

39 Mattei et al., "Chronic Diseases."

40 Michelle M. Hildago, "La diáspora en el liderato del MVC," *La Revista Ciudadana*, April 23, 2021; López Varona, "El premio."

41 Yarimar Bonilla and Marisol LeBrón, eds., *Aftershocks of Disaster: Puerto Rico before and after the Storm* (Chicago: Haymarket Books, 2019).

42 Antonio Flores and Jens Manuel Krogstad, "Puerto Rico's Population Declined Sharply after Hurricanes Maria and Irma," *Pew Research Center*, July 26, 2019.

43 Jaison R. Abel and Richard Deitz, "The Causes and Consequences of Puerto Rico's Declining Population," *Current Issues in Economics and Finance* 20, no. 4 (2014): 1–8.

44 Flores and Manuel Krogstad, "Puerto Rico's Population."

45 On the archipelago of racialized neglect, see Bonilla, "Coloniality of Disaster," 2.

46 Intergovernmental Panel on Climate Change, "Climate Change 2001: Synthesis Report," in *A Contribution of Working Groups I, II, and III to the Third Assessment Report of the Intergovernmental Panel on Climate Change* (Cambridge, UK: Cambridge University Press, 2001).

47 Anthony Oliver-Smith, "Anthropological Research on Hazards and Disasters," *Annual Review of Anthropology* 25, no. 1 (1996): 303–28.

CHAPTER 12

Hurricane Katrina, PTSD, Preparedness, Resilience, and Recovery

JANICE HUTCHINSON

This chapter addresses two questions: What are the multiple mental health impacts of a natural disaster like Hurricane Katrina? And how does one prepare for recovery from such a natural disaster? Studies point to the need for holistic and systems visions rather than separate, disconnected efforts.

In 2005 there was a warning that a storm of "biblical proportions" was barreling toward the Gulf Coast. According to the Free Dictionary, a hurricane is "a severe, rotating tropical storm with heavy rains and cyclonic winds exceeding 74 miles (119 kilometers) per hour."[1] Enter Hurricane Katrina, the third-largest hurricane to hit the United States at the time, striking the city of New Orleans as well as the Gulf Coast of Louisiana, Mississippi, Georgia, Florida, and Alabama.

ENVIRONMENTAL ANXIETY

Hurricane Katrina and other climatic events over the years have resulted in the emergence of a new climate-related vocabulary that addresses emotional and mental concerns. The term "eco-anxiety" describes heightened emotional, mental, or somatic distress in response to dangerous changes in the climate system. There is some confusion about the definition. "Eco-anxiety" is generally defined as the anxiety and fear people have regarding climate change and the future of the planet. The word is synonymous with terms such as "climate anxiety," "environmental anxiety," and "climate change anxiety." The American Psychological Association in a 2017 report embraced eco-anxiety as a "chronic fear of environmental doom."[2] New definitions and terms continue to appear. Some of the vocabulary used to describe environmentally related distress includes "ecological grief" (response to experienced or anticipated losses in the natural world,[3] "eco-angst" (despair of the fragile condition of the planet),[4] "environmental distress" (lived experience of desolation of one's home and environs),[5] and "solastalgia" (impact on people directly connected to their home environs).[6]

While "eco-anxiety" denotes fears related to the future, the popular term "solastalgia" is a homesickness you experience while that home has been damaged or changed in a negative way as a result of the environment having changed. In the context of the winds and rains of Hurricane Katrina, homes and communities were irrevocably altered in the face of climate extremes. The familiar smells of New Orleans' home-cooked creole and Cajun cuisine became the malodorous stench of dead animals and, sometimes, people. Homes and all their holdings were tumbled and sometimes not found. Domiciles and communities that were once a place of comfort, love, and support became places of inner disorientation, confusion, anger, and dissonance.

Eco-anxiety and solastalgia are not mental illnesses identified in the Diagnostic and Statistical Manual of Mental Disorders; they represent symptoms rather than researched, defined medical conditions. Rather, this terminology is used to capture real and appropriate concerns one may have regarding the state of ecology in relationship to our survival. Research has well established that high levels of depression and post-traumatic stress disorder (PTSD) occur in persons subject to large natural disasters like hurricanes.[7] The devastating and unpredictable physical and economic damage of climate change, as represented by Hurricane Katrina, resulted in an environment culture plate (similar to a culture of microbes) from which symptoms of mental distress emerged. These ranged from anger, anxiety, and depression to PTSD and sometimes to suicide. Symptoms were acute, transient, chronic, and/or life-changing.

New Orleans is a low-lying city that has withstood over ninety hurricanes since its European colonial founding in 1718.[8] Meteorologists often comment that New Orleans in essence sits in a bowl about six feet below sea level and surrounded by the waters of Lake Pontchartrain, the Gulf of Mexico, and the Mississippi River as well as by flood walls and levees to keep the water out. But on August 29, 2005, the US Army Corps of

Engineers, who was responsible for the design, development, and administration of the levees, received a report that water had breached the concrete flood wall between the 17th Street Canal and the city, followed by breaching of the Industrial Canal.[9] These breaches led to massive flooding of the Lower Ninth Ward. By late afternoon, with millions of Americans watching along with the citizens of the Gulf Coast, 80 percent of New Orleans was underwater, with depths variably reported as ten, fifteen, and twenty feet. The storm surge was as high as twenty feet. The water did not recede for weeks. A study authorized by the corps took some responsibility for the failures of the levee system.[10]

The New Orleans research organization called the Data Center estimated that the storm and flooding had displaced more than one million people, leaving many homeless.[11] Total damages (as calculated by the Federal Emergency Management Agency) were estimated at $125 billion, making Katrina the costliest US hurricane on record. The exodus from the Gulf Coast region and New Orleans was one of the largest and most sudden in US history. Before Katrina, African Americans made up 67.3 percent of the population of New Orleans and Whites 28.1 percent; 11.7 percent were sixty-five or older; and 28 percent lived in poverty (mostly African Americans). One month after the storm about six hundred thousand households had not returned. Baton Rouge became the second-largest city in Louisiana. In April 2000, the population in New Orleans was 484,674, and by July 2006, it had dropped to about 230,000. By 2020, the census reported a population of about 390,000 (80 percent of pre-Katrina numbers). Half of those who died in the storm were seventy-four years old and older, and a 2009 study showed that most of the deaths occurred near the failed levees near the Ninth Ward and the Lakeview area.

Ten long-term residents and survivors of Hurricane Katrina were interviewed for this chapter (and have been identified by their initials). The interviews were primarily unstructured and open to allow each interviewee to reveal their own personal experience and feelings during and after Katrina. However, each person was asked to share the most stressful aspect of their experience as well as whether and for how long they experienced nightmares or prolonged depression, insomnia, or other such feelings. They were also asked whether they stayed or left, why, how long they were away, and where they went; their financial, family, and physical status; and their terms of return. Survivors of Hurricane Katrina revealed other intangible, unquantifiable, and unimaginable but common and deleterious stressors. Most striking were the individuality and variability of responses, both in terms of personal experiences as well as personal stressors. Interviewees expressed emotions and observations that were consistent, though different, to those expressed in personal accounts of Hurricane Katrina in Douglas Brinkley's book *The Great Deluge*:[12]

A coworker said, "Hurricane Katrina turned to New Orleans that evening, be careful . . . so it hit me, I gotta go . . . ran home, packed a bag, just a few items—never thought the city was coming apart, never thought I couldn't return for six months, so I left with no personal papers, jewelry, few clothes . . . not prepared." —X

The stress could have cost us our marriage if not long together. . . . Coming back was no solution, and that was most devastating. —BR

There was four feet [of] water in the house. When water goes down, couldn't open the doors due to the swelling; once open couldn't close the doors, so people could come in and take stuff. —DD

No answers to anything, not knowing what was going on—the worst part and so much happening you didn't know about. . . . The airport was the scariest; lost my eighty-year-old mom in the confusion. . . . Ordered around by national guard, who wouldn't listen and were rough. There was a sea of people, dirty and standing in long, curved lines. . . . When I returned weeks later, there was nobody to help with redoing/restoring shop. —Q

Hurricane Katrina was a testing of whether you could survive the worst experience of your life. The hotel was shaking, the chandeliers, all the windows shattered due to the wind surge and pressure. . . . Whole families of children killed, people drowned in their houses, or crushed with debris, buses overturned. Homes knocked off their foundation, whole area like a bomb dropped. . . . People lost their roots, i.e., homes, history. . . . Amazing to see [a] fridge flipped over due to high water, mattresses off the beds and on the floor; [a] new appreciation of power and force of water, everything mold and mildew, turned over. —B

When I returned two weeks later, it was like a scene from an apocalypse: no power, maggots and flies, freezer turned over. . . . The stench was unbelievable, horrible. Can't describe it, mildew, sitting, wet, rotten smell through whole city. Looked like a sci-fi movie. Gray, no color at all. The trees were dead; everything. . . . Scariest thing was didn't know what [the] future held, if [the] city could rebound, could rebuild, relocate, not knowing. —X

Wind blowing hard, sounds like screaming, and sounds like someone being beat, so scary but we had to get out or drown since water kept coming in. We decided to walk, although the water covered our cars, with water up to our waist, falling and helping others to walk, walked up to I-10, where about a thousand people were walking towards the Superdome. . . . Scariest part of being in the convention center was the criminal-looking people, strangers lying next to us; afraid of us being hurt. Most embarrassing and heart-breaking was the bathroom . . . humiliating was urinating on self, had to keep on same clothes, had bowel movement on them. —Y

There was debris on the Pontchartrain expressway up to twelve feet high for three years . . . it was heartbreaking, never seen anything like it. —I.

Old people crying and reaching out, wore same clothes a week, everyone smelly, babies covered in feces, no toilet paper, people started fighting each other. —DA

Several of the interviewees described their mental state as follows:

I had sudden crying, not knowing why, just thinking about it. Spouse was totally puzzled, unproductive, unsupportive . . . had nightmares, sad. —I

Had depression, anxiety, upset things not happening quick enough. Angry about uncertainty. —Q

I walked through chest-deep water in the dark alone for one block to city hall . . . felt nervous, didn't want to see a snake or rat . . . high level of water so knew the covers of drainage might be off and I could fall into a hole, tricky to stay on solid ground. It makes me nervous to recall the experiences. . . . Really sad when I went to my house, saddest part was loss of family pictures, but I have kept even damaged pictures, I just couldn't get rid of them. . . . Makes me nervous to recall the experiences. . . . Difficulty with sleep. —B

Nervousness, depression, scared all the time, fear of getting hurt while in CC [convention center]. Son diagnosed with diabetes and depressed all the time, the whole family depressed all the time. —DA

We were wiped out, nothing to go back to, a mental thing of not knowing what to do next. —I

People with increased vulnerability to climate change often include the elderly, children, minorities, persons of lower socioeconomic groups, pregnant and postpartum women, the chronically ill, people with impaired cognition and mobility, and those with a previous history of mental illness.

A number of studies have documented the association between mental disorders and hurricane-related stressors. It was reported that one in six post-Katrina residents of New Orleans met the criteria for PTSD, and 49 percent of residents had an anxiety or mood disorder.[13] Outcomes of PTSD are consequential and include social, psychological, and health aspects: increased risk of unemployment, marital instability, homelessness, aggression, and criminality;[14] impaired social functioning;[15] secondary mental and substance use disorders;[16] increased risk of suicide;[17] and increased risk of cardiovascular, gastrointestinal, and musculoskeletal disorders.[18]

To determine the rates and predictors of PTSD in post-Katrina New Orleans researchers conducted a web-based survey six months after the storm. They surveyed 1,542 participants from the largest employer in New Orleans, Tulane University, and included faculty, staff and administrators. The prevalence of PTSD symptoms was 19.2 percent. Predictors of PTSD symptoms included being female, knowing someone who died in the storm, not having property insurance, undergoing a longer evacuation, having a much longer commute to work than before Katrina, and living in a newly purchased home, a rental house, or in a temporary trailer. Despite having insurance, only 28.5 percent had talked to a health professional about Hurricane Katrina or about issues they had experienced since the storm.[19] Given their key role in the economic redevelopment of the region, there is a tremendous need to identify those in the workforce with symptoms consistent with PTSD and to improve treatment options. The strong relationship between displacement from one's pre-Katrina residence and symptoms of PTSD suggests a need to focus resource utilization and intervention on individuals living in temporary housing.

Studies have consistently shown that Black people are more likely to suffer PTSD after a natural disaster than are Whites. One group of researchers contacted 279 Katrina survivors who reported that Hurricane Katrina was the most traumatic event in their lifetime.[20] The interviews were conducted about fifteen months after Katrina. At that time, Black people had twofold greater odds of screening positive for PTSD. The authors of the study suggested that the differential was due to that fact that Black people were more likely than White people to have experienced frequent mental distress prior to Hurricane Katrina (29 percent vs. 14 percent).

THE RISK STUDY

Unique to research on the mental health effects of Hurricane Katrina is the Resilience in Survivors of Katrina Project (RISK) study. In 2003, 1,019 low-income, mostly African American young women from New Orleans enrolled in the Opening Doors Louisiana study, which was designed to increase educational achievement among community college students. Researchers collected demographic, socioeconomic, and health information (including economic status, social support, and physical and mental health) from all 1,019 participants in the study. At the time Katrina struck, 492 of the 942 female participants had completed a twelve-month follow-up survey (since there were only seventy-seven men in the study, analysis was limited to the female participants). At the time of Katrina, the average age of the women was 25.2 years, average monthly earnings were $957.00, African Americans made up 85 percent of the study population, and 71 percent of the participants had received some form of public assistance. All were mothers, of whom 64 percent were single (not married or living with someone), and with an average 1.92 children. Most reported they were in excellent or good health. All had a high school degree or a GED.[21]

The RISK project researchers recognized the unplanned but rare opportunity to study the consequences of the storm and turned their attention to the Hurricane Katrina survivors. By 2006, they had telephonically surveyed 402 of the 492 survivors, for whom they had complete data, using the same questions as appeared on the predisaster survey. They also included measures of PTSD as well as information regarding the survivor's hurricane experiences. Additional relevant data includes the fact that 85 percent had evacuated and the other 15 percent were either in their homes or at the Superdome or the convention center. Of the survivors, one-third reported the death of a family member or friend in the storm, and 80 percent had severe home damage. Follow-up surveys were conducted in 2005–6 and again in 2010.[22]

The prevalence of mild to moderate mental illness or serious mental illness before Katrina was 23.3 percent. The initial studies after Katrina showed a 37.2 percent prevalence. Probable serious mental illness doubled (6.9 percent to 13.8 percent). After Katrina, 47.9 percent showed probable PTSD. People with higher baseline resources (higher personal income, more perceived social support, and ownership of a car) had fewer hurricane-associated stressors (e.g., unemployment and separation from family and friends). They were more able to evacuate and avoid many stressors. But those exposed to many hurricane traumas (e.g., being trapped in their houses, seeing people die, and sheltering at the convention center) had a markedly increased risk of developing serious mental illness or PTSD, regardless of baseline resources before the hurricane. Of those who reported significant mental health problems after Katrina, only 9.2 percent mentioned a visit with a psychiatrist, psychologist, or other mental health professional in the first year after Katrina, including the 15 percent with mild to moderate mental illness and the 16 percent with serious mental illness.[23] Participants who returned to New Orleans in 2006 reported less psychological distress, less perceived distress, and fewer health conditions than those who settled elsewhere or were still in interim housing.[24]

The post-RISK surveys in 2010 revealed that some survivors had improved but had still not recovered their prestorm mental health: about 30 percent (down from the 36 percent after Katrina but up from 24 percent before Katrina) still met criteria for mental illness, and one-third had symptoms of probable PTSD. Those who fully returned to prehurricane function had low levels of baseline psychological distress.

CHILDREN AND NATURAL DISASTERS

One of the many values of the Gulf Coast study was the information it provided on the especially vulnerable child population affected by Hurricane Katrina. An estimated 160,000-plus children were displaced. The effects were multiple, but primary factors included displacement, transiency, and social disruption in many aspects of their lives. The study group of David Abramson and colleagues evaluated the Gulf Coast Data to determine the long-term mental health effects of Hurricane Katrina exposure among youth and to show the path by which the disaster effect operated.[25] They assessed serious emotional disturbance among 283 school children in Louisiana and Mississippi. The study conducted interviews in four annual rounds; most of the data cited here was from the fourth round. More than one-third of the children received a clinical mental health diagnosis of depression, anxiety, or behavior disorder based on parent reporting. They were five times as likely as a pre-Katrina cohort to show serious emotional disturbance.

Children have a limited capacity to self-regulate and choose options for self-care. They depend on others to make choices (regarding their household, neighborhood, school, and social environs) that affect their well-being. The authors acknowledged the complexity of mental illness in children, noting both organic and social factors or both. The Abramson group considered that children's mental health recovery postdisaster can serve either as a bellwether of successful recovery or as a lagging indicator of system dysfunction and failed recovery. Hurricane Katrina may have affected children's sense of control, safety, or stability. All of this could be mediated by parental characteristics, schools, neighborhood, and displacement. It is noteworthy that pediatric mental health is a function of multiple systems. If any one of these systems is dysfunctional, limited, or absent, the child can experience heightened stress. It is noteworthy that many of the child-related studies of mental health post-Katrina did not directly address the role of development in children's responses to Hurricane Katrina. Developmental status is a fundamental determinant of a child's mental response, attitudes, and behaviors and must be considered in any evaluation of children's mental health.

TRAUMA AND DISPLACEMENT

A recurring theme associated with the development of mental illness among Hurricane Katrina survivors is that of displacement. The effects of the loss of a home and/or community can be profound. One's home is generally a source of safety, security, and protection. Survivors reported the following effects:

> The worst part was being displaced for two years; took my comfort zone, not what I was used to. —X

> Went to Monroe, Louisiana. And went to a center and wound up in Tampa. I worked at church in a day care center. FEMA paid the rent . . . returned when told could return, initially couldn't come back to city . . . didn't like Tampa. —Y

> Returned after two weeks, had to sneak in around detours, etc, . . . my house was demolished, destroyed . . . remained in Dallas, liked it, bought a house, still there, . . . was a lifelong resident of New Orleans with a good business but comfortable now in D[allas], knew people there. —MU

As noted previously, 80 to 90 percent of New Orleans residents evacuated as a result of Hurricane Katrina, and six hundred thousand had not returned a month after the storm. Some did not return for months or years, and some never returned, relocating to nearby places such as Baton Rouge, Dallas, and Houston as well as more distant places such as California, Florida, and South Carolina. The studies reviewed have often referenced mental illness in the context of displacement; factors often characterized as stressors included forced relocation, difficulty finding housing, and prolonged separation from community.

The psychiatrist Mindy Fullilove has described the psychological processes affected by geographic displacement.[26] She discusses components of the psychology of place, referencing the "good enough" environment that people require to live. Place-related psychological processes include attachment, familiarity, and identity. "Attachment" refers to a mutual caretaking bond between a person and a beloved place. "Familiarity" refers to the processes by which people develop detailed cognitive knowledge of their environs. "Identity" is the extraction of a sense of self based on places in which one passes one's life. All are threatened by displacement, allowing for the development of nostalgia, disorientation, and alienation and thereby creating the potential for mental health problems.

Researchers of natural disasters have identified four dimensions of displacement: geographic distance from the predisaster community, type of postdisaster housing, number of postdisaster moves, and time spent in temporary housing. A study group examined the impact of housing displacement on the mental health of low-income, non-Hispanic Black single mothers who survived Hurricane Katrina and completed pre- and postdisaster assessments. They created three profiles of displacement experiences: (1) returned to a predisaster community, (2) relocated to a new community, and (3) housed unstably as reflected in long periods in temporary housing and multiple moves. They assessed the relationship between the displacement profiles and three mental health outcomes: general psychological distress, post-traumatic stress, and perceived stress, controlling for predisaster characteristics, mental health, and hurricane-related experiences. "Perceived stress" refers to the degree to which someone perceives situations in their life in the past month as unpredictable, uncontrollable, and overloading. The Perceived Stress Scale measures these perceptions. Those in the relocated profile had slightly higher general psychological distress and perceived stress, and those in the unstably housed profile had significantly higher perceived stress compared to the returned profile. There are implications here for interventions and policies that reduce postdisaster housing instability and prioritize mental health services in communities of relocation.[27]

Still another study examines the relationship between displacement and mental health in the framework of geographic social network dispersion and post-traumatic stress.[28] Researchers found that network dispersion was associated with the following: likelihood of post-traumatic stress five years after Katrina, exposure to Katrina-related trauma, perceived social support, and New Orleans residency. Two social-psycho mechanisms that accounted for the effect of dispersion and post-trauma stress were a lack of deep "belonging" in the new location and a lack of "mattering," characterized as an inability to fulfill obligations to important distant ties. The study authors concluded that physical proximity to emotionally intimate network ties is important for long-term recovery.

SYSTEMS APPROACH TO TRAUMA RELIEF

The failures after Katrina that led to traumatic reactions are numerous, but hindsight offers a way to better employ foresight. The US Congress passed the Post-Katrina Emergency Management Reform Act in 2006 to preempt the scope of physical and mental damage and to support individuals and communities in recovery efforts. The act gave the Federal Emergency Management Agency (FEMA) the responsibility to lead and support efforts to reduce loss of life and property. This includes the ability to track the location of goods and services from FEMA to the affected state, the establishment of a pre-positioned equipment program in at least eleven locations nationwide to support local efforts, updated and improved technology systems, and assistance in the identification of illegal conduct or addressing public safety concerns, including those involving sex offender notification requirements. The act also authorizes search and rescue teams, and focuses on the reunification of families through a family Registry and Child Locator Center. The Post-Katrina Act called for a review of housing resources and existing housing stock, a compilation of housing resources for disaster victims, and attention to disabled and their housing needs as well as to low-income and special needs populations. There are also provisions for the repair of rental housing. In addition, the FEMA administrator must meet requirements for experience and background, and new positions of chief medical officer and advocate for state and rural health have been added.

During the Obama administration an initiative to establish a "whole community approach" was proposed that would allow the involvement of the private sector, community groups, and individual citizens in disaster preparedness. The goal is to draw attention to disaster resilience and to improve coordination. This allows the establishment of partnerships between the government and individuals in supportive responsive efforts. Both the government and individuals use social media to facilitate disaster response by reaching out through texts and tweets, identifying open and available shelters, and so on. Social and visual media can also reveal dangerous, difficult situations that prompt government response and provide quick, informed life-saving support and interventions.

Quality-of-life interventions that support physical and mental integrity should preferably occur before and not after a disaster. The public health/medical approach would be to focus on how to provide better sanitation, food, and clean water to victims of emergencies in coordination with other relief and government agencies. There should also be immediate provision in shelters for areas where people with infectious diseases such as the flu and the measles could be separated from healthy evacuees.

As one interviewee asked, How does one prepare for a rainy day that is different? Or, to rephrase, how does one address the trauma of climate disasters? The global surge of hurricanes, wildfires, and flooding is creating more traumatized individuals and communities. Derek Silove and colleagues have proposed a theoretical framework for the recovery of individuals and collectives known as the adaptation and development after persecution and trauma (ADAPT) model.[29] The model enumerates key psychosocial domains threatened by a disaster: safety and security, interpersonal networks, systems of justice, social identities and roles, and institutions that confer meaning. This is an important model because it provides multimodal and multidiscipline means of supporting the integrity of the individual and indirectly supporting a resilient response.

The Arkansas Model

Some consider the Arkansas model the prototype of the most efficient, effective intervention for the future. As described above, Hurricane Katrina affected several Gulf Coast areas, but Arkansas had undertaken predisaster planning. More than one thousand individuals displaced from Hurricane Katrina were relocated to Arkansas. Before the hurricane struck, there had been anticipatory disaster planning involving collaboration between the Arkansas Department of Health and the Arkansas Chapter of Social Workers. Accordingly, "A training manual was developed that outlined a mental health response that included the creation of seven mental health teams. These were interdisciplinary, composed of a team leader who was a licensed clinical social worker and five-to-seven team members who were licensed mental health providers e.g., social workers, psychologists, and psychiatric nurses."[30] The disaster mental health response to Hurricane Katrina's displaced survivors intervened at several levels. There was assessment for PTSD, family functioning, and loss of family support, with crisis interventions. These actions assured physical safety, confidentiality, nonjudgmental listening, and empathy; they also helped develop coping skills, connected services with families, gave information regarding assistance, and integrated survivors quickly as possible into the Arkansas community. There was also flexibility in providing services. The response, though not perfect, was organized and coordinated.[31]

This Arkansas model may be the model for the "intervention of the future." It includes four essential factors for effective intervention in the face of looming disaster. First, there is predisaster planning that includes state agencies and community organizations/personnel. Foresight usually facilitates prevention of negative, destructive outcomes. Second, there is organization and order rather than chaos and confusion, which impede recovery. Third, there is coordination and collaboration among critical personnel/agencies to provide more constructive and complete interventions. Fourth, there is identification of communitywide resources. Immediately, the survivors of a disaster are not isolated or left to their own devices. There is immediate guidance on housing, food, health providers, and the recovery of lost family members. The Department of Health in Arkansas served as the core of the response, and other health disciplines became the spokespersons. The Arkansas model has many strengths, but there are barriers to duplication and implementation in other jurisdictions that depend on the ability of participants to trust selected leadership and to respect the skills offered by each discipline.

RESILIENCE

The psychiatric researchers Dennis Charney and Steven Southwick have researched the neurobiological, psychological, sociological, and spiritual substrates of anxiety, stress, and trauma and have offered a new understanding of and approach to managing these emotions. They have codified these findings in the term "resilience." To be resilient is to "bend without breaking, to return to an original shape." Resilience is the ability to "bounce back" after encountering a difficulty. Charney and Southwick have outlined a "ten-step prescription for resilience" based on studies of adults that documents guidance for adult mastery of this skill:[32] (1) keeping a positive attitude (optimism is strongly related to resilience and is partly genetic but can be learned through cognitive behavioral therapy); (2) practicing cognitive flexibility through cognitive reappraisal (reframing the value and meaning of a traumatic event or failure, which is essential to growth); (3) embracing a personal moral compass (developing unshatterable core beliefs, having faith in conjunction with strong spiritual beliefs, practicing altruism); (4) finding a resilient role model; (5) facing one's fears (increasing self-esteem, practicing skills to move through fear); (6) developing active coping skills (active versus passive skills, embracing positivity about oneself, seeking the support of others); (7) establishing and nurturing a supportive social network (having a safety net, getting emotional strength from close relationships); (8) attending to physical well-being (physical exercise); (9) training regularly and rigorously in multiple areas (emotional intelligence, moral integrity, and physical endurance); and (10) recognizing, using, and fostering signature strengths to deal with difficult/stressful situations.

However, at least two interviewees offered a clue to resilience that involved their faith:

> There was overwhelming destruction; but now we can celebrate the resilience and perseverance in terms of character that was formed post–Hurricane Katrina. It gives me a headache to talk about it but in faith we can endure . . . faith can endure . . . faith sustained me many a night. —C

> We survived by the Lord; prayed, some joined them in prayer. —Y

As one survivor explained,

> I did not feel helpless, "Don't put all your eggs in one basket" is how my mom raised me. So I was not hurt and I could weather the events, made a difference

emotionally. . . . Makes a difference in response to catastrophe. I could look
forward to how to get to next step . . . my son-in-law noted I was not worried
and not suffering like others. —BR

Hurricane Katrina was not a trauma that resulted in damage only to the physical and mental well-being of individuals but also to whole communities.

Community resilience typically involves a set of attributes that are linked
to "a positive trajectory of functioning and adaptation after a disturbance."[33]
In the framework developed by Fran Norris and her colleagues, these attributes include economic development, social capital, information and communication, and community competence.[34] Accordingly the government
and private sector embrace the notion of resilience as an organizing concept
for disaster management. A foundational commitment to resilience allows
actions beyond rebuilding. It encourages investment in flood prevention, as
in levees and canals, and limits construction in low-lying, flood-prone areas.

One study considered data from two population-based cohort studies
of the long-term effects of Hurricane Katrina on the health and well-being
among minority populations. One study population consisted of only
Vietnamese Americans and was named the Katrina Impacts on Vietnamese Americans study, and the second was the Gulf Coast Child and Family
Health study of African American and White residents of Katrina-affected
regions. Comparison between the two groups in 2007 and again in 2018
showed that African Americans faced disadvantages in mental health two
years after Katrina but that these disappeared after twelve years. Analysis suggested that poor employment prospects and nonmarried status
(single income, no companion) accounted for the poorer mental health
in African Americans in 2007. African Americans had significantly longer displacement periods than Whites or Vietnamese Americans and were
therefore in unfamiliar environs for longer periods. There was also evidence of clear preferences for hiring Vietnamese Americans immediately
after Katrina by local employers.[35]

Negative and positive stereotypes were considered factors as well as
social forces, but employment did increase for African Americans over the
years (from 69 percent in 2007 to 84 percent in 2018) while employment
for Vietnamese Americans remained steady. Unfortunately, postdisaster
trauma continues to be influenced by socioeconomic status, gender, and
race. Interactions between individuals and police, insurance companies,

contractors, housing authorities, and others may be still be influenced by socioeconomic class, race, age, and other subjective factors. People with a lack of resources or social connections face chronic obstacles and limited support. In terms of policy, programs that consider the challenges of race, poverty, gender, and age may facilitate the acquisition of employment, housing, medical sources, and other resources for African Americans and other disadvantaged minorities soon after a disaster. These programs may hasten the recovery of mental health.

A Katrina survivor gave voice in her interview to the resilience that came during the recovery process. Her comments reflect some resolution of fear and the presence of hope for the future:

> Immediate evacuation and evacuation policies, voluntary to mandatory, have changed. Now [we] have to evacuate nursing homes, hospitals at Cat 2[,] and have to have relationships with other communities to receive evacuees. It is so disruptive for those in fragile states to travel to get to another destination, to environs unfamiliar, and have uncomfortable rides. There are some nursing homes where many seniors died, but new policies ensure better outcomes. It is a question of personal economics, independence, ability to move, transportation systems to get people out, ways to care for those who can't get going on their own. Now there is an excellent system for disabled, a 311 call system where you register in advance and let your location [be] known. Others can go to a designated area (stick man sculpture) where you can wait for a ride. Communities have to be wise, thoughtful. —B

And what does resilience mean for children? There is wide recognition in the pediatric/child psychiatry community of traumatic events known as "adverse childhood events." These have been identified as physical abuse, neglect, parental mental illness, divorce, and exposure to violence. Natural disasters like wildfires, floods, and hurricanes can also be adverse childhood events, especially when a child experiences the loss of their home, family member, or friend. The result can be social, emotional, and cognitive impairment and can give rise to risk-taking behaviors. Helping children to develop resilience may be the best counter—a blend of optimism, flexibility, problem solving, and motivation. Providing a stable school structure and accessible mental health services to children and parents in a school space, a church, or other places of worship, through mobile mental health units or other structures, can facilitate and support positive mental health in traumatized children.

The notion of "unconditional love" applies to adults as well as children. Everyone needs to feel safe and supported, no matter their age, actions, or choices predisaster. So instead of berating an adult about their decision to shelter in place, not having an adequate supply of food and medication, or not having prepared relocation and job plans, we simply offer solutions, reassurance, and resources to facilitate recovery. In other words, we meet people in the state and place they are at and do not judge where they might be if they had made other choices. It shows them a way forward during what might feel like an overwhelming disaster. They can express their wants and needs without fear of judgment or negative responses or consequences. This helps them have a sense of control and competence to overcome.

As demonstrated thus far, climate change can bring anxiety and stress, but resilience can improve mental health. Parents and caretakers can maintain involvement in their child's life and promote healthy risk taking in a responsible way. In other words, they can help a child step outside their comfort zones, try new activities, and meet new people. This builds self-confidence. They can show the child the value of persistence in overcoming obstacles by sharing their own stories of overcoming. Communities can also promote resilience to climate-related trauma by providing high-quality early education, reducing crime, training teachers to help children with trauma-related experiences, improving parks and making them accessible, and providing more green spaces.

Climate resilience recognizes the disproportionate impact of climate disasters on marginalized groups: communities of color, the poor, females, children, seniors, and others. Black and Brown people already face profound discrimination in access to housing, transportation, food, and jobs. Climate change of any type tends to exacerbate inequalities. Women are more likely to live in poverty than men. Persons of limited income are less able to purchase supplies and overcome displacement due to limited finances. The cost of repairing and restoring homes, cars, books, clothes, and other elements of lost lives is most challenging for minorities, women, and the poor. The young and the old have a limited ability to fight and gather resources for themselves. These marginalized groups have little political representation. Trauma itself is complex, with interracial and intergenerational considerations. Equity and the inclusion of all people must be part of the strategy and actions that fuel climate action planning.[36]

Some people went beyond resilience and actually reported post-traumatic growth. They felt that they were stronger, more spiritual, and had a greater appreciation for life as a result of having survived the disaster.[37] The psychologists Richard Tedeschi PhD and Lawrence Calhoun PhD developed the concept of post-traumatic growth in the mid-1990s. In some ways, it is the flip side of PTSD. Survivors of trauma, while having many features of PTSD, may ultimately find a new sense of self; a new understanding of themselves, the world they live in, and how to relate to others; and a better understanding of how to live. The New Orleans politician Karen Carter, in Spike Lee's film *When the Levees Broke*, commented in the aftermath, "Out of brokenness we began to mend." Some feel that post-traumatic growth is a variation of resilience, but others argue that it is a different construct:

> I used to collect antique dolls, music box collections from trips and Sunday hats. Since Katrina, I have never bought another hat, music box, doll... not sure why, maybe the heartbreak, perspective of loss, time and money spent collecting things destroyed... better to invest in people. —B

Similarly, another survivor noted that their habits had changed after Hurricane Katrina:

> I collected carousels in the past, all destroyed but I was not trying to save anything... never bought another carousel... just not that important. —X

SUMMARY

Hurricane Katrina is one of the few events in the geohistory of the world that has affected the citizens of an entire city (New Orleans) and region of the country (the Gulf Coast). Katrina drowned a city and part of a region and caused destruction and devastation that were not only external but also internal. Interviews with survivors of Katrina revealed an experience that far exceeded the scope of normal sight and sound. They portrayed Hurricane Katrina as "the worst experience of your life," "like a scene from the apocalypse," and "like a sci-fi movie." These are stark, sobering images that primarily resulted in short- and long-term anxiety, depression, and PTSD, depending on one's social status, race, age, gender, and pre-Katrina mental health. Personal humiliation and loss of one's roots (home and history) are primal in nature and therefore more threatening to one's sense of self.

Predisaster planning that focuses on provisions for basic human needs, that is, food, shelter, and clothing; the establishment of mental health resources; legislation that supports FEMA response to limit the loss of life and property; and provision of a multimodal, multidisciplinary approach that protects several psychosocial domains and connects multiple issues is key to mitigating effects of climate trauma. Perhaps most significantly, psychiatric researchers have shown that almost everyone, irrespective of demographics, can develop skills to effectively manage stress and trauma. The ultimate lesson is that the climate has deteriorated to the point that extreme weather events threaten every aspect of our lives. No aspect of life is sustainable. The best solution to protect our mental health in the face of increasing natural disasters is to protect our climate.

NOTES

1   *American Heritage Dictionary of the English Language,* 5th ed. (Boston: Houghton Mifflin, 2016).

2   Understanding and Coping with Eco-Anxiety, Mental Health Commission of Canada, April 21, 2023, https://mentalhealthcommission.ca/resource/understanding-and-coping-with-eco-anxiety/.

3   Ashlee Cunsolo Willox and Neville R. Ellis, "Ecological Grief as a Mental Health Response to Climate Change–Related Loss," *Nature Climate Change* 8, no. 4 (2018): 275–81.

4   Daniel Goleman, "The Age of Eco-Angst," *New York Times,* 2009.

5   Nick Higginbotham, Linda Connor, Glenn Albrecht, et al., "Validation of an Environmental Distress Scale," *Ecohealth* 3, no. 4 (2006): 245–54.

6   Glenn Albrecht, "'Solastalgia': A New Concept in Health and Identity," *PAN: Philosophy, Activism, Nature* 3 (2005): 41–55.

7   Fran H. Norris, Matthew J. Friedman, Patricia J. Watson, Christopher M. Byrne, Eolia Diaz, and Krzysztof Kaniasty, "60,000 Disaster Victims Speak: Part I: An Empirical Review of the Empirical Literature 1981–2001," *Psychiatry* 65, no. 3 (2002): 207–39, https://doi.10.1521/psyc.65.3.207.20173.

8   US Corps of Engineers, "Hurricane Study, New Orleans," released June 2006, H. Rpt. 109–377, "A Failure of Initiative: Final Report."

9   William Morrow, *The Great Deluge: Hurricane Katrina, New Orleans and the Mississippi Gulf Coast* (New York: Harper Collins, 2006).

10   Morrow, *Great Deluge,* 7.

11   Allison Plyer, "Facts for Features: Katrina Impact," The Data Center, 2018, https://www.datacenterresearch.org/data-resources/katrina/facts-for-impact/.

12 Douglas Brinkley, *The Great Deluge* (New York: Harper Perennial, 2007).

13 Sandro Galea, Chris R. Brewin, Michael Gruber, et al., "Exposure to Hurricane-Related Stressors and Mental Illness after Hurricane Katrina," *Archives of General Psychiatry* 64, no. 12 (2007): 1427–34, https://doi.10.1001/archpsyc.64.12.1427.

14 Rajeev Ramchand, Rena Rudavsky, Sean Grant, et al., "Prevalence of, Risk Factors for, and Consequences of Posttraumatic Stress Disorder and Other Mental Health Problems in Military Populations Deployed to Iraq and Afghanistan," *Current Psychiatry Reports* 17, no. 5 (2015): 37, doi:10.1007/s11920-015-0575-z.

15 Henry L. Lew, Malissa Kraft, Terri K. Pogoda, et al., "Prevalence and Characteristics of Driving Difficulties in Operation Iraqi Freedom/Operation Enduring Freedom Combat Returnees," *Journal of Rehabilitation Research and Development* 48, no. 8 (2011): 913–25, doi:10.1682/jrrd.2010.08.0140.

16 Naomi Breslau, "Epidemiologic Studies of Trauma, Posttraumatic Stress Disorder, and Other Psychiatric Disorders," *Canadian Journal of Psychiatry* 47, no. 10 (2002): 923–29, doi:10.1177/070674370204701003; Naomi Breslau, G. C. Davis, and Lonnie R. Schultz, "Posttraumatic Stress Disorder and the Incidence of Nicotine, Alcohol, and Other Drug Disorders in Persons Who Have Experienced Trauma," *Archives of General Psychiatry* 60, no. 3 (2003): 289–94, doi:10.1001/archpsyc.60.3.289.

17 Ronald C. Kessler, "Posttraumatic Stress Disorder: The Burden to the Individual and to Society," *Journal of Clinical Psychiatry* 61, suppl. 5 (2000): 4–14.

18 Melissa Farley and B. M. Patsalides, "Physical Symptoms, Posttraumatic Stress Disorder, and Healthcare Utilization of Women with and without Childhood Physical and Sexual Abuse," *Psychol Reports* 89, no. 3 (2001): 595–606, doi:10.2466/pro.2001.89.3.595; C. Zayfert, A. R. Dums, R. J. Ferguson, and M. T. Hegel, "Health Functioning Impairments Associated with Posttraumatic Stress Disorder, Anxiety Disorders, and Depression," *Journal of Nervous and Mental Disease* 190, no. 4 (2002): 233–40, doi:10.1097/00005053-200204000-00004.

19 Karen B. DeSalvo, A. D. Hyre, Danielle C. Ompad, et al., "Symptoms of Posttraumatic Stress Disorder in a New Orleans Workforce following Hurricane Katrina," *Journal of Urban Health* 84, no. 2 (2007): 142–52, doi:10.1007/s11524-006-9147-1.

20 Adam C. Alexander, Jeanelle Ali, Meghan E. McDevitt-Murphy, et al., "Racial Differences in Posttraumatic Stress Disorder Vulnerability following Hurricane Katrina among a Sample of Adult Cigarette Smokers from New Orleans," *Journal of Racial and Ethnic Health Disparities* 4, no. 1 (2017): 94–103, doi:10.1007/s40615-015-0206-8.

21 Mary C. Waters, "Life after Hurricane Katrina: The Resilience in Survivors of Katrina (RISK) Project," *Sociol Forum* 31, suppl. 1 (2016): 750–69, doi:10.1111/socf.12271.

22 Patricia Salber, "Mental Health after a Tragedy: Lessons from Katrina," *The Doctor Weighs In*, July 4, 2020, https://thedoctorweighsin.com/mental-health-after -disaster/.

23 Jean Rhodes, Christina Chan, Christian Paxson Christina, et al., "The Impact of Hurricane Katrina on the Mental and Physical Health of Low-Income Parents in New Orleans," *American Journal of Orthopsychiatry* 80, no. 2 (2010): 237–47, doi:10.1111/j.1939–0025.2010.01027.x.

24 Elizabeth Fussell and Sarah R. Lowe, "The Impact of Housing Displacement on the Mental Health of Low-Income Parents after Hurricane Katrina," *Social Science & Medicine* 113 (2014): 137–44, doi:10.1016/j.socscimed.2014.05.025.

25 David M. Abramson, Yoon Soo. Park, Tasha Stehling-Ariza, and Irwin Redlener, "Children as Bellwethers of Recovery: Dysfunctional Systems and the Effects of Parents, Households, and Neighborhoods on Serious Emotional Disturbance in Children after Hurricane Katrina," *Disaster Medicine and Public Health Preparedness* 4, suppl. 1 (2010): S17–S27, doi:10.1001/dmp.2010.7.

26 Mindy T. Fullilove, "Psychiatric Implications of Displacement: Contributions from the Psychology of Place," *American Journal of Psychiatry* 153, no. 12 (1996): 1516–23, doi:10.1176/ajp.153.12.1516.

27 Elizabeth Fussell and Sarah R. Lowe, "Impact of Housing Displacement."

28 Katherine A. Morris, and Nicole M. Deterding, "The Emotional Cost of Distance: Geographic Social Network Dispersion and Post-traumatic Stress among Survivors of Hurricane Katrina," *Social Science Medicine* 165 (2016): 56–65, doi:10.1016/j.socscimed.2016.07.034.

29 Derek Silove and Z. Steel, "Understanding Community Psychosocial Needs after Disasters: Implications for Mental Health Services," *Journal of Postgraduate Medicine* 52, no. 2 (2006): 121–25.

30 Chris Lang, "Climate Change and Mental Health," *Psychiatric Times*, October 26, 2015, https://www.psychiatrictimes.com/view/climate-change-and-mental -health.

31 Natalie Timoshin, "In This Issue: Volume 32, no. 10," *Psychiatric Times*. November 16, 2020, https://www.psychiatrictimes.com/view/issue-volume-32-no-10.

32 Steven M. Southwick and Dennis Charney, *Resilience: The Science of Mastering Life's Greatest Challenges* (Cambridge: Cambridge University Press, 2018).

33 Norris et al., "60,000 Disaster Victims Speak," 6.

34 Norris et al., "60,000 Disaster Victims Speak," 6.

35 Mengxi Zhang, Mark VanLandingham, Yoon Soo Park, et al., "Differences in Post-disaster Mental Health among Vietnamese and African Americans Living in Adjacent Urban Communities Flooded by Katrina," *PLoS One* 16, no. 8 (2021): e0255303, doi:10.1371/journal.pone.0255303; L. Peek, "Weathering

Katrina: Culture and Recovery among Vietnamese Americans," *Contemporary Sociology* 48, no. 2 2019): 225–27, https://doi.org/10.1177/0094306119828696pp.

36 Emily Wright, "Climate Resilience Requires Dismantling Internalized White Supremacy," Medium, September 7, 2019, https://emily-wright.medium.com/climate-resilience-requires-dismantling-internalized-white-supremacy-c4a2d-b11acd4.

37 Patricia Salber, "Mental Health after a Disaster: Lessons from Katrina," Medium, July 4, 2020, https://docweighsin.medium.com/mental-health-after-a-disaster-lessons-from-katrina-1edfa87f3a06.

Part 4

**Acquiring Understanding:
Two Case Studies of
Long-Term Impacts**

CHAPTER 13

Consequences of and Responses to Compounded Vulnerabilities Rooted in Colonialism

The Case of Vieques, Puerto Rico

LORENA M. ESTRADA-MARTÍNEZ, ROSALYN NEGRÓN, JORGE L. COLÓN, CRUZ MARÍA NAZARIO, HELEN POYNTON, VALERIA HERNÁNDEZ-TALAVERA, ADRIANA OYOLA-VIVAS, AND ANA ROSA LINDE-ARIAS

The archipelago of Puerto Rico in the Caribbean consists of the main island (La Isla Grande), Vieques (La Isla Nena), Culebra, and several minor uninhabited keys. The US government has used Puerto Rico, long coveted as a strategic commercial and military outpost, for numerous military training and testing activities (e.g., Agent Orange and napalm in the protected rainforest).[1] Vieques is an island-municipality located approximately six miles southeast of the main island (figure 13.1). From 1941 to 2003, the US Navy and NATO allies used Vieques for target practice with live and inert ammunition and ordnances, storage and processing of supplies, and waste disposal.[2] Vieques also has elevated risks of chronic disease compared to Puerto Rico's main island, including a higher frequency of cancer; cardiovascular, renal, and neurological diseases; and diabetes, as well as poor infant and maternal health outcomes. Environmental and epidemiological concerns in Vieques are part of a broader socioecological context. Residents continue to be concerned about contaminant exposure and related stress, distrust government assessments, and experience inadequate healthcare access.[3] Climate change presents a significant risk to small island communities because of the intersection of climate hazards (e.g., coastal flooding and impacts on infrastructure and agriculture), exposure (e.g., temperature increases and droughts), and vulnerability (e.g., food and water insecurity and minority and low-income populations).[4] Among the consequences of climate change is its potential to alter the environmental distribution, cycling, and risks of chemical pollutants.[5] The result is increased chronic diseases due to elevated temperatures and degraded air quality, high levels of PM2.5 particles, greater risk of outbreaks of existing and emerging infectious diseases,[6] and, in the case of Vieques, enhanced contaminant toxicity stemming from US Navy military practices in addition to new exposure pathways.[7]

In January 2021, our team, which includes a transdisciplinary group of researchers, including Puerto Rican scientists based in Puerto Rico and Massachusetts, and a group of Vieques community leaders, partnered to establish a multiyear collaboration called Vieques, Ambiente, Salud y Acción Comunitaria (Vieques, Environment, Health, and Community Action). Through community meetings, interviews, focus groups, citizen science, and contaminant analysis, this community-based participatory research project aims to understand better the human health and environmental impacts of the US Navy's presence in Vieques.

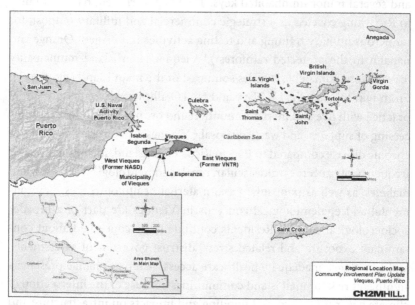

FIGURE 13.1. Regional location map. Source: US NAVFAC, 2015, Community Involvement Plan Update, Vieques, PR, CH2MHILL.

Our purpose in this chapter is threefold. First, using the social determinants of health framework and the concept of cascading disasters, we analyze Vieques' political, military, and socioeconomic history.[8] Specifically, we address the social and health implications of land expropriation and occupation and environmental exposure after military activities intensified in the 1970s. We emphasize grassroots movements' role and public health's integration into the protests and eventual removal of the US Navy in 2003. Next, we explain critical public health issues in the post-Navy era, including ongoing health concerns, new forms of land displacement, and recent exposure pathways considering extreme climate events. Finally, we describe health, environmental, and social community-based initiatives and provide policy and research recommendations.

Theoretical Frameworks: Social Determinants of Health and Cascading Disasters

The social determinants of health (SDoH) are the nonmedical conditions in which we are born, grow, live, work, play, and age.[9] The SDoH framework highlights the importance of economic stability, access to quality healthcare systems, and social and community contexts. These broad domains are intertwined, mutually reinforcing, and incorporate historical contexts and long-term exposures. In Puerto Rico, colonialism, militarism, and unchecked capitalism have shaped all SDoH dimensions, magnifying their impacts in Vieques for the past eighty years.

Given the multiple, compounded, and continued threats to the health and well-being of Viequenses, the concept of cascading disasters provides a valuable framework for understanding Vieques' past, present, and future. Disasters are major events that significantly overwhelm a community's coping capacity. As a result, they impair systemic functions and lead to human suffering, loss, and environmental damage.[10] For Vieques, a historical chain of hazards has slowly but relentlessly unfolded since Puerto Rico's military invasion by the United States in 1898. Hazards alone do not create disasters. Instead, vulnerability is at the root of all disasters.[11]

Cascading disasters result from a chain of causally related events. The impacts of each disaster compound, thus rendering communities vulnerable to future hazards, both human- and nature-caused.[12] Cascading disasters begin from a triggering disaster that leads to secondary and tertiary effects and may follow nonlinear and unpredictable trajectories.[13] Vieques' geographical location results in a marginalization from Puerto Rico and the US mainland, with clear consequences for accessibility to essential materials, services, and administrative support. Ultimately, this has rendered Vieques especially vulnerable to environmental contaminant-linked health disparities.

The nonlinearity of cascading disasters means that an initial hazard has branching effects in parts of the system not directly touched by the triggering event. In this case, the sustained presence of the US Navy in Vieques and its bombing exercises had severe direct effects, such as environmental contamination. The Navy's presence also had some indirect effects. For example, land expropriation and the subsequent contamination of land, air, and water affected human health in ways not yet thoroughly investigated. Moreover, land expropriation made areas in Vieques uninhabitable and unavailable to local economic development. This forced the island's economic activity to cater to the military personnel stationed there. These vulnerabilities have compounded and placed Vieques at an increased risk of climate change–linked extreme weather events in the Caribbean region.

Taking a holistic account of the case of Vieques, we can point to at least five disasters: colonialism (Disaster #1), land expropriation and occupation (Disaster #2), intensification of military activities (Disaster #3), socioeconomic obstruction (Disaster #4), and persistent exposure to environmental injustice (Disaster #5). Because Vieques has a limited healthcare infrastructure, these disasters have unfolded in the context of a weak healthcare system. In addition, we can declare the responses from the state and federal governments as a disaster of institutional neglect.

While the five disasters are specific to Vieques, they play out against the backdrop of global climate change. For example, recent climate-related events, such as Hurricane Maria in 2017, destroyed critical infrastructure, including the only hospital; dredged up dormant contaminant and ordnance deposits; and heightened inconsistent accessibility to and from the main island. While Hurricane Maria was an exceptional storm, tropical cyclone activity is expected to increase in the Caribbean as climate change progresses.[14] Vieques is vulnerable to other climate change effects, such as droughts, extreme heat events, and sea level rise, further stressing its health care system and social services.[15] Therefore, as the five cascading disasters unfold in Vieques, it is in the context of a climate justice community already vulnerable due to its Caribbean location and limited infrastructure to mitigate the effects of climate change.

Disaster #1: Colonialism

The triggering disaster in our analysis is the annexation of Puerto Rico, Vieques, and Culebra in 1898 as a colony of the United States to serve its economic and military interests. In the 1940s, the United States established itself as a world superpower. The military and economic role of Puerto Rico shifted to provide American investors with a captive market, showcase military might, and provide sites for different types of biomedical, chemical, and weapons testing. The US government installed twenty-five military bases in Puerto Rico in the twentieth century, including Roosevelt Roads, which became the most extensive US naval base globally.[16] The base included Vieques as its firing practice and amphibious landing site. The US Navy effectively controlled Vieques' land, water, and air by managing all nautical routes, flight paths, aquifers, and zoning laws in the civilian territory as well as ownership of much of the "resettlement tracts" where most residents lived.[17]

One resident described the uniquely marginalizing power dynamics that have troubled Vieques for decades: "Vieques es una colonia de una colonia" (Vieques is a colony of a colony). On the one hand, the island municipality is subject to the US government's colonial prerogatives. On the other hand, it cannot rely on protection and prioritization by Puerto Rico's state government. Puerto Rico's financial woes are magnified in Vieques, where the Navy's presence irrevocably altered the course of its development as an agricultural and fishing economy.

Colonialism is the triggering disaster we can trace to Vieques' present vulnerabilities. In its most basic sense, resource extraction leaves occupied communities with few resources to build capacity absent colonial power. For example, land extraction, and the attendant uprooting from land, create deep psychological wounds. Such traumas interact with the stressors of economic insecurity that come with the loss of financial resources. Scarce access to land may also set up resource competition, such as land disputes, that can fracture familial and community bonds.

Moreover, Puerto Rico's colonial status meant that Viequenses had little say in whether their land could be appropriated and used as a training and testing ground for military operations, manifesting as Disasters #2 and #4. Disaster #3 clarifies that aggressive land appropriations devastated Vieques' demography, natural resources, and economy. Ultimately, this cascade of disasters has culminated in Disaster #5, or the contamination of Vieques' land with consequences for human health yet to be fully understood.

Disaster #2: Land Expropriation and Occupation

The broad context of colonialism set the stage for US Public Law 247 in 1941, which allowed the US Navy to expropriate 78 percent of the total land in Vieques to build a military base (figure 13.2). As a result, most Viequenses were forcibly removed and relocated to the middle of the island with little notice and no compensation.[18] They received no help to relocate and had to leave valuable livestock and crops behind.[19] The US

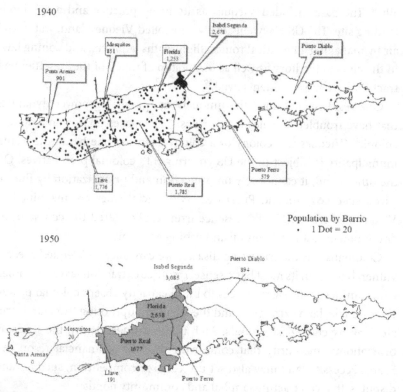

FIGURE 13.2. Population distribution by barrio in 1940 (pre-Navy) and 1950 (post-Navy). Source: Ayala and Bolívar, *Battleship Vieques*, 63–64.

Navy also expropriated four large sugar mills, significant employers on the island. Finally, after one year of economic boom due to the construction of the base, unemployment forced many to leave.[20]

With thousands of soldiers participating in military maneuvers in Vieques, prostitution and riots flourished. Several people died in fights between Viequenses and Navy troops.[21] Military personnel frequently got drunk at local bars, sexually assaulted women, and urinated in front of people's houses. As a result, Viequenses learned to remain locked up in their houses when the Navy troops were allowed to leave the base.

In 1961, the Department of Defense made multiple attempts to abolish the Vieques municipality and use the entire island for military training. However, after pressure from the Puerto Rican government, the Vieques mayor and population, and the Catholic Church, they ceased their repeated attempts.[22]

Disaster #3: Socioeconomic Obstruction

The expropriation of land had cascading disaster effects on economic development in Vieques. First, hundreds of acres of land were made unusable for agricultural activity. This was particularly detrimental given that Vieques's economy was primarily agriculture-based before the Navy's presence on the island. Consequently, 40 percent of the population lost their income, and the island underwent an unemployment and hunger crisis.[23] Second, the expropriation of land led to the displacement of farmers, who could not use the restricted lands for grazing their animals.[24] Moreover, the Navy actively discouraged economic development because more civil travel to Vieques would negatively affect their military exercises. As a result, thousands left Vieques in search of work, reducing the number of working-age Viequenses.[25]

Third, the US government expropriated the most fertile land in Vieques.[26] Today, approximately 20 percent of its land is available for development. However, the open land is suboptimal for agriculture or tourism, given its aridity and barrenness. These economically unjust conditions are exacerbated by a lack of investment in sustainable economic development and a disinvestment in essential social services (transportation, education, food, water, and electricity access).[27] The expropriation of land was a triggering event that obstructed opportunities to create an economy and community that was self-sufficient and self-sustaining. However, grassroots community groups (including churches) have tried to fill this gap without institutional investment.

Disaster #4: Intensification of Military Activities

The context of colonialism and the expropriation of lands meant that the US military had a location to conduct ongoing military training that would otherwise have been illegal. The first large-scale military practice in Vieques occurred on January 16, 1948. It included sixty ships, 350 airplanes, and fifty thousand troops from all branches of the US Armed Forces. In 1949, military practices involved over thirty-five thousand troops, several hundred planes and ships, and napalm-incendiary bombs.[28] Then, in the mid-1970s, after thirty-five years of military exercises on the nearby island of Culebra ceased, those exercises were added to the existing military practices on the island of Vieques, intensifying adverse environmental and health effects.[29]

By the early 1980s, Viequenses were exposed to two hundred days of air-to-ground combat exercises, 158 days of naval bombardment, twenty-one days of amphibious landings, and 3,400 bombs annually. During the most intense training periods, 7,600 bombs were dropped each month (around 253 daily). Meanwhile, the ten thousand Viequenses inhabitants resided only eight miles away, downwind from the military ranges. In addition, the Navy rented the Vieques facility for eighty million dollars a year for training military maneuvering and weapons testing among NATO allies.

According to Navy records and local testimonies, the bombs and other military munitions used in Vieques contained toxic substances such as TNT, RDX, HMX, Tetryl, HBX, PETN, heavy metals (e.g., lead, cadmium, arsenic, mercury), perchlorate, phosphorus and other pyrophoric materials, napalm, Agent Orange, chaff, and residues of organic and inorganic chemical components, among other unknown contaminants.[30] In recent years, the US Navy and Department of Defense have also admitted to using illegal depleted uranium munitions and biological and chemical warfare agents, including trioctyl phosphate (a simulant of the nerve agent VX), during exercises in Vieques.[31]

On April 19, 1999, a F/A-18C Hornet fighter jet negligently dropped two MK-82 five-hundred-pound bombs near a marked observation post in the live impact area of the bombing range in east Vieques, killing the civilian security guard David Sanes Rodríguez. This tragedy sparked a new phase of the protracted struggle to end the bombing in Vieques, including massive protest and civil disobedience. The international community became aware of the risks to the Viequenses of continued bombing exercises. Supported by mainland Puerto Ricans, the Puerto Rican diaspora, and others worldwide, local Viequenses engaged in massive civil disobedience, forcing the Navy to end the bombing on May 1, 2003.[32]

Disaster #5: Environmental Injustice

Even though the US Congress in 1980 had concluded that the Navy's activities were detrimental to Viequenses, it was not until 2003 that the Navy stopped military practices. On February 11, 2005, the Environmental Protection Agency (EPA) placed the Vieques bombing range and surrounding waters on its Superfund National Priorities List under the designation Atlantic Fleet Weapons Training Area–Vieques.[33] The National Priorities List is the EPA's list of sites throughout the United States and its territories with known or potential releases of hazardous substances, pollutants, or contaminants. Nevertheless, Viequenses continue to be exposed to toxic contaminants that infiltrated and tainted the water and soil during the occupation and that are also produced by the open detonation of unexploded ordnances as part of the cleanup efforts of the current Superfund site. Viequenses have objected to these alleged cleanup efforts, since they mainly involved removing bombs and exploding many of them in the open air.

In 1996, responding to a request from the Committee for the Rescue and Development of Vieques, researchers from the Graduate School of Public Health at the University of Puerto Rico analyzed cancer data from the Puerto Rico Department of Health.[34] They confirmed that (1) the incidence rate of cancer in Vieques increased from the early 1970s until 1989; (2) the rate followed the pattern of increasing days and intensity of US Navy use of the Vieques Bombing Range after the Navy closed the Culebra Bombing Rage in 1970s and transferred its maneuvers to Vieques, with a lag time of twenty years; (3) by the early 1980s the cancer incidence rate in Vieques had surpassed the one in Puerto Rico; and (4) the cancer risk for 1985–89 was 27 percent higher for the population of Vieques compared to that of Puerto Rico. Moreover, the risk was exceptionally high among young Viequenses. For example, children ages zero to nine years and ten-to nineteen-year-olds were at 117 percent and 257 percent higher risk of cancer than children of the same ages in Puerto Rico.[35]

High levels of arsenic, aluminum, copper, iron, HMX and RDX explosive compounds, and cadmium have been detected in foods, including fish, biota, and produce (e.g., pigeon peas).[36] Researchers have found heavy metals (i.e., mercury) in hair samples,[37] urinary samples (i.e., nickel, lead, arsenic, uranium, and cadmium), and blood samples (i.e., mercury, uranium, and lead).[38] High levels of lead and arsenic have been found in residential soil samples, and traces of TNT, HMX, RDX, and metabolites in live impact area samples.[39] In December 2020, the US Navy indicated that drinking water on the island was contaminated with six different per- and polyfluoroalkyl (PFAS) compounds,[40] and fourteen sites within the restricted areas were identified with potential PFAS contamination.[41] Previous studies in Vieques evaluated the association between seafood consumption and inorganic arsenic in adults' nail, hair, and urine samples. Results indicated a positive association with fish consumption, particularly for men, increasing with years of residency in Vieques.[42]

The population in Vieques is aware of the health impacts of multiple environmental contaminants from military exercises in the air, water, and soil. However, Viequenses refused to actively participate in the Restoration Advisory Board meetings that are part of the Comprehensive Environmental Response, Compensation, and Liability Act cleanup process. The meetings are conducted in English (with interpreters that use technical vocabulary), and are simply informative, rather than consultative, and therefore do not consider questions and suggestions from the community, thereby limiting valuable community engagement.

CURRENT CONCERNS

Viequenses continue to grapple with the legacies of colonialism, contamination, and economic stagnation due to the US Navy's presence. The following section provides an overview of Vieques' current health and healthcare concerns and the complex and synergistic relationships between climate change and environmental exposure to contaminants used by the US Navy.

Socioeconomic

According to the American Community Survey, in 2019 there were 8,300 residents and 2,260 households in Vieques.[43] The median age of Viequenses was 44.5 years. Children made up 20 percent of the population. The 2019 American Community Survey indicators show a challenging economic context.

After the Navy expropriated land, including acres usable for agriculture, the Vieques economy became dependent on their presence. Since the Navy's exit in 2003, Vieques economic development has been hampered by the limited use of former military land for agricultural and housing, even in noncontaminated areas. Having agricultural lands available for local food is especially important in Vieques because of its distance from the main island. Currently, 46 percent of the population lives in poverty, including 81 percent of children and 77 percent of seniors.

Health and Health Care

Vieques has many critical health and healthcare concerns, including limited access to healthcare services, medications, and air transport for medical emergencies; lack of reliable transportation to and from the main island; and inaccurate morbidity and mortality data. The fact that Hurricane Maria destroyed the only hospital in Vieques highlights how the island is uniquely vulnerable to extreme weather events spurred by climate change. The hospital has yet to be reconstructed. The current urgent care facility consists of a few trailers. It offers essential prenatal services, has an X-ray machine, and houses a dialysis unit that often does not work. In addition, several general practitioners and a pediatrician provide services several days a week but no other specialist care. There is one private pharmacy on the island.

Patients needing additional care must obtain a referral for a provider in La Isla Grande. To get there, some Viequenses arrive at dawn at the ferry dock and try to come back on the last trip of the day. Others leave the day before the appointment, secure housing nearby, and return by ferry after receiving services from their healthcare provider. For many, this means taking days off work, arranging childcare, and paying for additional housing and transportation. For those who become severely ill, it often means finding long-term housing near their healthcare providers.

New Forms of Land Displacement

Gentrification unfolds at the intersection of multiple SDoH contexts: economic, neighborhood, built environment, and social and community. American investors and expatriates—many former US Navy personnel—have moved to the island in the last two decades. They now own most restaurants and bars and cater primarily to tourists and the upper class.[44]

In recent years, real estate developers have purchased many land parcels and houses from Viequenses, transforming them into hotels or high-end properties that cater to foreigners.[45] Vieques is a beautiful island with clear blue-water beaches, unique natural resources (e.g., the bioluminescent bay), and precious scenery. Notably, the out-migration of Viequenses has consequences for social support networks and community cohesion in the neighborhoods they leave behind.

The Synergistic Risks with Climate

The Intergovernmental Panel on Climate Change has stated that the archipelago of Puerto Rico may be especially vulnerable to global warming. Impacts include increases in air and ocean temperatures, ocean acidification, precipitation, sea level rise, and erratic weather events (severe storms and droughts). One of the most significant impacts of climate change is the shift in precipitation patterns.[46] The US Fifth National Climate Assessment concluded that global carbon emissions would likely reduce average rainfall in this region, constraining freshwater availability.[47] Higher temperatures will also increase evaporation rates, further intensifying droughts. The climate forces responsible for reducing average rainfall will also create conditions for intensified storm events.[48] For example, tropical cyclone activity is also expected to increase in frequency and intensity, such that when it does rain, it occurs as an extreme storm event that causes widespread flooding.[49] As a result, extreme events pose significant risks to life, property, and the economy in the Caribbean.

The most concerning impacts of climate change on Vieques include storm surges and extreme hurricanes. The frequency and extent of storm surges have increased with the rise in sea level. Hurricanes' wind and rain forces can cause substantial damage to biological resources, including wetlands and coastal environments. In addition, climate change can alter the distribution and biological effects of chemical contaminants, potentially exacerbating the effects of pollution on Vieques.[50] The fate and exposure of contaminants in soil and water depend on the chemical properties and the environment in which they are found. Both may be altered by global climate change impacts such as heating, increased rainfall rates, extended dry periods, erosion, and increased water level.[51] Even temperature changes can modify, alter, and increase the toxicity of some contaminants.[52] Significant exposure risk to humans arises when contaminants are mobilized from the soil into groundwater and are spread downstream, potentially contaminating drinking water.[53] Increased contaminant mobility may cause higher bioaccumulation within food webs or decrease the quality of crops through plant uptake.[54] For example, cadmium, observed in local vegetation,[55] is more rapidly accumulated in vegetation when temperatures increase, facilitating its entry into the food chain and risk to human health.[56] Finally, massive water movements during extreme weather raise the potential to redistribute contaminants throughout the affected areas (soil, groundwater, etc.). Following a storm event, resuspended contaminants settle into topsoil and sediments, potentially exposing the local community and coastal ecosystems for years. Given the contamination challenges in Vieques, it is critical to understand how climate change may influence the mobilization and bioavailability of heavy metal and military-related contaminants. This emphasizes the need to identify and create solutions to mitigate its effect on the community and the ecosystems that are most vulnerable.

CURRENT COMMUNITY INITIATIVES

Social activists have become a critical component of local health, environmental, and social service networks responding to the adverse health, ecological and social impacts of the US Navy's presence in Vieques. Direct actions by civic groups and organizing networks, such as the Comité Pro Rescate y Desarrollo de Vieques (Committee for the Rescue and Development of Vieques), continued to serve Viequeses after the Navy left. Today, Vieques has grassroots organizations building capacity and serving needs across multiple domains of life, including agriculture, health, youth development, environmental justice, and small-business development.

Many community groups have focused on providing direct services to residents (e.g., Vieques en Rescate, Vieques Love, Iglesia Fé que Transforma, COREFI), environmental cleanup, and decontamination (e.g.,

Vidas Viequenses Valen). Specifically, Vidas Viequenses Valen has created awareness among domestic and international audiences. It works through educational forums, peaceful protests, media publications, meetings with public officials, and public appearances. Vidas Viequenses Valen insists on the demilitarization, decontamination, and devolution of the lands still controlled by the US federal government.[57] In addition, it works on issues related to deficient transportation and the lack of adequate health facilities on the island.[58] For example, several children have died because they could not be transported by ferry or airplane from Vieques to a hospital facility in Puerto Rico, about forty miles away.[59]

POLICY RECOMMENDATIONS

Vieques continues to suffer from policies of neglect by federal and state governments. We outline a few recommendations that address immediate needs and concerns for Viequenses' health.

Undoubtedly, building a quality hospital in Vieques needs urgent attention. The Federal Emergency Management Administration approved $43.5 million to rebuild the hospital in Vieques.[60] However, hospital rebuilding efforts have not begun in earnest as of the writing of this chapter, more than three years after the funds were obligated. Even when the hospital rebuilding is complete, Vieques will still need a more robust preventive care system. To address the shortage of healthcare professionals in rural areas, a program to recruit and retain physicians and other healthcare professionals is necessary for Vieques, similar to other initiatives on the US mainland.[61] Vieques currently has one federally qualified health center but it is insufficient and would benefit from other programs.

Policy recommendations that address SDoH include economic, transportation, and land management domains. Puerto Rico's and Vieques' economic development policies should consist of strategies that encourage employment for residents of Vieques and compensate them adequately to promote retention. Similarly, recruitment initiatives to encourage highly qualified personnel to live and work in Vieques should offer proper compensation. Ferry service between Vieques and Puerto Rico's main island should ensure access and on-time departures and arrivals, prioritizing transportation for Vieques residents.

More significant investment is needed to diversify the Vieques economy, building on local knowledge and experience. For example, Vieques once had a productive agricultural economy, based mainly on pineapple and sugarcane. Several local farms, such as La Semillera, Isla Nena Composta, and Finca Conciencia, contribute to Vieques' efforts at ecologically sustainable farming. Additional financial, knowledge-sharing, and technology transfer support from local and federal agencies such as Puerto Rico and the US Departments of Agriculture or the EPA would further their efforts. Local farms can provide sustainable food sources for Viequenses. Suitable lands should be available for ecologically sustainable agriculture, especially after decontamination.

That said, making more land available for civilian use raises concerns about real estate investors outside of Vieques acquiring land for redevelopment or resale. Such activities would mean profits are going primarily to non-Viequenses while displacing residents. Existing antigentrification policy models from rapidly gentrifying cities could offer pathways for the municipality to follow.[62] These include establishing a community land or investment trust to purchase and conserve properties for local use, including affordable housing, and establishing a locally controlled and transparent oversight committee to review and approve land and property purchase and development proposals. Furthermore, community benefits agreements could also be implemented. These contracts between the community and developers detail the benefits to community members of the proposed development. Such benefits could include guaranteed employment for residents, local health infrastructure investments, and incentive programs to draw and retain new potential residents who can contribute to Vieques' growth, community cohesion, and long-term vitality. This would be the road to their empowerment.

NOTES

1 US Government Accountability Office, "Agent Orange: Actions Needed to Improve Accuracy and Communication of Information on Testing and Storage Locations," *U.S. Government Accountability Office: Report to Congressional Addresses*, GAO-19-24, November 2018, https://www.gao.gov/assets/gao-19-24.pdf; Albin L. Young, *The History, Use, Disposition and Environmental Fate of Agent Orange* (New York: Springer, 2009).

2   Katherine T. McCaffrey, *Military Power and Popular Protest: The U.S. Navy in Vieques, Puerto Rico* (New Brunswick, NJ: Rutgers University Press, 2002); Amílcar Antonio Barreto, *Vieques, the Navy, and Puerto Rican Politics* (Gainesville: University Press of Florida, 2002).

3   Katherine T. McCaffrey, "Environmental Remediation and Its Discontents: The Contested Clean-up of Vieques, Puerto Rico," *Journal of Political Ecology* 25, no. 1 (2018): 80–103, https://doi.org/10.2458/v25i1.22631.

4   US Environmental Protection Agency, *Climate Change and Social Vulnerability in the United States: A Focus on Six Impacts,* EPA 430-R-21–003, 2021, https://www.epa.gov/cira/social-vulnerability-report.

5   S. H. Schneider et al., *Climate Change 2007: Impacts, Adaptation and Vulnerability: Contribution of Working Group II to the Fourth Assessment Report of the Intergovernmental Panel on Climate Change,* ed. M. L. Parry, O. F. Canziani, J. P. Palutikof, P. J. van der Linden, and C. E. Hanson (Cambridge: Cambridge University Press, 2007), 779–810; Henrique Cabral, Vanessa Fonseca, Tânia Sousa, and Miguel Costa Leal, "Synergistic Effects of Climate Change and Marine Pollution: An Overlooked Interaction in Coastal and Estuarine Areas," *International Journal of Environmental Research and Public Health* 16, no. 15 (July 31, 2019): 2737, https://doi.org/10.3390/ijerph16152737.

6   David R. Reidmiller, Christopher W. Avery, David R. Easterling, et al, "Impacts, Risks, and Adaptation in the United States: Fourth National Climate Assessment, Volume II," US Global Change Research Program, 2017.

7   Pamela D. Noyes, Matthew K. McElwee, Hilary D. Miller, et al., "The Toxicology of Climate Change: Environmental Contaminants in a Warming World," *Environment International* 35, no. 6 (2009): 971–86.

8   On the social determinants of health framework, see the US Department of Health and Human Services, "Healthy People 2030: Social Determinants of Health," accessed on August 8, 2021, https://health.gov/healthypeople/objectives-and-data/social-determinants-health. On the concept of cascading disasters, see Gianluca Pescaroli and David Alexander, "A Definition of Cascading Disasters and Cascading Effects: Going beyond the "Toppling Dominos" Metaphor," special issue on the 5th IDRC Davos 2014 *Planet@Risk* 3, no. 1 (March 2015): 56–87, http://citeseerx.ist.psu.edu/viewdoc/download?doi=10.1.1.874.4335&rep=rep1&type=pdf.

9   US Department of Health and Human Services, "Healthy People 2030"; World Health Organization Commission on the Social Determinants of Health, *Closing the Gap in a Generation: Health Equity through Action on the Social Determinants of Health Final Report of the Commission on Social Determinants of Health* (Geneva: CSDH, 2008); Richard C. Palmer, Deborah Ismond, Erik J. Rodriquez, and Jay S. Kaufman, "Social Determinants of Health: Future Directions for Health Disparities Research," *American Journal of Public Health* (1971) 109, no. S1 (2019): S70–S71, https://doi.org/10.2105/AJPH.2019.304964.

10 Pescaroli and Alexander, "Definition of Cascading Disasters"; Susan L. Cutter, "Compound, Cascading, or Complex Disasters: What's in a Name?," *Environment: Science and Policy for Sustainable Development* 60, no. 6 (October 26, 2018): 16–25, https://doi.org/10.1080/00139157.2018.1517518.

11 David Alexander, *Natural Disasters* (London: University College of London Press, 1993).

12 Shlomo Mizrahi, "Cascading Disasters, Information Cascades and Continuous-Time Models of Domino Effects," *International Journal of Disaster Risk Reduction* 49 (October 2020): 101672, https://doi.org/10.1016/j.ijdrr.2020.101672; Anthony Webster, "Multi-Stage Models for the Failure of Complex Systems, Cascading Disasters, and the Onset of Disease," *PLoS One* 14, no. 5 (May 20, 2019): e0216422, https://doi.org/10.1371/journal.pone.0216422; Amir AghaKouchak, Laurie S. Huning, Omid Mazdiyasni, et al., "How Do Natural Hazards Cascade to Cause Disasters?," *Nature* 561, no. 7724 (2018): 458–60, https://doi.org/10.1038/d41586-018-06783-6.

13 Robert D'Ercole and Pascale Metzger, "La vulnérabilité territoriale: Une nouvelle approche des risques en milieu urbain," *Cybergeo: European Journal of Geography* 447 (2009): 87–96.

14 Intergovernmental Panel on Climate Change (IPCC), *Climate Change 2021: The Physical Science Basis: Working Group I Contribution to the Sixth Assessment Report of the Intergovernmental Panel on Climate Change* (Cambridge: Cambridge University Press, 2023), https://doi.org/10.1017/9781009157896.

15 US Environmental Protection Agency, *Climate Change and Social Vulnerability.*

16 On Roosevelt Roads, see and Héctor R. Feliciano Ramos, "Las bases e instalaciones militares de Estados Unidos en Puerto Rico y su impacto en la sociedad puertorriqueña," *Cátedra: Revista de los Investigadores de Humanidades* (2011): 10–11, *Universidad de Panamá*, "Cátedra: Revista Especializada en Estudios Culturales y Humanísticos." https://centroinvestigacionhumanidades.up.ac.pa/sites/fachumanidades/files/revista10_11/Hector%20Feliciano.pdf.

17 McCaffrey, "Environmental Remediation."

18 Arturo Meléndez López, *La batalla de Vieques* (Río Piedras, Puerto Rico: Editorial Edil, 2000); César Ayala, "Recent Works on Vieques, Colonialism, and Fishermen," *Centro Journal* 15, no. 1 (Spring 2003): 212–25; César Ayala and José L. Bolívar, *Battleship Vieques: Puerto Rico from World War II to the Korean War* (Princeton, NJ: Marcus Wiener, 2011); Miguel Ángel Santiago Ríos, *Militarismo y clases sociales en Vieques: 1910–1950* (San Juan, Puerto Rico: Ediciones Huracán, 2007).

19 R. Rabin, M. Sobá, and Carlos Zenón, "Cátedra UNESCO de Educación para la Paz de Vieques a la Universidad: Lecciones y necesidades del pueblo de Vieques en lucha por la paz y el desarrollo," in *Tercera Lección Magistral de la Cátedra UNESCO de Educación para la Paz de la Universidad de Puerto Rico, Río Piedras* (San Juan, Puerto Rico: Universidad de Puerto Rico, 2003); Grupo

de Apoyo Técnico y Profesional para el Desarrollo Sustentable de Vieques, *Guías para el Desarrollo Sustentable de Vieques* (San Juan, Puerto Rico: Publicaciones Gaviota: 2002).

20 Meléndez López, *La batalla de Vieques*; Tere Villegas Pagán, *Taso: Un pedazo de Vieques, Puerto Rico* (Caguas, Puerto Rico: Impresos Taínos, 2001); Carlos ('Taso) Zenón, *Memorias de un pueblo pobre en lucha: Primer libro, 1978–1998* (San Juan, Puerto Rico: Editorial El Antillano, 2018); Ana M. Fabián Maldonado, *Vieques en mi memoria: Testimonios de vida* (San Juan, Puerto Rico: Ediciones Puerto, 2003); John Wargo, *Green Intelligence: Creating Environments That Protect Human Health* (New Haven, CT: Yale University Press, 2009).

21 Meléndez López, *La batalla de Vieques*; Zenón, *Memorias*.

22 Meléndez López, *La batalla de Vieques*; Evelyn Vélez Rodríguez, *Proyecto V-C, Negociaciones Secretas entre Luis Muñoz Marín y la Marina: Plan Drácula* (Rio Piedras, Puertro Rico: Editorial Edil, 2002).

23 Miguel Ángel Santiago Rios, *Militarismo y clases sociales en Vieques: 1910–1950* (San Juan, Puerto Rico: Ediciones Huracán, 2009).

24 Meléndez López, *La batalla de Vieques*.

25 Barreto, *Vieques*.

26 Mario A. Murillo, *Islands of Resistance: Puerto Rico, Vieques, and U.S. Policy* (New York: Seven Stories Press, 2001).

27 Data USA, accessed July 11, 2023, https://datausa.io/profile/geo/vieques-pr/.

28 Hanson W. Baldwin, "Island Bombarded in Navy War Game: Live Shells, Bombs, and Rockets 'Soften' Vieques, Near Puerto Rico, for Landing Today," *New York Times*, March 2, 1949, 11, https://www.nytimes.com/1949/03/02/archives/island-bombarded-in-navy-war-game-live-shells-bombs-and-rockets.html.

29 McCaffrey, "Environmental Remediation."

30 John Wargo, *Green Intelligence: Creating Environments that Protect Human Health* (New Haven, CT: Yale University Press, 2010); Arturo Massol-Deyá and Elba Díaz de Osborne, *Ciencia y Ecología: Vieques en Crisis Ambiental* (Adjuntas, Puerto Rico: Publicaciones Casa Pueblo, 2013).

31 US Government Accountability Office, "Report to the Senate and House Committees on Armed Force, Chemical and Biological Defense, D.O.D. Needs to Continue to Collect and Provide Information on Tests and Potentially Exposed Personnel" (May 2004) https://www.gao.gov/new.items/d04410.pdf. On illegal depleted uranium munitions, see Naval Facilities Engineering System Command Atlantic Public Affairs Office, *Vieques Environmental Restoration Fact Sheet: Depleted Uranium in the Live Impact Area of Vieques* (Vieques, Puerto Rico: NAVFAC, 2018), https://www.navfac.navy.mil/content/dam/navfac/Environmental/PDFs/env_restoration/vieques/Depleted_Uranium_FS_English.pdf.

32 McCaffrey, "Environmental Remediation"; Zenón, *Memorias*; José M. Atiles-Osoria, "Environmental Colonialism, Criminalization and Resistance: Puerto Rican Mobilizations for Environmental Justice in the 21st Century," *Revista crítica de ciencias sociais*, no. 100 (May 1, 2013): 131–52, https://doi.org/10.4000/rccsar.524.

33 US Environmental Protection Agency, "National Priorities List for Uncontrolled Hazardous Waste Sites," *Environmental Protection Agency Federal Register*, 70, (2005).

34 Cruz María Nazario and Erick Suárez-Pérez, "Critical Analysis of the Puerto Rico Department of Health's Report Cancer Incidence in Vieques," March 10, 1998, Department of Biostatistics and Epidemiology, Graduate School of Public Health, Universidad de Puerto Rico, presented to the Comité Pro Rescate de Vieques, CDC, and Puerto Rico Department of Health.

35 Nazario and Suárez-Pérez, "Critical Analysis."

36 Imar Mansilla-Rivera, Cruz María Nazario, Farah A. Ramírez-Marrero, et al., "Assessing Arsenic Exposure from Consumption of Seafood from Vieques-Puerto Rico: A Pilot Biomonitoring Study Using Different Biomarkers," *Archives of Environmental Contamination and Toxicology* 66 (2014): 162–75; Elba Diaz and Arturo Massol-Deyá, "A Trace Element Composition in Forage Samples from a Military Target Range, Three Agricultural Areas, and One Natural Area in Puerto Rico," *Caribbean Journal of Science* 39 (2003): 215–20; Prabhat Kumar Rai, Sang Soo Lee, Ming Zhang, et al., "Heavy Metals in Food Crops: Health Risks, Fate, Mechanisms, and Management," *Environment International* 125 (2019): 365–85, https://doi.org/10.1016/j.envint.2019.01.067.

37 Carmen Ortíz-Roque and Yadiris López-Rivera, "Mercury Contamination in Reproductive-Aged Women in a Caribbean Island: Vieques," *Journal of Epidemiology and Community Health* 58, no. 9 (September 2004): 756–57, https://saludparavieques.org/images/pdfs/SALUD/Ortiz-Roque%20&%20Lopez-Rivera%20-%20Mercury%20contamination%20in%20reproductive%20age%20women%20in%20a%20Caribbean%20island%20-%20Vieques.pdf.

38 Juan Alonso-Echanove and Luis Manuel Santiago, "Executive Summary of the Prevalence of Heavy Metals in Vieques," *An Evaluation of Environmental, Biological, and Health Data from the Island of Vieques, Puerto Rico* (March 19, 2013): A63–A64, Agency for Toxic Substances and Disease Registry, Atlanta GA, https://www.atsdr.cdc.gov/hac/pha/vieques/Vieques_Summary_Final_Report_English_2013.pdf.

39 Díaz and Massol-Deyá, "Trace Element Composition."

40 B. D. Weiss, "Official Navy Letter to Building Occupants," December 7, 2020, Jacksonville, FL, available at https://www.govinfo.gov/content/pkg/CHRG-117hhrg46634/pdf/CHRG-117hhrg46634.pdf.

41 Naval Facilities Engineering Command Atlantic, *Preliminary Assessment Report for Per- and Polyfluoroalkyl Substances Atlantic Fleet Weapons Training Area—Vieques, Former Naval Ammunition Support Detachment and Former Vieques Naval Training Range, Vieques, Puerto Rico*, prepared by CH2M HILL, Virginia Beach, VA, accessed August 8, 2021, accessed from https://www.navfac.navy.mil/niris/ATLANTIC/VIEQUES/N69321_004181.pdf.

42 Imar Mansilla-Rivera et al., "Assessing Arsenic Exposure."

43 US Census, "American Community Survey 2019 5-Year Estimates Data Profile—Vieques," 2021, https://data.census.gov/cedsci/table?g=0400000US72_1600000US7285971&tid=ACSDP5Y2019.DP05&hidePreview=true.

44 Zorrilla Lassus and María del Carmen, "La puesta en valor del paisaje a través de la educación, propuesta para el desarrollo de La Isla de Vieques (Puerto Rico)" [The Enhancement of Landscape through Education Proposal for the Development of the Island of Vieques (Puerto Rico)], *Espacio, Tiempo y Forma*, Revista de la Facultad de Geografía, Serie 6, Geografía, no. 6–7 (2015): 281–314, https://doi.org/10.5944/etfvi.6-7.0.14857.

45 Lassus and del Carmen, "La Puesta"; Luis Galanes Valldejuli and Yolanda Rivera Castillo, "Tourism and Language in Vieques: Ethnography of the Post-Navy Period," *Centro Journal* 32, no. 2 (2020): 157–63.

46 Kevin E. Trenberth, "Changes in Precipitation with Climate Change," *Climate Research* 47, no. 1/2 (2011): 123–38, https://doi.org/10.3354/cr00953.

47 Méndez-Lazaro, P.A., P. Chardón-Maldonado, L. Carrubba, N. Álvarez-Berríos, M. Barreto, J.H. Bowden, W.I. Crespo-Acevedo, E.L. Diaz, L.S. Gardner, G. Gonzalez, G. Guannel, Z. Guido, E.W. Harmsen, A.J. Leinberger, K. McGinley, P.A. Méndez-Lazaro, A.P. Ortiz, R.S. Pulwarty, L.E. Ragster, I.C. Rivera-Collazo, R. Santiago, C. Santos-Burgoa, and I.M. Vila-Biaggi, 2023: Ch. 23. US Caribbean. In: Fifth National Climate Assessment. Crimmins, A.R., C.W. Avery, D.R. Easterling, K.E. Kunkel, B.C. Stewart, and T.K. Maycock, Eds. U.S. Global Change Research Program, Washington, DC, USA. https://doi.org/10.7930/NCA5.2023.CH23.

48 Trenberth, "Changes in Precipitation."

49 US Global Change Research Program, "Fourth National Climate Assessment."

50 Protusha Biswas, Tania Bhattacharya, Abhra Chanda, et al., "Urban Wetlands–CO_2 Sink or Source? A Case Study on the Aquaculture Ponds of East Kolkata Wetlands," *International Journal of Recent Scientific Research* 9 (2018): 24158–65.

51 Karina Acevedo-Whitehouse and Amanda L. J. Duffus, "Effects of Environmental Change on Wildlife Health," *Philosophical Transactions: Biological Sciences* 364, no. 1534 (November 27, 2009): 3429–38; Biswas et al., "Urban Wetlands."

52 Doris Schiedek, Brita Sundelin, James W. Readman, and Robie W. Macdonald, "Interactions between Climate Change and Contaminants," *Marine Pollution Bulletin* 54, no. 12 (2007): 1845–56; Noyes et al., "Toxicology of Climate Change."

53 Jerker Jarsjö, Yvonne Andersson-Sköld, Mats Fröberg, et al., "Projecting Impacts of Climate Change on Metal Mobilization at Contaminated Sites: Controls by the Groundwater Level," *Science of the Total Environment* 712 (April 2020): 135560–135560, https://doi.org/10.1016/j.scitotenv.2019.135560.

54 Li-Qiang Ge, Long Cang, Hui Liu, and Dong-Mei Zhou, "Effects of Warming on Uptake and Translocation of Cadmium (Cd) and Copper (Cu) in a Contaminated Soil-Rice System under Free Air Temperature Increase (FATI)," *Chemospher* 155 (2016): 1–8, https://doi.org/10.1016/j.chemosphere.2016.04.032.

55 Kumar Rai et al., "Heavy Metals."

56 Majeti Narasimha Vara Prasad and Marcin Pietrzykowski, *Climate Change and Soil Interactions* (Amsterdam: Elsevier, 2013).

57 Torres Gotay Benjamín, "Muerte de bebé vuelve a acentuar la falta de un hospital en Vieques," *El Nuevo Día*, May 27, 2021, https://www.elnuevodia.com/noticias/locales/notas/muerte-de-bebe-vuelve-a-acentuar-la-falta-de-un-hospital-en-vieques/.

58 United Nations General Assembly *Special Committee on Decolonization Approves Text Calling upon the United States to Promote Puerto Rico's Self-Determination, Eventual Independence,* resumed session 5th meeting (AM), Special Committee on Decolonization, GA/COL/3346, June 18, 2021, https://www.un.org/press/en/2021/gacol3346.doc.htm.

59 Valeria Pelet, "Puerto Rico's Invisible Health Crisis: The Island of Vieques Has Some of the Highest Sickness Rates in the Caribbean. Is the U.S. Navy Responsible?" *Atlantic*, September 3, 2016, https://www.theatlantic.com/politics/archive/2016/09/vieques-invisible-health-crisis/498428/.

60 Federal Emergency Management Agency, *FEMA Obligates over $39.5 Million for the Vieques Community Health Center,* FEMA DR-4339-PR NR 413, January 21, 2020, https://www.fema.gov/press-release/20231115/reconstruction-begins-vieques-health-facility

61 Daniel G. Mareck, "Federal and State Initiatives to Recruit Physicians to Rural Areas," *AMA Journal of Ethics* 13, no. 5 (2011): 304–9, https://doi.org/10.1001/virtualmentor.2011.13.5.pfor1-1105.

62 Juliana Broad, "Fighting Gentrification and Displacement: Emerging Best Practices," The Next System Project, February 19, 2020, https://thenextsystem.org/fighting-gentrification-best-practices.

CHAPTER 14

Climate Change and Long-Term Impact on Women's Reproductive Health

An Unfolding and Untold Social Crisis in the South Coast Region of Bangladesh

REAZUL AHSAN, PROKRITI NOKREK, AND AFRIDA ASAD

Individuals in the Asia-Pacific region are four to twenty-five times more susceptible to natural disasters than their counterparts in Africa or North America/Europe. Asia experienced 75 percent of the over two million deaths caused by natural disasters between 1974 and 2003, contributing to 32 percent of global extreme climatic events and a significant number of deaths in the 1990s. Women in the Asia-Pacific region are more vulnerable to natural disasters than men because of social and cultural setup. Global initiatives such as the Sustainable Development Goals and the Paris Agreement acknowledge this issue in light of the connection between climate change and women's health. The socioeconomic conditions of women in South Asia are considerably affected by climate change, particularly in regions such as southwest Bangladesh. Bangladesh, situated in the vulnerable Asia-Pacific region, faces a significant number of cyclone-related fatalities and challenges resulting from the sea level rise caused by climate change. Women of the marginal community in the southwest coastal areas in Bangladesh are the most exposed to climate change and its long-term effects.

In their positions as family caregivers and domestic managers, women are exposed to perilous circumstances, such as waterborne diseases and malnutrition. Salinity intrusion disrupts the fresh water supply, requiring NGOs to provide alternatives. Due to climatic effects, pregnant women face mobility restrictions and increased hazards of unhygienic reproductive health conditions. Women in the southwest of Bangladesh are exposed to elevated levels of salinity, which negatively affects their health, as a result of climate change and salt intrusion, according to research. Climate change disrupts women's livelihoods by causing crop and livestock losses, reducing the availability of fresh water, and necessitating lengthier journeys to obtain potable water. Migration of males exacerbates the vulnerability of women by imposing additional responsibilities and financial obstacles.

Despite the large number of studies investigating the relationship between climate change and women's health, relatively few examine the effect of water salinity on reproductive health. This chapter investigates the reproductive health of Bangladeshi women in the southwest littoral region, which has struggled for thirty years with soil and water salinity due to sea level rise. Profit-driven shrimp aquaculture exacerbates the salinity problems in this historically agriculturally productive region. The purpose of this chapter is to examine the interconnected issues of climate change, soil salinity, displacement, means of subsistence, and social and environmental justice from the perspectives of health specialists, climatologists, and social scientists.

INTRODUCTION

Climate change is a significant threat to global health, and exacerbated by the ongoing issue of gender inequality. Globally, approximately 1.3 billion people live below the poverty line in low- and middle-income countries, with 70 percent of those being female. Coastal communities, particularly in developing countries, are particularly vulnerable to climate change, which has immediate and long-term consequences for community health.[1] Following natural disasters and extreme weather events caused by climate change, women frequently face unequal access to economic and technical resources.[2]

According to the 2010 Asia Pacific Disaster Report, people in the Asia-Pacific region are four to twenty-five times more vulnerable to natural disasters than people in Africa or North America/Europe. Asia was responsible for 75 percent of the over two million deaths caused by 6,367 natural disasters between 1974 and 2003. In the 1990s, the Asia-Pacific region accounted for 32 percent of global extreme climatic events, 84 percent deaths, and 88 percent of people affected.[3]

The role of women in combating climate change, in general, has been prioritized in many recent global agendas, including the Sustainable

Development Goals, the Paris Agreement on Climate Change, and the United Nations Framework Convention on Climate Change, all of which recognize the link between climate change and women's health.[4] Bangladesh is one of the countries in the Asia-Pacific region that is highly exposed to extreme climate events and sea level rise. Bangladesh is hit by only 1 percent of the world's cyclones, however, approximately half of all cyclone-related deaths occur in Bangladesh, where over 70 percent of the country and eighty million people are flood-prone.[5]

The southwest coastal region of Bangladesh is the most disaster-prone and is extremely vulnerable to climate change–induced sea level rise. For the last thirty years, the region's sea level has been rising at a rate of three to four millimeters per year.[6] Such sea level rise poses a threat to the coastal environment, ecology, and livelihood and health in the short and long term. Extreme weather events put women and children at risk the most. Climate change has significant socioeconomic impacts on women. It disproportionately affects them in a variety of domains, such as agriculture, livelihood, food security, physical and mental health, water, and sanitation in South Asia.[7] In the southwest region of Bangladesh, women are also bread earners for their family. In addition to their household duties, they also work in saltwater fishing farms locally known as GHER.[8] Due to these responsibilities, women cannot avoid being exposed to hazardous living conditions such as waterborne diseases and malnutrition.[9] The fresh water supply in coastal areas is also affected by water salinity. Rainwater is the only available fresh water source, and the local nongovernmental organizations provide fresh water to the villagers once a week. Rainwater harvesting is a costly endeavor, as it requires a water tank, water treatment, purification, preservation, and a large investment of time and money.[10] A few families can afford rainwater tanks; others rely on other sources. Once per day, local NGOs deliver portable drinking water containing fresh water, but the villagers have to pay for the fresh water provided by the local NGOs. Others who cannot afford to pay for fresh water from NGOs rely on deep-tube wells for their daily drinking water requirements. However all the households rely on the nearby canals and ponds for their daily domestic water needs (cooking and cleaning). However, salinity intrusion has made these water sources saline. Consequently, the women are constantly exposed to saltwater at work and at home. On average, each individual is exposed to saltwater for twelve to fifteen hours per day.

Pregnant women have limited mobility and are often trapped in slick conditions, forcing them to stay indoors and fall prey to unsanitary reproductive health conditions. Women in southwest coastal districts such as Shatkhira are routinely afflicted by more frequent cyclones, salinity intrusions, increased tidal surges, coastal flooding, extreme temperatures, droughts, coastal erosion, and waterlogging. Indeed, climate change and extreme events have had a long-term negative impact on the socioeconomic status of women in the southwest coastal communities.[11]

Two distinct research organizations conduct different soil and water salinity studies in Bangladesh's southwest region to assess the health consequences of climate change and salinity. Bangladesh's Center for Environmental and Geographic Information Services, a government agency, conducted one independent study in 2003 and another in 2011. According to the findings, women in the southwest region consumed and were exposed to eighteen to twenty times the amount of saline water recommended by the World Health Organization. Furthermore, between 2012 and 2016, the Bangladesh-based research organization International Center for Diarrheal Disease conducted a field-based study on 12,867 pregnant women who had been exposed to soil and water salinity due to climate change. The research found that women who lived in coastal areas (within twenty kilometers of the sea) were more likely to miscarry than those who lived in the highlands, as they were more exposed to water salinity.[12]

Climatic hazards affect women's livelihoods as well, due to the destruction of crops and livestock, which are the primary means of female livelihood at the household level in the coastal area. Salinity intrusion reduces freshwater supply options, forcing women to travel long distances to collect drinking water. Extreme heat, a lack of rain, and drought-like conditions annihilate homestead gardens and any vegetable production, so coastal erosion and flooding force coastal residents to relocate. When male family members migrate, the entire family becomes socially and economically vulnerable, particularly the women, who are forced to take on additional responsibilities and find additional funds for the household.[13]

Work in saline water affects coastal women's reproductive health and family structure, as their reproductive organs are exposed to saltwater for upward of eight to ten hours a day. Over time, as soil salinity increases due to sea level rise, saline water shrimp farming has become the region's only source of livelihood and revenue.

After two concurrent cyclones hit the southwest coastal area it become a focal point for climate and social researchers. Cyclones Sidr in 2007 and Aila in 2009 drew researchers from around the world, including public health scientists, climatologists, and social scientists, to study climate change, soil salinity, displacement, livelihoods, and social and environmental justice issues.

LITERATURE

According to reports from the Soil Resources Development Institute, the total amount of salinity-affected land in Bangladesh increased from 83.3 million hectares in 1973 to 105.6 million hectares in 2009.[14] The situation is still worsening due to the reduced freshwater flow from the upstream region. Salinity in the country has increased by roughly 26 percent in the last thirty-five years, and is spreading to noncoastal areas. Local agriculture and livelihood practices have been drastically transformed, forcing marginalized individuals to pursue a narrower range of occupations. Shrimp and crab aquaculture, for example, use saltwater to trap and store saline water in the ground for longer periods, and locals blame these farming practices for the soil salinity and the destruction of other crop productions. The livelihoods of many people are associated with salt shrimp farming called chingri gher in southwest Bangladesh. Moreover, due to the extreme poverty, women of all ages are largely associated with day labor work in those ghers.

Between 2012 and 2016, studies were undertaken by the World Bank, the Institute of Water Modeling, and World Fish, Bangladesh, to assess the consequences of increasing salt in river waters in coastal Bangladesh, including the Sundarbans. Researchers from the International Centre for Diarrhoeal Disease Research Bangladesh have discovered an unusually high rate of miscarriage in Chakaria, a small village near Cox's Bazaar on Bangladesh's east coast. After further investigation, scientists concluded that climate change could be to blame.

Both surface water and groundwater in Bangladesh's coastal areas are contaminated by various levels of salinity, which can harm the health of the thirty-five million people who live there through direct or indirect use of water resources.[15] In low-lying areas, the population is sensitive to climate change and sea level rise, and they frequently drink untreated water and eat unprocessed food. They are also likely to be exposed to high sodium intake. According to the World Health Organization, a considerable portion of the population in the coastal region of Bangladesh is exposed to more sodium than is recommended for daily consumption (more than 5 gallons per day).[16] The consumption of high-salt-containing foods and potable water is positively associated with hypertension, kidney disease, and skin disease in the studied population over twenty years of age in the rural coastal area of Bangladesh.[17] Cardiovascular disease, abdominal pain, gastric problems, ulcers, dysentery, skin disease, and typhoid are also reported to be largely related to high salt intake.[18]

RESEARCH METHOD AND STUDY AREA

Each member of the research team (Reazul Ahsan, Prokriti Nokrek, and Afrida Asad) has previously worked on Bangladesh's southwest coast. Climate change and the plight of low-lying coastal villages has occupied a significant portion of our professional work lives. Women in the south coast region are particularly vulnerable when it comes to natural disasters and postdisaster consequences. Furthermore, societal and religious structures prevent them from speaking out. Women in the southern coastal region have no choice but to work on shrimp farms, where they are exposed to hazardous conditions with no protection. After witnessing and working in this region's coastal villages, we felt it important to study this particular area, where respondents have little or no say, in order to understand what types of interventions were possible to address the reproductive health of women and how it is affected by climate change. This particular section is missing in mainstream climate change and public health research.

A qualitative research approach has been adopted to carry out this study. A total of two focus group discussions and two in-depth interviews were conducted with women laborers who engaged in livelihood activities in shrimp ghers. The study participants were selected following purposive and snowball sampling techniques. To recruit participants from the communities, the researchers supported local fixers to find and recruit study participants.

The team began with a brief explanation of the purpose of the study. We got verbal consent to conduct focus group discussion and to use participants' photographs. We also took into consideration participants' comfort in talking with the facilitator and their level of distraction. Both the facilitator and the notetaker were female.

Furthermore, after visiting the fields, we checked and extended our data collection by talking to the interviewees over phone calls. To protect the confidentiality and anonymity of the participants, we also used pseudonyms.

Shyamnagar is between 21.360 and 22.240 degrees north latitude and 89.000 and 89.190 degrees east longitude. It is a remote subdistrict within the Satkhira District situated fifty kilometers south of the city of Satkhira. The total population of the subdistrict is 313,781.[19] Fishing, agriculture, shrimp farming, salt farming, and tourism are the main economic drivers in this area. However, farmers have abandoned agricultural lands because of increased soil and water salinity. Two sample villages (Bon Bibitola and Abad Chandipur Pankhali) in Burigoalini union, Shyamnagar subdistrict, Sathkhira district have been selected considering the highest number of shrimp ghers and highest number of the women in the labor force.

On the one hand, a random snowball sampling procedure was used to select participants for focus group discussion from a pool of potential candidates. Focus group discussion is a common qualitative method for gaining a comprehensive understanding of social issues. The method seeks to collect information from a purposefully selected group of individuals instead of a statistically representative sample of the larger population.[20] Therefore, researchs consider the focus group discussion as a best data collection approach because the samples were purposefully selected. On the other hand, the researchers wanted to find the area where shrimp farm/gher was the only alternative source of income. The Shyamnagar subdistrict was selected because there the only livelihood for the local people (men and women) is on the gher. They have no other alternative livelihoods because of the water and soil salinity, a long-term effect of climate change. The gher is the primary source of income for the people of Shyamnagar, and researchers purposefully choose the respondents from the villages in Shyamnagar.

Respondents' Profiles

Of the thirty-one focus group discussion respondents, 100 percent were female. The median age of respondents was forty-one (minimum twenty-

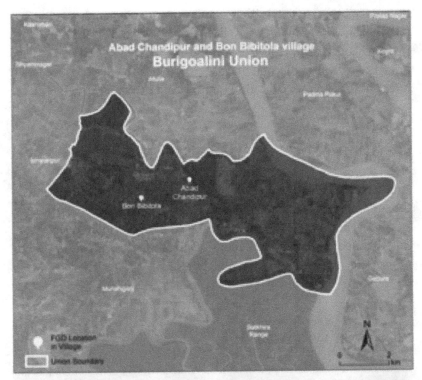

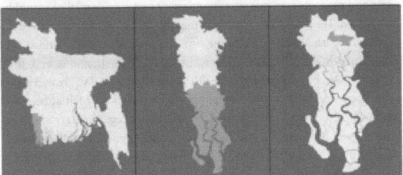

Satkhira District, Bangledesh   Shyamnager Upzila, Satkhira   Buri Goatini Union, Shyamnager

FIGURE 14.1. Study area: two selected villages in the southwest coastal region of Bangladesh. The researchers (authors of this chapter) prepared the maps in this figure using GIS (Geographical Information System) and Google Earth. Google Earth is a free access source.

five and maximum sixty). The sample was predominantly of a lower-income class. The team interviewed participants from two religious groups (Muslim and Hindu). Most did not go to school; few had completed primary-level education; and only three participants had completed their junior secondary-level education (grades eight and nine).

Limitations of the Study

Due to the COVID-19 pandemic, the team conducted only two focus group discussions and two in-depth interviews. It would have been good if the team could have conducted some interviews with community members to properly understand the context and circumstances of these two areas. Despite these challenges, the team has been able to draw some conclusions about the current situations of climate change and its impact on womens' reproductive health in this study area.

FINDINGS AND DISCUSSION

Initially, we established rapport with the focus group discussion participants and made it clear that their anonymity and confidentiality would be maintained. The team also mentioned that this discussion might not directly benefit the participants at that time but that their experiences and stories would help to identify the current climate change situations and their impact on women's reproductive health in the study area. We created a welcoming environment for the women participants, so they felt at ease and were willing to share their experiences and stories regarding genital infections. A local NGO worker was hired to establish rapport with the community and also facilitate the venue. Also, he was responsible for ensuring that no men were present during the focus group discussion and in-depth interviews. The participants also inquired about the most effective treatment for their vaginal infections. This was their first time attending a discussion regarding reproductive health. During the focus group discussion, the participants addressed the fact that they had been interviewed at various times by numerous other organizations and individuals performing research on water salinity and other related concerns. But no one had ever asked them about their reproductive health or their work schedule in saline water. They had been suffering for a long time, but due to societal and cultural norms, they had never shared their sorrow with others.

During an in-depth interview, the team primarily focused on sensitive issues, such as unhappiness in the relationship between the husband and wife due to a vaginal infection. As stated previously, we did select participants from the focus group discussion for in-depth interviews. Therefore, participants in in-depth interviews felt more at ease sharing details of their stories and sufferings.

Work Pattern

As a ground laborer in shrimp ghers, all focus group discussion participants stated that they spend seven to eight hours removing weeds and applying inorganic fertilizers such as TSP, UREA, and lime to maintain water quality. They had no other options but to take these positions within the research area. In the absence of any other options or a relevant educational background, they were left with no other options. As a result, they had no choice but to work in the gher as laborers. Soil salinity due to climate change supports only saline water shrimp farming, which is the only source of income for both men and women in that area.

Reproductive Health Issues

All the participants in two focus group discussions reported that they were suffering urinary and vaginal infections, for example, a burning sensation when urinating, pimples, or itching in the genital area. They have to spend long hours in saline water. Findings revealed that participants who have worked longer in shrimp ghers (twenty-five to thirty years) have been suffering from urinary and vaginal infections for a long time. Some of them also mentioned their uterus problems and a regular white discharge (leukorrhea). Another study by carried out in two different villages in Bangladesh similarly found that women who used water with excessive saline content were prone to uterine inflammation and uterine ulcer.[21]

Women participants also mentioned that they used inorganic fertilizers such as TSP, UREA, and lime to maintain water quality. Most participants noted that they knew about the health hazards from these toxic substances/waste and water pollution, but they did not maintain any health safety measures. They do not have any dress code, and they do not use gloves.

The majority of women workers wear a maxi dress. A maxi dress is a kind of nightgown from head to toe without any underwear. Some women dress in sarees. Women in rural Bangladeshi villages generally wear a saree, which is a long piece of cloth wrapped around the body in a unique style. The saree is typically worn with a blouse and a petticoat (a woman's light, loose undergarment worn beneath a skirt or dress). Because they do not wear underwear, all of these pollutants and harmful substances have a direct negative impact on their genital organs and reproductive health.

The focus group discussion participants from two villages noted that they had to take extra precautions during their menstrual period. They needed to work underwater since they had to remove weeds, snails, and pebbles. Therefore, they tried to put the stress on their waists and lean toward the water during this period. It is not easy to stand up straight after long working hours. This stress causes lower back pain. One respondent shared that it had caused a gap in the bones of her spine. Despite this physical stress, the women have to work to make a living. Only a few had taken time off from their work during their menstrual period.

Women working on shrimp farms in coastal Bangladesh often face poor working conditions, which have been highlighted by various studies and reports. Here are some key points regarding their working conditions:

Long working hours: Shrimp processing women endure eight- to twelve-hour shifts, standing on their feet for long periods of time, and their bodies are exposed to toxic water.

Working circumstances are harsh: These women must labor in extremely low temperatures, which can be physically difficult and uncomfortable.

Lack of sufficient safety precautions: The working environment on shrimp farms frequently lacks proper safety measures, putting the workers' health and well-being in danger. Working in deep water also exposes their reproductive organs to saline water.

Inadequate perks and protections: These women's bad working conditions are exacerbated by a lack of social protections and benefits.

Long-Term Health Issues

The team also asked about the trends of puberty. Focus group discussion participants who are now fifty years old mentioned that their puberty had begun at the age of fourteen or fifteen, but nowadays puberty begins early, at the age of ten. The majority of participants ages twenty-five to forty-five shared that they took birth control injections rather than the pill, and for them, birth control methods led to early menopause. The elderly participants argued it didn't used to be like that. This had started in the last fifteen to twenty years, since the women had been working in gher and were experiencing these challenges in their children and grandchildren. Such concerns raise a real question in this research: Is climate change limited to greenhouse gas emissions and temperature rise, or does it also have long-term effects on the health and social-cultural structure of vulnerable communities? In this study, our emphasis is on women's reproductive health in southwest coastal communities, aiming to highlight that the repercussions of climate change extend beyond immediate impacts. It is crucial to recognize climate change's broader, long-term effects and assess how extensively they could influence various aspects, including reproductive health.

Women's Health-Seeking Behavior

Health-seeking behavior is one of the important determinants of women's health. In this patriarchal society, women are disadvantaged by discrimination rooted in sociocultural factors. We asked respondents about their health-seeking behavior when they suffer from skin and genital diseases.

The focus group discussion data shows that the respondents initially treat their genital infections at home using local home remedies, not medicine prescribed by doctors. They clean their vaginas with warm water and apply antiseptic cream and antibacterial ointment. Some respondents shared that they also put burnt tobacco in their vaginas (as prescribed by the local elders). They also boil water that they use for cleaning shrimp and put it in their vaginas. They believe these traditional practices help them to cure their vaginal infections. Some respondents noted that they had gone to a traditional healer to get an amulet (tabij). When these traditional home practices do not work, they seek homeopathic medicine. Rina Begum is one of those who sought homeopathic treatment:

Rina Begum is working in shrimp gher for about thirty to thirty-five years. She has to choose this profession as her husband cannot work anymore for his sickness. Furthermore, Rina does not have her own homestead land, and lives on other people's land where she has to pay rent. She is a day laborer and there is only verbal agreement about her daily wage. Sometimes the owner of the gher pays weekly or sometimes even a bit later. At that time, she has to borrow from others. Daily wage 150 to 160 [Bangladeshi Taka] equals one and a half to two US dollars per day. She works seven hours a day [from 6 a.m. to 1.00 p.m.], and she does not have any written contract for this work. On average, she earns 2,250 to 3,000 BDT equals thirty to thirty-five US dollars per month. There is no dress code; she wears maxi [locally called maxi dresses] with a petticoat. She had suffered from genital rashes and fungal infections. One of her neighbors suggested putting tobacco on it, and that was the most horrible experience in her life. In addition to this, she also put earth vermilion [mete Sindur], but those harmful practices did not cure her health problems. Therefore, she has to go to the homeopath doctor. She feels better after taking homeopathic medicine, but it triggers itchy skin and nasty rash when she works in saline water.

Most participants admitted that they do not share such health problems (they consider the vagina a secret organ and do not feel comfortable talking to others about it). Very few went to see the allopathic doctor in the town. They also added that the women participants did not want to visit doctors due to the distance between the study area and the community clinic/health care provider.

Focus group discussion participants shared that they have limited access to health services, resources, or information related to sexual and reproductive health and rights. They also stated that this was first time researchers had asked about this problem. Everyone asked about saline water drinking habits and skin problems. They are suffering a great deal but cannot open up because it is socially not very acceptable:

Fahima [forty years old] has been working in shrimp gher since 2000 [about twenty years]. She stopped working for a couple of years, but since 2010 she is continuing her work regularly. She works seven hours a day [from 6 a.m. to 1 p.m.]. She has been suffering from urinary tract infections for the last twenty years. To her, they drink less water during their work and there is no restroom in their working areas. So they have to wait until they get into their home. She also shared that it is very painful to have an active UTI and have intercourse with her husband. Saying no to her husband also caused family and social problems.

Such changes in women's livelihood due to saline water intrusion affects their health, income, family, and conjugal life and they have no place to discuss or share this.

Wage Discrimination and Social Vulnerability

During the focus group discussion participants shared that their wages were lower than those of men. The owners paid women 160 to 180 BDT and men 300 to 350 BDT. Despite their unequal wages/pay, they do not protest the unfair pay practice. Due to their minimal skills and resources, they can not change their occupation.

Most focus group discussion participants shared that they have been suffering from leucorrhoea due to exposure to a high level of saline water. Few participants discussed their unhappy relationship with their husband due to the leucorrhoea. One participant shared that her husband had left her because of her uterus. Furthermore, N. Kulkarni and P. M. Durge also confirmed that leucorrhoea is regarded with serious concern by both men and women. It also leads to depression, verbal abuse, sexual violence, concern about a husband's extramarital affairs, low social integration, and autonomy.[22]

Government doctors are available in the local union and subdistrict (Upazila); however, most local doctors are male. There are a few female doctors, but the local women do not feel comfortable sharing sensitive issues with outsiders. Moreover, social and religious norms discourage women from seeking care from male physicians, particularly regarding reproductive health matters. The respondents shared that all government doctors are male and live far from the village. They are not at ease sharing their private illnesses or showing the doctors their private parts. Additionally, they believe it is socially inappropriate to be treated by a male physician, particularly when it concerns a female issue.

Review of the literature and other supporting climate change and health research indicates that Bangladesh's southwest region is extremely vulnerable to climate change and sea level rise. Women in the southwest region are particularly vulnerable to climate change due to their exposure to salinity. This study discovered the same facts through field research. Researchers such as Reaz Haider and UN Women address climate change, pregnancies, and miscarriage. Additionally, this study discovered that women's reproductive health is jeopardized by water salinity, which has a direct result of climate change.[23] Additionally, Aneire E. Khan and colleagues and Mashura Shammi and colleagues concluded that increased sodium levels in local water as a result of sea level rise pose an immediate health risk to the local community.[24] Climate change has also affected local livelihoods and social structures, as Umme Kulsum and colleagues and Reazul Ahsan noted in their research.[25] This study concurs and discovers similar facts via focus group discussion. However, an intriguing finding from this study is that prolonged exposure to saline water directly affects the reproductive health of indigenous women. Climate change has altered the local people's livelihood patterns; they are now forced to work in ghers and spend extended periods of time in saline water. There is a dearth of research on this subject.

In addition, the social and religious beliefs of these coastal communities prevent them from discussing such sensitive matters with others. They consider it a social disgrace to discuss reproductive health issues with others, especially male family members or male doctors. As a result, they do not discuss such health concerns with physicians or others and seek their own solutions, but women continue to endure prolonged suffering. This reproductive health issue is inextricably linked to climate change and has historically been overlooked. This study aims to shed light on this issue and establish that climate change has a long-term impact on women's reproductive health, which will contribute to disruptions in Bangladesh's social and economic structure.

CONCLUSION

Drinking and everyday use water in the southwest region of Bangladesh is frequently salted and contains high salt concentrations. The effects of climate change on the coastal environment and saline intrusion go far beyond the immediate impact on the physical environment; they have long-term effects on human health and the local economy. For the most part, women are exposed to water salinity in their livelihoods and everyday lives.

The health impacts of high salinity–contaminated potable water on local people's health, maternal health, hypertension, and infant mortality in Bangladesh's coastal population have been studied and analyzed by researchers and scientists.[26]

Unfortunately, the impact of high salinity–contaminated water on coastal women's reproductive health and reproductive organs has never or seldom been discussed. The reasons could be that this issue is not widely discussed by the community or that it does not garner much interest from researchers. The majority of the research focused on the environmental effects of climate change and the difficulties of agricultural production and water salinity. Also, the sociocultural norms of the local community make them feel uncomfortable discussing female diseases with others. Furthermore, government health facilities are mostly serviced by male doctors, and those women do not feel comfortable discussing their genital or reproductive health issues with them. There are female doctors, yet the social structure and norms do not allow them to open up and discuss their problems with doctors. Those local women also do not know much about sexual diseases or the health of their reproductive organs. Therefore they can not explain their problems or seek help. The most important thing to them is that they are afraid of losing their jobs or income. They feel that if they discuss these types of health issues, they might lose their job and their family will suffer.

This study demonstrates that women reproductive health consequences brought about and accelerated by climate change are often overlooked or buried beneath cultural norms, ignorance, and a lack of research interest. This study tries to determine the effect of saline water exposure on women's reproductive organs and health. Additionally, the study provides a forum for women to discuss how their livelihood affects their reproductive health, and how these are connected with climate change. The Focus group discussion reveals that women are forced to work long hours in the gher and are exposed to chemically polluted saline water. Some researchers may argue that this is unrelated to climate change. However, soil salinity is a long-term and indirect effect of climate change, and saline intrusion that affects marginal coastal communities' livelihood patterns.

The public health sector in Bangladesh and international organizationshave highlighted water salinity and associated health issues in Bangladesh. Several published research and medical findings have proven that water salinity causes a number of health issues in the coastal areas and that women and children are the most vulnerable to health problems resulting from water salinity. Several research studies have been published on climate change sea level rise, salinity, and women's health. This chapter refers to some of those studies. However, none of this research addresses salinity and women's reproductive heath or reproductive organs.

The Bangladesh government has a policy on climate change strategy and action plan and adaption, however, these policy documents do not cover climate change and health issues, particularly reproductive health. The two major policy documents prepared by the Bangladesh government are the Bangladesh Climate Change Strategy and Action Plan and the National Adaptation Program of Action.[27] The Bangladesh Climate Change Strategy and Action Plan addresses six pillars of the climate change action plan: (1) food security, social protection, and health; (2) comprehensive disaster management; (3) infrastructure; (4) research and knowledge management; (5) mitigation and low carbon development; and (6) capacity building and institutional strengthening. The National Adaptation Program of Action is the development of a countrywide program that encompasses the immediate and urgent adaptation of activities that address the current and anticipated adverse effects of climate change, including extreme events.

None of the policy papers address women's health and salinity. Local nongovernmental organizations and international donor agencies are trying to provide safe drinking water but not a safe working environment in the saline water.

More in-depth study is required to address this little-known problem in women's health. Climate scientists should collaborate with health care providers, local government officials, development organizations, and members of the communities they serve on this research, which should be conducted in a multidisciplinary manner. The local population must also be educated on maintaining good health in their reproductive and sexual organs. There must be proper job security, paid sick leave, and a water-resistant uniform for the women who work in the gher. The study's findings show a significant discrepancy between what people know and what they believe about women's reproductive health how the reproductive health is affected due to climate change. Climate change has a long-term impact on the environment, public health, social justice, and the livelihoods of the most vulnerable communities in Bangladesh's coastal areas, and needs to be addressed nationally and internationally.

NOTES

This article is dedicated to the marginalized coastal women's communities whose tales of climate hardship have never been told.

The authors of this article wish to express their gratitude to the University of Utah Asia campus for seed funding this research and encouraging them to raise the voice of Bangladesh's marginal coastal women, who are disproportionately affected by climate change and sea level rise and have very little to say in their defense.

1 Zalak Desai and Ying Zhang, "Climate Change and Women's Health: A Scoping Review," *GeoHealth* 5, no. 9 (August 18, 2021): e2021GH000386, https://doi.org/10.1029/2021GH000386.

2 Ana Langer, Afaf Meleis, Felicia M. Knaul, "Women and Health: The Key for Sustainable Development," *Lancet* 386, no. 9999 (2015): 1165–1210, https://doi.org/10.1016/S0140-6736(15)60497-4.

3 Kazuyuki Uji, *The Health Impacts of Climate Change in Asia-Pacific*, UNDP (New York: UNDP, 2012/16), https://cdn.who.int/media/docs/default-source/climate-change/the-health-impacts-of-climate-change-in-asia-pacific-c65a37137-3449-4936-a711-7fcaa7c1b4ae.pdf?sfvrsn=8358390a_1&download=true.

4 Cecilia Sorensen, Virginia Murray, Jay Lemery, and John Balbus, "Climate Change and Women's Health: Impacts and Policy Directions," *PLoS Medicine* 15, no. 7 (2018): e1002603; Langer et al., "Women and Health."

5 Uji, *Health Impacts*.

6 David Freestone and Clive Schofield, "Sea Level Rise and Archipelagic States: A Preliminary Risk Assessment," *Ocean Yearbook Online* 35, no. 1 (2021): 340–87, https://doi.org/https://doi.org/10.1163/22116001_03501011.

7 Sangram Kishor Patel, Gopal Agrawal, Bincy Mathew, et al., "Climate Change and Women in South Asia: A Review and Future Policy Implications," *World Journal of Science, Technology and Sustainable Development* 17, no. 2 (2020): 145–66, https://doi.org/10.1108/WJSTSD-10-2018-0059.

8 Gher farming is a traditional agriculture system in Bangladesh in which a pond is dug into a rice field to use for fish farming. WorldFish, Bangladesh, https://www.worldfishcenter.org/content/gher-farming-bangladesh.

9 Reazul Ahsan, "Climate-Induced Migration: Impacts on Social Structures and Justice in Bangladesh," *South Asia Research* 39, no.2 (2019): 184–201, https://doi.org/10.1177/0262728019842968.

10 Gulsan Ara Parvin, Nina Takashino, Md Shahidul Islam, et al., "Disaster-Induced Damage to Primary Schools and Subsequent Knowledge Gain: Case Study of the Cyclone Aila Affected Community in Bangladesh," *International Journal of Disaster Risk Reduction* 72 (2022): 102838, https://doi.org/https://doi.org/10.1016/j.ijdrr.2022.102838.

11 UN Women: Asia and the Pacific, *Women's Livelihoods in the Coastal Belt of Bangladesh* (Dhaka: UN Women Bangladesh, 2022), https://asiapacific.unwomen.org

/en/digital-library/publications/2015/1/women-s-livelihoods-in-the-coastal
-belt-of-bangladesh.

12 Reaz Haider, "Climate Change–Induced Salinity Affecting Soil across Coastal
Bangladesh," *Reliefweb*, January 15, 2019, https://reliefweb.int/report/bangla
desh/climate-change-induced-salinity-affecting-soil-across-coastal-bangladesh.

13 UN Women, *Women's Livelihoods*.

14 Minsitry of Agriculture, Soil Resource Development Institute, ed., *Saline Soils
of Bangladesh*, (Dhaka: Minsity of Agriculture, 2010).

15 Aneire E. Khan, Wei W. Xun, Habibul Ahsan, and Paolo Vineis, "Climate
Change, Sea-Level Rise, and Health Impacts in Bangladesh," *Environment: Science and Policy for Sustainable Development* 53, no. 5 (2011): 18–33, https://doi.
org/10.1080/00139157.2011.604008.

16 Khan et al., "Climate Change"; Aneire E. Khan, Andrew Ireson, Sari Kovats, et
al., "Drinking Water Salinity and Maternal Health in Coastal Bangladesh: Implications of Climate Change," *Environ Health Prespect* 119, no. 9 (2011): 1328–32.

17 Khan et al., "Climate Change"; Radwanur Rahman Talukder, Shannon Ruth-
erford, and Cordia Chu, "Salinization of Drinking Water in the Context of
Climate Change and Sea Level Rise: A Public Health Priority for Coastal Bangladesh," *International Journal of Climate Change: Impacts and Responses* 8, no.
1 (2015): 21–32; Mashura Shammi, Md M. Rahman, Serene E. Bondad, and Md
Bodrud-Doza, "Impacts of Salinity Intrusion in Community Health: A Review
of Experiences on Drinking Water Sodium from Coastal Areas of Bangladesh,"
Healthcare 7, no. 1 (2019): https://doi.org/10.3390/healthcare7010050.

18 Rishika Chakraborty, Khalid M. Khan, Daniel T. Dibaba, et al., "Health Implications of Drinking Water Salinity in Coastal Areas of Bangladesh," *International Journal of Environmental Research and Public Health* 16, no. 19 (2019):
3746, https://doi.org/10.3390/ijerph16193746.

19 Bangladesh Bureau of Statistics, ed., Bangladesh Population Census (Dhaka:
Ministry of Planning, 2012).

20 Tobias O. Nyumba, Kerrie Wilson, Christina J. Derrick, and Nibedita Mukher-
jee, "The Use of Focus Group Discussion Methodology: Insights from Two
Decades of Application in Conservation," *Methods in Ecology and Evoluation* 9,
no. 1 (2018): 20–32.

21 Faruk Zaman, "Impact of Salinity on Poor Coastal People's Health: Evidence
from two Coastal Villages in Bangladesh," *Journal of the Asiatic Society of Bangladesh* 62, no. 1 (2017): 1–14.

22 R. N. Kulkarni and P. M. Durge, "A Study of Leucorrhoea in Reproductive
Age Group Women of Nagpur City," *Indian Journal of Public Health* 49, no. 4
(2005): 238–39.

23 Haider, "Climate Change–Induced Salinity."

24 Khan et al., "Climate Change"; Shammi et al. "Impacts of Salinity Intrusion."

25 Umme Kulsum, Jos Timmermans, Marjolijn Haasnoot, et al., "Why Uncertainty in Community Livelihood Adaptation Is Important for Adaptive Delta Management: A Case Study in Polders of Southwest Bangladesh," *Environmental Science & Policy* 119 (2021): 54–65, https://doi.org/https://doi.org/10.1016/j.envsci.2021.01.004; Ahsan "Climate-Induced Migration."
26 Shammi et al., "Impacts of Salinity Intrusion."
27 "Bangladesh Climate Change Strategy and Action Plan" (Dhaka: Ministry of Environment and Forests, 2009); Ministry of Environment and Forests, ed., "National Adaptation Program of Action" (Dhaka: Ministry of Environment and Forest, 2005).

Centering Vulnerable Groups, Adopting Humility, and Embracing Holistic Systems

RAJINI SRIKANTH AND LINDA THOMPSON

The fact that you can think of seven generations back and seven generations forward is a big lesson for us. Who can think that way? You can. The way you can take care of the land and the way you take care of one and other, we can learn from that. And so your gift to us we now have to carry forward and do.

—Cellist Yo-Yo Ma speaking with the Wabanaki people

Vulnerability to climate events manifests in multiple ways and among diverse populations. This volume has presented the climate-exacerbated health vulnerabilities of several groups of people—pregnant women, infants, the elderly, people of color, indigenous communities, and people experiencing homelessness, to name a few. These vulnerabilities have roots, in most instances, in multiple generations of deprivation, discriminatory practices, and intentional withholding of resources. Addressing these long-term injustices requires radical new approaches in thinking and living and the jettisoning of decades of fragmented responses to problems. Fundamentally, as the authors of this volume argue, there is overwhelming evidence of paucity in our frameworks of understanding, rooted in self-centeredness in our ways of engaging with one another and with the natural resources of the world. If humans could be honest in assessing our actions, we would recognize that humility, gratitude, and recognition of the interdependence of all living beings and the natural world are urgent perspectives we must adopt.

HOLISTIC SYSTEMS

Acts of reimagining are challenging. But this volume has also provided examples of situations in which precisely such crucial changes were initiated, and collaborations across disparate spheres of activity were adopted to begin the process of rebuilding broken structures and creating nimble and responsive frameworks. Barbara Sattler's chapter features a number of visionary nurses who took the initiative to bring about meaningful change in their hospitals and communities by thinking capaciously to encompass all aspects of a hospital's operations—including its large-scale purchasing policies—and reaching out to agricultural and other underserved communities with which they were connected. Nursing organizations recognized their influence and, particularly following Hurricane Katrina, began to articulate at both the local and global levels the types of interventions that had to occur to sustain health.

Laura Peters and colleagues in their chapter focus on two practices that can be mobilized to create positive health outcomes: participatory action research (involving communities that are negatively affected by climate change as equals in framing "solutions") and remote sensing (the gathering of climate-focused satellite data from forests, agricultural fields, buildings, and roadways). Rather than embracing satellite data as "objective," Peters and colleagues explain how analysis of this data, with care and attention to its limitations (particularly with respect to small tracts of land that the satellite misses), ensures responsible public health interventions. Their chapter looks at three dimensions of justice—space, time, and relationships—to discuss how ethical decisions can be made.

Systems thinking—not discrete and disconnected spheres of activity—is required. Nurses, policymakers, researchers, scientists, epidemiologists, mental health professionals, and disaster-recovery agencies are slowly beginning to acknowledge the long-term consequences (the "slow violence") of colonialism, racism, extractive and unregulated production, and untempered anthropocentric thinking, and they are beginning the attitudinal changes to repair the wounds of injurious practices. In their chapter on the impact on maternal and infant health outcomes of excessive heat, Lisa Heelan-Fancher and Laurie Nsiah-Jefferson note that policymakers are typically unable to think beyond polarities: they articulate situations as antipodes of "sameness and difference" and "black and white" rather than conceptualizing complex circumstances as akin to the multiple pieces of a kaleidoscope that can fall in a range of patterns, depending on how you frame the challenge to be addressed. They call for educating future nurses, social workers, and policymakers on the interconnectedness among their seemingly disparate various spheres of activity.

CENTERING VULNERABILITIES

Climate justice cannot be rooted in an economic framework of carbon credit trading. Rather, as Surili Sutaria Patel and Adrienne Hollis note in their chapter, climate justice must begin with the acknowledgment of fault and the recognition of injury by those responsible for the discriminatory policies and actions that affect poor communities and communities of color. This recognition of complicity in causing the injury must be followed by ethical and material investment in correcting those wrongs.

Lorena Estrada-Martínez and colleagues' chapter reveals that these reparative efforts can fall short if they don't address the multiple facets of long-term deprivation caused by colonialist and imperialist decisions. Her chapter underscores that involvement by local residents as equal partners with the US military is a crucial step in identifying necessary reparative actions.

Clair Cooper, writing about nature-based solutions in Europe, observes that without authentic participation by groups typically marginalized—the elderly, for instance—urban renewal projects can lead to unintended consequences, such as creating upscale locations that attract better-educated residents and diminish attention to the needs of the elderly. Kandel and Raciti's chapter about mitigating excessive heat in Boston also makes the same point: the perspectives and voices of those who are most affected and who live with the day-to-day consequences of extreme climate events must be at the center of policy and decision making. José Martínez-Reyes and Camille C. Martinez, writing of the danger to traditional agricultural practices of the Maya of Mexico, explain how green land-grabbing intended to protect nature to provide a carbon offsetting mechanism in fact threatens land tenure and climate justice for the Maya.

The neglect of the elderly by public health departments, first responders, health care providers, and society in general is a central focus of the chapter by Caitlin Connelly and colleagues; they observe that it is not until an extreme climate event occurs that we remember the circumstances and needs of the elderly. They argue for heightened recognition of the conditions suffered by the elderly—"social isolation, socioeconomic disadvantage, and substandard housing"—and they call for the education of long-term care givers of the elderly and the commitment of policy framers to address the impacts of climate change on this vulnerable population.

HUMILITY

In the opening chapter of this volume, Deborah McGregor and her coauthors draw on United Nations reports to make the claim that while Indigenous peoples "make up less than 5 per cent of the world's population . . . they speak an overwhelming majority of the world's estimated 7,000 languages." Furthermore, while they "own, occupy or use resources on some 22% of the global land area, [this land] in turn harbours 80% of the world's biological diversity." McGregor and colleagues remind us that language is not commonly considered a protective factor in the face of climate change, and yet language retention and revitalization initiatives are in fact critical to adaptation and resilience. Language is intricately tied to consciousness and ways of thinking and engaging the world. Their chapter establishes the connections among language revitalization, biodiversity, and climate change, underscoring that cultural priorities are inextricably linked to environmental and climate justice.

Deborah McGregor is of the Anishinaabe First Nations people of Canada. Also from Canada, Diana Lewis (not a contributor to this volume) and two coauthors from across the globe in New Zealand, Lewis Williams and Rhys Jones (also not contributors to this volume), offer a keen insight about health: settler colonists came to occupy others' lands, found themselves in places of extraordinarily rich resources, proceeded to wipe out the indigenous population, and then started to use the physical landscape to yield returns. Indigenous peoples' relationship to the land is a crucial facet of their physical, emotional, and cultural health and well-being. When connections to the land are severed or disrupted, then Indigenous peoples suffer for multiple generations as they lose the traditional practices that kept their communities healthy. To undo or repair centuries of oppressive policies and practices, the authors contend that one must center Indigenous knowledge and Indigenous perspectives.[1]

The Cherokee poet Marilou Awiakta published a poem in 1991 titled "Mother Nature Sends a Pink Slip." The poem is written in the form of a corporate memo, the recipient of which is the human species. The subject is straightforward: "Termination." Humans are being pink-slipped because they have failed to abide by the rules of teamwork, says Mother Nature. Despite the warnings given to humans—"I have . . ./—made the workplace too hot for you/—shaken up your home office/—utilized plagues to cut back personnel"—there has been no change or correction to behavior, and therefore Mother Nature has no other option than to terminate the human species. There can be no appeal. Today, three decades after Awiakta's poem was published, the situation is dire, and the planet is in the throes of an emergency in terms of long-term survival. Awiakta plunges into her Cherokee tradition to ask if there is any way that we can avert the pink slip. The answer that she offers is "respect." It is through respect for Mother Nature that we can salvage and then repair a planet that we wrongly assumed we had control over and could use without any consequence to life—not just our human lives but the lives of all living creatures and the physical landscapes of the planet. But respect is not an attitude or sentiment that we can withdraw from some safe repository, like withdrawing cash. Respect comes from careful nurturing, early cultivation, and continuous introspection. Without this respect for our ecosystem, our relationship to the living beings and natural spaces of our planet can be only transactional and exploitative.[2]

Affective Attachment to Place/Our Planet

Indigenous peoples have long underscored the centrality of place to knowing, being, and enduring. Western epistemologies and European ways of thinking, by contrast, see place as separate from the humans who inhabit it, as something to be acted on. Jay T. Johnson (not a contributor to this volume) writes, "By detaching our histories, our stories and our science from place, Western science has developed an arrogance which seeks to elevate it above other knowledge systems, particularly those knowledge systems which have remained more attached to place."[3] He makes the powerful observation that

> when we are engaged with place, we are carrying out an act of remembrance, a retelling of the stories written there, while also continually rewriting these stories. Being-in-place is continually an act of engaged/active learning. Each place name acts as a mnemonic device, helping us to remember the story associated with that toponym. Each story is a text within the metanarrative of a particular culture, aiding through its remembrance, the continual re/creation of that society. We can look out across our landscape, seeing a series of place names, remembering the stories associated with creating and recreating our culture; I would argue that this storied landscape is the equivalent of a library. Its placed knowledge serves as a repository for the narratives, which[,] through an oral literacy, are employed to access the knowledge produced within a knowledge system which values being-in-place.[4]

David Gruenewald (not a contributor to this volume) makes a similar claim: "As centers of experience, places teach us about how the world works and how our lives fit into the spaces we occupy. Further, places make us: As occupants of particular places with particular attributes, our identity and our possibilities are shaped."[5]

Given the central role of housing in providing stability in people's lives, it should come as no surprise that people experiencing homelessness are among the most vulnerable to the disruptive tumult of extreme climate events. Kim Flike and colleagues' chapter discusses the pressure-and-release model to draw attention to a "progression of vulnerabilities, including systemic inequalities that deny access to power and resources," that have contributed to the conditions of those experiencing housing instability. In order to respond to the needs of this population when extreme weather events occur (such as excessive heat and cold, flooding, and wildfires), the authors stress that the activities of climate change researchers, affordable housing policy analysts, and health care professionals must intersect.

Janice Hutchinson's chapter on the psychological stressors caused by Hurricane Katrina reminds us of the centrality of place and home in the capacity of New Orleans' displaced residents to recover from the ravages of the hurricane. Connection to place and social relations within familiar places give meaning to our lives, and the longer that residents remain in temporary housing, the less likely that they will be able to satisfactorily rebuild their lives and restore their well-being. Affective attachment to place provides us with the source of emotional wellness.

Climate-induced displacement and the emotional turmoil it causes is the focus of Rosalyn Negrón and colleagues' chapter on decisions made by Puerto Ricans after Hurricane Maria struck their island. The authors offer a nuanced and complex analysis of the role of affect/emotion in people's justifications for whether to leave or stay when a climate event strikes unexpectedly. Negrón and colleagues write, "Puerto Ricans who left the archipelago after Maria carried the traumas of the disaster and the heartache of saying goodbye. But many also have survivor's guilt, grounded in the knowledge that their loved ones continued to struggle." It would appear that understanding the role of emotions (and their influence on mental health) must become an essential part of policy regarding climate justice and public health.

Thousands of miles away from Puerto Rico lies another climate-buffeted region: Bangladesh. The closing chapter of this volume by Reazul Ahsan and colleagues focuses on women working in the shrimp farms of southwest coastal Bangladesh. Their stories are seldom told; Ahsan and colleagues dedicate their chapter to these women, whose reproductive organs (and therefore reproductive health) are negatively affected by the saltwater in which they are immersed for several hours a day as they gather shrimp, which offers a lucrative export market. The increase in salinity in many of the water bodies in Bangladesh is a direct consequence of sea-level rise. Sea-level rise has been shown to be a direct consequence of global warming, which is attributed to human activity, specifically the burning of fossil fuels. Ahsan and colleagues' chapter places the health and lives of these women within the global context of UN Sustainable Development Goals and other global interventions addressing the impact of climate change. The health of female shrimp farmers in Bangladesh is not disconnected from the energy decisions made in the Western world.

Mia Motley (not a contributor to this volume), the prime minister of Barbados and one of the most assertive leaders of the developing world, makes it very clear: it is the responsibility of developed nations as well as international monetary organizations to help poorer nations transition to renewable energy sources and to create defenses against extreme climate events. At the Paris Climate Summit in June 2023, she said, "The reason why these institutions [global lending and energy agencies] exist is that they were created to help the world in the reconstruction effort after World War Two. We are in a moment that is equal to World War Two with respect to climate."[6] A concrete solution she proposes is that a lending agency such as the World Bank offer loans at low interest to developing countries rather than charging them higher rates than developed countries to borrow money. She stresses the relationship between biodiversity and the fate of humanity and emphasizes that developing nations must recognize that the countries of the globe are collectively facing one of the most dire threats in history. As we read in the introduction to this volume, Motley has been delivering her consistent and powerful message for the past several years.

Globally and locally, increasing numbers of scientists, activists, policymakers, and educators are casting off their restraint and pointing to the destructive impulses that have brought us to the present situation. The anthropologist, essayist, and novelist Amitav Ghosh (not a contributor to this volume) observes, "The muting of a large part of humanity by European colonizers cannot be separated from the simultaneous muting of 'Nature.' Colonization was thus not merely a process of establishing dominion over human beings. It was also a process of subjugating, and reducing to muteness, an entire universe of beings that was once thought of as having agency, powers of communication, and the ability to make meaning—animals, trees, volcanoes. . . . These mutings were essential to the processes of economic extraction—because . . . in order to see something as mere resource, 'we first need to see it as brute, as something that makes no normative demands of practical and moral engagement with us.'"[7]

Ghosh is justifiably critical of "experts" who don't recognize the gaps in their own thinking:

> When issues like climate migration and climate justice are discussed by experts in international conferences, one of the guiding assumptions is that the negotiations are intended to produce fair and just outcomes. Another such assumption . . . is that structures of governance, national and international, exist to promote people's welfare and to advance the causes of equality, security, and justice. What migrants . . . know, on the other hand, is that every aspect of their plight is rooted in unyielding, intractable, and historically rooted forms of class and racial injustice. They know that if they were wealthy or White they would not have to risk their lives on rickety boats. They know that the processes that have displaced them are embedded in very old and deeply entrenched social relationships of power, national and international.[8]

We conclude this volume as we began it: with an exhortation to recognize the connectedness of all facets of life—humans, animals, plants, water, land; to decenter the human and to practice humility in our interactions with all beings on the planet; to fulfill our responsibility to the next generation of holistic thinkers and doers by rejecting destructive and fragmented ways of thinking. We urge an interrogative stance on the certainties that researchers and policymakers have taken for granted. The Canadian indigenous researcher Julie Bull (not a contributor to this volume) challenges us to consider why our accepted ways of thinking about research and policy can and should be open to the stance of indigenous researchers who come to research with a mindset that asserts that "all of my relations are the focus of all my research." Narrowly focused research questions and policy work must give way to expansive and interconnected endeavors.[9] The future health of our planet urgently depends on it.

NOTES

1   Diana Lewis, Lewis Williams, and Rhys Jones, "A Radical Revision of the Public
    Health Response to Environmental Crisis in a Warming World: Contributions
    of Indigenous Knowledges and Indigenous Feminist Perspectives," *Canadian
    Journal of Public Health* 111, no. 6 (December 2020): 897–900; see also Edwin
    Ogar, Greta Peci, and Tero Mustonen, "Science Must Embrace Traditional and
    Indigenous Knowledge to Solve Our Biodiversity Crisis," *One Earth* 3, no. 2
    (August 2020): 162–65.

2   Marilou Awiakta, "Mother Nature Sends a Pink Slip," *Appalachian Heritage* 19,
    no. 1 (Winter 1991): 5. The poem was subsequently re-published in Awiakta,
    Selu: Seeking the Corn Mother's Wisdom (Wheat Ridge, CO: Fulcrum, 1993).

3   Jay T. Johnson, "Place-Based Learning and Knowing: Critical Pedagogies
    Grounded in Indigeneity," *GeoJournal* 77, no. 6 (2012): 829–36.

4   Johnson, "Place-Based Learning and Knowing," 833.

5   David A. Gruenewald, "Foundations of Place: A Multidisciplinary Framework
    for Place-Conscious Education," *American Educational Research Journal* 40, no. 3
    (2003): 619–54, 621 http://www.jstor.org/stable/3699447; David Gruenewald and
    Gregory A. Smith, eds., *Place-Based Education in the Global Age: Local Diver-
    sity* (London: Routledge, 2014); See also the speakers and the topics they cover
    at the Environmental Humanities Initiative's climate change conference at the
    University of California, Santa Barbara, held May 3–24, 2016. The conference
    was almost entirely digital in format and, therefore, "nearly carbon-neutral."
    The conference foregrounded the "View from the Humanities" on climate
    change. http://ehc.english.ucsb.edu/?page_id=12613. The journal *Humanities*
    published a special issue in 2020 titled "Environmental Humanities Approaches
    to Climate Change." The submissions are accessible at https://www.mdpi.com/
    journal/humanities/special_issues/environmental_humanities.

6   Justin Rowlatt, "Barbados PM Fights for Shake-Up of Global Climate Finance," BBC
    News, June 22, 2023, https://www.bbc.com/news/science-environment-65962997.

7   Amitav Ghosh, *The Nutmeg's Curse: Parables for a Planet in Crisis* (Chicago:
    University of Chicago Press, 2021), 190.

8   Ghosh, *Nutmeg's Curse*, 158.

9   Julie Bull, "Research Is Relational: From Principles to Practice in Reconcilia-
    tion," keynote address at Carleton University Institute on Ethics in Research on
    Indigenous Peoples, June 6, 2017, https://www.youtube.com/watch?v=uqkqlb
    -9EQk.

Creating Hope and Systems Thinking in the Next Generation of Climate Stewards and Public Health Workers

PATRICK BARRON, PAUL KIRSHEN, KELLY LUIS, RAJINI SRIKANTH, LEONARD VON MORZÉ, AND CEDRIC WOODS, WITH MARC ALBERT

A team of faculty members (Patrick Barron, Alan Christian, Paul Kirshen, Cheryl Nixon, Rajini Srikanth, Leonard von Morzé, and Cedric Woods) and the evaluation consultant Elena Stone, at University of Massachusetts Boston, came together in 2017–18 to apply for a grant from the National Endowment for the Humanities (NEH). The NEH grant, titled "Humanities Connections," spotlights connections between the humanities and the natural and social sciences and is meant to encourage the cultivation of new and transdisciplinary ways of thinking in students.

The UMass Boston team submitted "Living with the Urban Ocean: Inquiring, Imagining, Embracing," a project focused on Boston Harbor and the islands in it, bringing together members from anthropology, English, ethnic studies, environmental sciences, and indigenous studies. In addition, the National Park Service in Boston—specifically Marc Albert, the director of Natural Resource Partnerships—has been an integral partner with the UMass Boston team. The application to the NEH was grounded in place-based learning and experiential pedagogy. The proposed program of activities features the effectiveness of humanities methodologies in knowing about, caring for, and acting to preserve and protect our natural resources.

In the executive summary of the grant proposal we submitted, we wrote,

The campus's humanities departments, including English, Philosophy and History, as well as Anthropology, have had a long-standing collaboration with the university's School for the Environment (SFE), a unique academic home that integrates environmental sciences with humanistic concerns. Our project team, drawn from diverse institutional locations, is proposing to implement a cluster of three courses that focus on Boston Harbor, the Boston Harbor Islands, and the diverse coastal communities surrounding the harbor that many of our students call home. The course cluster is designed to incorporate humanistic methods of inquiry such as literary analysis, archival research, storytelling, writing and reflection, and interpretive exercises, with the goal of increasing students' awareness of the close and millennia-long interaction between human consciousness and activity and the natural world. This increased awareness provides the framework within which students engage in experiential learning that strengthens their appreciation for the harbor's impact on the city, its history and its future.[1]

Graduate students Kelly Luis and Sarah Shapiro became an essential part of the team and were instrumental in designing and implementing two of the new courses—"The Urban Ocean" (a seminar for first-year students) and "Writing and the Environment" (an intermediate seminar), respectively— that we crafted to feature interdisciplinarity and underscore the value of systems-level thinking that foregrounds the connections among diverse spheres of knowledge and experience. Both Shapiro and Luis have since graduated. Luis currently works for the National Aeronautics and Space Administration's Jet Propulsion Lab, and Shapiro teaches creative writing. Two other graduate students—Sara Bistany and Tania Ploumi—continued teaching the courses originally taught by Luis and Shapiro. (Three of the original participants—Alan Christian, Cheryl Nixon, and Elena Stone—are no longer with the university and left in the grant's second year.) We mention the involvement of our graduate students because one of our objectives is to cultivate the next generation of climate justice stewards. Our graduate students are conscious of their pedagogical role in furthering systems-level and holistic thinking among the undergraduate students they come into contact with and teach/guide/mentor in classrooms, field sites, and labs, as well as of how they themselves will prioritize systems thinking in their professional trajectories.

INDIGENOUS PERSPECTIVES FROM THE OUTSET

A significant insight that the team members gained from working together was how crucial it is to center indigenous perspectives in conceiving, designing, and implementing activities related to Boston Harbor and its islands. Team member Cedric Woods, director of the Institute for New England Native American Studies, underscored for us that inviting the participation of indigenous peoples *after* the project has been designed is not respectful. If we are serious about recognizing the importance of traditional ecological knowledge in helping us move toward climate justice and reciprocity with the land and sea, then we cannot treat indigenous ways of knowing as an afterthought or even an embellishment. They must be an integral core of the project. Indigenous peoples were traveling between the many islands of Boston Harbor long before the settlers arrived in the seventeenth century, and the deep relationships between indigenous communities and the harbor and its islands must be positioned as central to any project that we undertake as an academic institution. The introduction to this volume spotlights the planetary damage that has ensued from humanity's exploitation of indigenous peoples and natural resources, and from our devaluing the deep connections between humans and nonhuman living creatures and natural landscapes. In looking to the future, we, as educators and researchers, acknowledge that our responsibility is to reestablish the reciprocity between humanity and the natural world and to ensure that the next generation of policymakers, community organizers, educators, scientists, artists, storytellers, healers, and inventors lives and acts with humility toward and gratitude for (echoing Robin Wall Kimmerer) what our planet has provided.

LOCATING THE BODY

Graduate student Sarah Shapiro's "The Situated Body (in) Place" activity helped us experience the embodied dimension of place. She invited students and faculty to walk along the periphery of Boston Harbor that borders the UMass Boston campus and to be fully conscious to the multiple ways in which one's body absorbs and makes meaning from the sensory stimulations of walking along the water. She frames the activity thus:

> We will shake off the day, breathe in the salt air and check in with our bodies. We will take note of our extended senses (outside of our bodies—sight, sound, touch, taste, smell, and in our bodies—heat/cold, pain, balance/ gravity, general body awareness).

We will move along the path in a series of irregular movements (short steps, long steps, dance steps, backwards) which allow us to slow down and notice the smaller details in our surroundings and ourselves. As we move along the Harbor Walk, I will point out moments and details which I notice as encouragement for the new movers to consider what it is that they usually notice along the Harbor Walk, if that is somewhere they frequent at all.

At each of the stops there will be time to compile notes and reflections of our experiences and complete writing exercises (these generative exercises are aimed to facilitate the articulation of the movers Harbor Walk experiences) and/or have discussions which explore the movers' connections to the Harbor Walk environment (this is more of an ad hoc comparative conversation, what do you usually notice, if anything, and what do you notice now?). These exercises will provide structures which will help incorporate our notes and reflections into creative pieces.

Questions we will consider throughout this workshop are:
Do you have time to slow down? How do you take time to slow down?
How do you to walk, watch and write in a city/urban ocean environment?
How do you bring "the sublime" into the city?[2]

The questions that Shapiro poses to the walkers ("movers," as she calls them) are designed to generate appreciation for place, to recognize the value in how we see, hear, and use all of our senses.

A COMMUNITY OF PLACE, A COMMUNITY OF INTEREST

The Living with the Urban Ocean project created a community of place (Boston Harbor and the Boston Harbor islands) and a community of interest—bringing together faculty and students who are committed to embracing the natural resource of Boston Harbor and its islands and understanding their importance to the city of Boston. Dear Harbor Radio, an art installation that was commissioned by the Dartmouth campus of University of Massachusetts system, is intended to generate a "vocabulary of care and mutual connection" to the harbor. A "pirate radio" cart, attached to a bicycle and propelled by a bicyclist, invites participants to express their appreciation for and attachment to the "nonhuman" in the form of a "love letter." The "letters, observations, recordings, songs, [and] conversations [serve as] a series of calls of love from humans to non-humans."[3] Patrick Barron, one of our team members, incorporated a visit to the Dear Harbor Radio installation when it came to the UMass Boston campus in his course "Literature,

Culture, and the Environment." Students visited the installation at the UMass Boston art gallery, wrote a love letter to the harbor, and recorded their letters for Dear Harbor Radio. The act of expressing in words one's attachment to the harbor and the nonhuman creatures and objects in it heightens the importance and foregrounds the value of these natural resources to our own physical and emotional well-being.[4]

Barron, who is deeply passionate about the pedagogical value of outdoor activities and is an accomplished kayaker, has taken his students to two other islands in the harbor—Peddocks and Thompson—combining the experiential physicality of being on the islands with reading and aurally absorbing texts that describe nature. He speaks eloquently of how deeply absorbed his students were as they passed around a text by Rachel Carson and together read aloud from it on their visit to Thompson Island. Barron draws from the *Norton Book of Nature Writing*, which encompasses a broad swath of history. Carson's piece is titled "The Marginal World," and is excerpted from her book *The Edge of the Sea*.[5] It is not difficult to understand why the students were so affected by reading Carson's words aloud: "The shore is an ancient world, for as long as there has been an earth and sea there has been this place of the meeting of land and water. Yet it is a world that keeps alive the sense of continuing creation and of the relentless drive of life. Each time that I enter it, I gain some new awareness of its beauty and its deeper meanings, sensing that intricate fabric of life by which one creature is linked with another, and each with its surroundings."[6]

PREPARING FOR SEA LEVEL RISE

Sea level rise is an inevitable phenomenon that is at the forefront of the City of Boston's planning for a resilient future.[7] Paul Kirshen, a member of our team, is one of the principal scientists working with the City of Boston's resilience plan to determine how best to mitigate the impact of sea level rise and flooding of coastal communities—many of them with socioeconomically vulnerable residents who are primarily people of color. Kirshen and his expert colleagues from universities and consulting firms conducted a feasibility study for constructing a concrete seawall barrier, and they came to the conclusion that it is an impractical approach and economically prohibitive. Instead, their recommendation is to design eco-friendly solutions, such as raised shoreline greenspaces that will act as barriers to the rising tides.[8]

Kirshen is also the director of the Stone Living Lab, a research endeavor set up on one of the harbor islands to study sea level rise and nature-based responses to contain it. The lab engages the following questions:

- How can we ensure nature-based solutions will work in weather conditions that are more volatile and unpredictable than we've ever experienced?
- How can the public help lead in implementing resilience solutions, and how can we ensure these solutions help address issues of climate justice and inequality?
- What are the best next-generation solutions we should start investing in across the world?
- How do we engage and collaborate with the regulatory and development community to streamline permitting and financing so that we solve these problems before it's too late?[9]

The Stone Living Lab's steering committee includes key partners and representatives from the Massachusett Tribe at Ponkapoag and the Woods Hole Group; its multidisciplinary advisory board has over forty members from diverse backgrounds, such as engineering, planning, regulatory, and environmental justice; Native American Tribes; municipalities; environmental groups; and many others. In addition, there is a scientific advisory committee.

REJECT SILOS AND TRANSFORM THINKING

Kirshen and a team member from the English department, Leonard von Morzé, cotaught a course in the spring of 2020 that served as a capstone (culmination) for majors in environmental science. Typically, this course engages a community environmental need, and the students are tasked with coming up with engineering and scientific responses to address this need. In the cotaught model, however, Kirshen and von Morzé infused history, culture, and indigenous perspectives to show the importance of this type of holistic thinking when coming up with solutions to environmental challenges. They were ably supported by Debra Butler, who received her doctoral degree at the School for the Environment and is now executive director of the American Society of Adaptation Professionals. Butler brought to the cotaught course her deep knowledge of grassroots environmental activism involving indigenous and African American communities.

Von Morzé identifies as a faculty member who focuses on earlier phases of US cultural history. This position affords him the opportunity to discuss episodes from the past that will surprise a present-day student. While such surprises are sometimes valuable in themselves—a reminder of the insight that studying the past can be as rich a source of new experiences as traveling to a foreign country—he finds himself designing courses in environmental literature that affirm a different truism, which is that the past continues to weigh on the present in ways that cause suffering for marginalized communities. The question he poses is, What can we learn from historical case studies about climate justice and public health today?

One of the most powerful realizations that came to von Morzé is that students deeply appreciated learning about the cleanup of Boston Harbor. This cleanup campaign, which took more than fifty years, provides a powerful example of what relentless and determined advocacy across multiple sectors of society can accomplish. For over one hundred years Boston Harbor had been used as a literal dump site for the city's waste. It was a court-ordered directive by the judge David Mazzone that set in motion the restoration of Boston Harbor as a natural resource that every person in Boston and Massachusetts could enjoy. There is a memorial to Judge Mazzone on Deer Island. Von Morzé's students saw in the success story of the Boston Harbor cleanup evidence that even complex problems that have long historical roots can be solved through strategic citizen participation. This example reaffirms for them the possibilities of civic engagement and democratic citizenship, and it reminds them of their own power to shape the societies in which they live.

The papers of Judge Mazzone are housed in UMass Boston's archival collections in the campus' Healey Library.[10] In fact, UMass Boston has the archival holdings related to many of the harbor islands. Students thus have the opportunity to acquire important skills in archival research as well as an appreciation of the connections between the past and the present. Knowing how "ordinary" citizens were able to engage government officials and other individuals in power to bring about changes that improved the quality of and access to their natural resources provides today's students—the next generation of climate justice stewards—with hope and confidence.

Von Morzé also asks students to look at cases of environmental activism from the more distant past. While individuals then did not have the scientific instruments now available to present-day students, Bostonians in the eighteenth and nineteenth centuries acted to respond to local conditions. While they did not understand the relationship between climate and public health as a meliorable condition, because no one recognized that human activity changes the climate, they still identified environmental problems that had human causes and could therefore be remedied. Yet this limitation on environmental understanding can also illustrate for students the power of communal action, as we have noted above. These were *local* actions motivated by topophilia and concern for public health.[11] Concerned Bostonians of the past responded to public health emergencies by identifying environmental problems and working to solve them. In a series of case studies, von Morzé's class examined responses to the pollution of Boston Harbor from the eighteenth century to the present day.

Landmaking projects that led Boston to grow exponentially in geographic extent between 1620 and 1980 reflected responses to public health problems, even as they also exacerbated them. Bostonians "wharfed out" the city by constructing ropewalks through which sewage (which derived from the term "seaward") could be dumped in the water. By 1800 the prevailing theory that disease was transmitted by smells known as "miasma" held that new neighborhoods, inevitably constructed on land that had not been completely drained of water, were epicenters of contagion. Urban problems caused by dirty water were fundamentally, at that time, treated as problems of air pollution.

As the case studies von Morzé and his students examined reveal, it was not until the later nineteenth century that water pollution was recognized as a threat in its own right. The students looked at the construction of the Calf Pasture Pumping Station, which is situated on the UMass Boston campus, as a site that attempted to bring the city's raw sewage outflow under centralized control. The construction of an airport and the rise of the military-industrial complex in the twentieth century, however, brought new threats to the harbor, this time chemical. The students studied selected parts of Rachel Carson's history-changing *Silent Spring* (1962) because the book shows the increasing biotoxicity of even tiny amounts of chemicals as they move up the food chain, from plankton to birds to human beings. Neal Stephenson's eco-thriller *Zodiac* (1988) offers a localized novelization of this phenomenon, as its hero turns to media to refute the popular PATEOTS theory that the ocean is an appropriate dumping ground for even the most toxic chemicals following the idea that the chemicals in Boston Harbor are no larger in relation to the ocean than the concluding period is in relation to the rest of this sentence. Humorously evoking the local professional sports franchise, the hero uses local TV shows and newspapers to point out that the PATEOTS theory fails to grasp that even if the pollutant occupies only a space the size of a banana in relation to a football field it can have disastrous consequences for the players who might slip and fall on it. Written just as the cleanup of Boston Harbor was beginning, Stephenson's novel represents local public spheres along the Atlantic seaboard as effective platforms for persuading citizens to take collective action on public health emergencies. With the shift from pollution to global warming as our major environmental concern today, the global scale of our climate justice problem can make students feel helpless. The reminder of the ability of historical actors such as Carson to effect change in popular attitudes can provide students with inspiration.

A final source of hope to be found in our history is traditional ecological knowledge. Von Morzé teaches Wampanoag cosmology not only because it holds that islands had their own value, and can be enjoyed seasonally without needing to be bridged through acts of landmaking, but also because it teaches that land and water are animate, and thus need to be treated responsibly as fellow agents in a shared environment.[12] This pedagogical shift comes with its own challenges. Native histories of New England's waterways require that we learn how to access the archive within the environment rather than only in a library, since "texts" in the traditional sense are not the only evidence of history and culture, especially for the early perspectives of Native Americans in the Boston area who did not have a writing system. Involving Indigenous authorities in discussions of questions of problems of environmental justice offers a way to recognize expertise, structure participation, and reflect our conviction that learning from the first stewards of the environment is critical to making all students responsible future actors. Thus, our approach to history takes us "beyond the text" to engage students with contemporary voices and with the physical environment.

NOTES

1   This is taken from "Living with the Urban Ocean," the executive summary of a grant submitted by University of Massachusetts Boston to the National Endowment for the Humanities in October 2017. The grant PI was Rajini Srikanth.

2   Shapiro, document shared with authors, August 2021.

3   See Dear Harbor Radio at http://www.plotformplot.org/project/dhr/.

4   The limited playlist of audio recordings at Dear Harbor Radio can be found at https://soundcloud.com/user-50501571-570280616/dhr?utm_source=www.plot formplot.org&utm_campaign=wtshare&utm_medium=widget&utm_content =https%253A%252F%252Fsoundcloud.com%252Fuser-50501571-570280616 %252Fdhr.

5   Rachel Carson, "The Marginal World," in Norton Book of Nature Writing, ed. Robert Finch and John Elder (New York: W. W. Norton, 1990), 480–85. Excerpted from Carson, 1955?

6   Carson, "The Marginal World," 481.

7   Climate Ready Boston, "Climate Change and Sea-Level Rise Projections for Boston: The Boston Research Advisory Group Report," 2016, https://www.boston.gov/sites/default/files/document-file-12-2016/brag_report_-_final.pdf.

8   Lisa Allen, "Turning the Tide," UMass Boston: News, January 15, 2021, https://www.umb.edu/news/detail/turning_the_tide.

9   UMass Boston Office of Communications, "Stone Living Lab Launches as New Hub for Nature-Based Resilience Solutions," October 28, 2020, https://www.umb.edu/news/detail/stone_living_lab_launches_as_new_hub_for_nature_based_resilience_solutions; See also https://www.stonelivinglab.org.

10 Judge Mazzone Chambers Papers on the Boston Harbor Cleanup Case (1985–2005), https://openarchives.umb.edu/digital/collection/p15774coll8/id/187; Steven Rudnick, "Remaking Boston Harbor: Cleaning Up After Ourselves," in Remaking Boston: An Environmental History of the City and Its Surroundings, ed. Anthony N. Penna and Conrad Edick Wright (Pittsburg: University of Pittsburgh Press, 2009), 56–74.

11 On topophilia, see Yi-Fu Tuan, Topophilia: A Study of Environmental Perception, Attitudes, and Values (Edgewood Cliffs, NJ: Prentice Hall, 1974).

12 The class discusses the Wampanoag stories about how the islands were formed in Christine DeLucia, Memory Lands: King Philip's War and the Place of Violence (New Haven, CT: Yale University Press, 2018), 32–33.

CONTRIBUTORS

REAZUL AHSAN, PhD, is an associate professor of City and Metropolitan Planning at the University of Utah. Dr. Ahsan received his PhD from the University of South Australia in climate migration and urban challenges. His primary research interests are climate-related social and urban concerns and climate-adaptive urban planning. This work highlights his and his team's interest in tackling climate change and its long-term implications for the social structures of the marginal population in Bangladesh's coastal region.

MARC ALBERT is the director of Science and Stewardship Partnerships for the National Parks of Boston, which includes the Boston Harbor Islands National and State Park, Boston National Historical Park, and Boston African American National Historic Site. Marc facilitates the study and management of natural areas and cultural landscapes of the parks through partnerships and community engagement. Marc studied plant ecology at Tufts University and UC Berkeley, and has more than twenty years of experience linking park management to science and community in both San Francisco and Boston.

DANIEL ALDRICH, an award-winning author, has published five books, including *Building Resilience* and *Black Wave*; more than eighty peer-reviewed articles; and written op-eds for the New York Times, CNN, Huff-Post, and many other media outlets. He has spent more than five years in India, Japan, and Africa carrying out fieldwork, and his work has been funded by the Fulbright Foundation, the National Science Foundation, the Abe Foundation, and the Japan Foundation, among other institutions. Aldrich was the 2021 Klein Lecturer at Northeastern University.

SHOSHANA V. ARONOWITZ, PhD, MSHP, is a family nurse practitioner, community-engaged health services researcher, and assistant professor in the Department of Family and Community Health at Penn Nursing. Her research examines innovative delivery models to promote equitable access to substance use treatment and harm reduction services. Dr. Aronowitz holds a PhD in nursing from Penn Nursing, and earned her master's in nursing from the University of Vermont, and master's in health policy from the University of

Pennsylvania. She completed her postdoctoral fellowship at the National Clinician Scholars Program University of Pennsylvania site.

TERI ARONOWITZ is a professor in the Tan Chingfen Graduate School of Nursing at UMass Chan Medical School. She is the editor in chief of the Journal of American College Health. Her seminal contribution has been to advance the science of evidence-based care for sexual health promotion and healthy adolescent development. She has completed dozens of community-based participatory research projects with African American, Asian American, and Native American families to promote youth resilience against risk behaviors. She is a fellow of the American Academy of Nursing.

AFRIDA ASAD is a commonwealth scholar on public health and environmental issues. She is also one of the pupils of Reazul Ahsan's postgraduate research team. Afrida's key research interest is in public health (particularly the health of women) and environmental issues, and she has more than five years of experience working in this area.

PATRICK BARRON is professor of English at University of Massachusetts Boston. He received his PhD from the University of Nevada, Reno. His areas of expertise include environmental literature and ecocriticism, Native American literature, literary translation, and twentieth-century American and Italian poetry. His publications include *The Agropastoral Landscape of the Majella National Park / Il paesaggio agro-pastorale del Parco Nazionale della Majella* (author and translator) and *Italian Environmental Literature: An Anthology* (editor).

CASEY BAS, BHA (honors), is a research associate at York University. Her research focuses on social determinants of health, health equity, and gendered aspects of environmental and climate justice and Indigenous peoples.

KATHRIN BOERNER, PhD, is professor at the Department of Health Services Research, Carl von Ossietzky University of Oldenburg, Germany, as well as fellow of the Gerontology Institute at the Manning School of Nursing and Health Sciences, University of Massachusetts Boston. Her research expertise is in adult development and aging, with a focus on social relationships in late and very late life, as well as coping with chronic illness, end of life, and bereavement. Dr. Boerner has a strong publication and funding record (over one hundred journal articles and book chapters, with US and international funding).

KESHBIR BRAR, JD, BS, is a graduate of Osgoode Hall Law School and the University of Waterloo. Her research focuses on the intersection between the built environment, law, and population health.

NATASHA BRYANT, MA, is senior director of Workforce Research and Development at the LeadingAge LTSS Center @UMass Boston, an independent applied research center within LeadingAge. Her work focuses on the center's efforts to develop, test, and disseminate workforce improvement initiatives to attract and retain quality staff at all levels—from managers to nurses to direct care workers. Ms. Bryant also works on improving racial and ethnic diversity in the long-term services and supports career pipeline, with a focus on mid-level and senior-level positions.

JORGE L. COLÓN, PhD, is a professor of inorganic chemistry at the University of Puerto Rico at Río Piedras. He is PI of the NSF-PREM Center for Interfacial Electrochemistry for Energy Materials (CIE2M), a collaboration between the UPR, the Universidad Ana G. Méndez, and Cornell University's synchrotron facility, CHESS. He is coauthor of *Guías para el Desarrollo Sustentable de Vieques* (2002), *El proyecto de explotación minera en Puerto Rico 1962 a 1968: Nacimiento de la conciencia ambiental moderna* (2014), and coeditor of *Tailored Organic Inorganic Materials* (2015). His research interests are in nanomaterials with applications in solar energy, electrocatalysis, and anticancer drug delivery.

CAITLIN CONNELLY, MS, is a PhD candidate and research assistant in gerontology at the Manning School of Nursing and Health Sciences, University of Massachusetts Boston. Her research focuses on the impact of natural disasters on older adults with physical and cognitive impairments.

CLAIR COOPER, PhD, is a postdoctoral research fellow with Trinity College Dublin and recently completed a PhD at Durham University. Using geometric data analysis techniques, thematic mapping, and quantitative text analysis, Clair's thesis explores the nexus between nature-based solutions and structural conditions that influence poor health in cities. Clair's wider research interests are influenced by her experience in the water industry in the UK, where she was responsible for designing community-based programs for water-use behavior. These interests include nature-based solutions and novel ecosystems, behavior change, nature connectedness, and mixed-method approaches.

LORENA M. ESTRADA-MARTÍNEZ, PhD, MPh, is a social epidemiologist and associate professor in the School for the Environment at University of Massachusetts Boston. Her research uses life-course and ecological frameworks and community-based participatory research principles to address racial and ethnic health inequities. She is the PI of Vieques, Ambiente, Salud y Acción Comunitaria. She has also led studies on how communities of color understand and experience climate change, and mental health implications of relocation decisions among Puerto Ricans in the wake of Hurricane María.

KIM FLIKE, PhD, MSN, RN, is a postdoctoral fellow at the Center for Healthcare Organization and Implementation Research (CHOIR) at VA Bedford Healthcare System. Her research focuses on improving the health and well-being of people experiencing homelessness through recovery-oriented whole person care. Dr. Flike received her PhD in nursing from University of Massachusetts Boston and her masters of science in nursing at Pacific Lutheran University.

LISA HEELAN-FANCHER, PhD, is an associate professor in nursing at University of Massachusetts Boston and a board-certified family and adult nurse practitioner whose program of research seeks to improve childbirth outcomes. In collaboration with colleagues from other disciplines, her research focuses on decreasing the high number of unnecessary cesarean deliveries, advancing birth equity, and examining the impact of climate change and environmental stressors on maternal and infant health outcomes.

VALERIA HERNÁNDEZ-TALAVERA, MS, is a UMass Boston School for the Environment doctoral student. She conducts research using ecotoxicological tests on individual substances or contaminant mixtures in sediment to determine their effects and create risk assessments. Valeria has experience collecting data and conducting experiments for environmental, molecular, and public health studies. Valeria also works as an online science teacher for middle and high school students, promoting science education to the younger generations. Valeria hopes to increase her professional experience in laboratory techniques and academic career and use her acquired knowledge to assist vulnerable communities.

ADRIENNE L. HOLLIS, PhD, JD, Vice President for Environmental Justice, Health and Community Revitalization at the National Wildlife Federation, leads the environmental justice team to advance climate justice policy and programs. With more than twenty years of experience in the environmental justice and public health arena as a toxicologist and an attorney, Hollis is at the intersection of environmental justice, health disparities, and climate change. She works to identify priority health concerns related to climate change and other environmental assaults and to evaluate climate and energy policy approaches for their ability to effectively address climate change and benefit underserved communities.

JANICE HUTCHINSON, MD, MPh, is a board-certified pediatrician and adult and child psychiatrist. She attended Stanford University and University of Cincinnati Medical School. Her international medical practice includes missions to Liberia, Thailand, and Haiti. Her work has focused on all forms of child abuse, including sex trafficking, HIV and AIDS education and prevention, mental illness in children and minorities, neuropsychiatric substrates of mental illness, juvenile forensics, and health in special populations, for example, women and low-income groups. She is the author of several articles and chapters and is a coauthor of the book *Losing Control: Loving a Black Child with Bipolar Disorder.* Dr. Hutchinson has consulted to the military regarding PTSD, TBI, and other issues. She is currently the chairperson of the Washington Psychiatric Society "Psychiatrists in Society," which addresses issues of climate change, structural racism, and other biopsychosocial elements that affect psychiatric medical health.

SAJANI KANDEL received her PhD in environment sciences at the School for the Environment, University of Massachusetts Boston. Her scholarly work spans the intersection of urban planning, environmental science, and environmental justice and she uses a transdisciplinary approach to redistribute expertise in climate resiliency decision making. She also leads an experiential summer learning program that engages youth of color from environment justice communities in Boston in intensive field-based research on the extreme heat challenges faced by their communities.

ILAN KELMAN is professor of disasters and health at University College London and a professor II at the University of Agder, Kristiansand, Norway. His overall research interest is linking disasters and health, integrating climate change into both. His three main areas of study are disaster diplomacy and health diplomacy, island sustainability involving safe and healthy communities in isolated locations, and risk education for health and disasters.

PAUL KIRSHEN, PhD, is professor of climate adaptation in the School for the Environment at University of Massachusetts Boston. His areas of expertise include water resources engineering and management, climate change vulnerability assessment, and climate change adaptation planning. He is one of the principal scientists sought out by the City of Boston for guidance on building resilience to sea-level rise.

ANA ROSA LINDE-ARIAS, PhD, is a graduate of biology and has a master's in genetics from Oviedo University, Spain; a PhD in environmental sciences from Hokkaido University, Japan; a doctorate in biology from Oviedo

University, Spain; and is a postdoctoral fellow at Merck Research Laboratories. She was the scientific coordinator of a research collaboration between the Oswaldo Cruz Foundation, Brazil, and the Kitasato Institute, Japan, with the research group led by professor Satoshi Omura, 2015 Nobel Laureate of Medicine. Linde-Arias is currently a senior researcher at the Getulio Vargas Foundation, Brazil, and senior research fellow at the Gaston Institute, UMass Boston. Her fields of expertise include global, public, and environmental health.

KELLY LUIS was born and raised on Maui, Hawaii. She received her BA in environmental science from Columbia University and her MS and PhD in marine science and technology from University of Massachusetts Boston. Kelly joined the Water and Ecosystems Group as a NASA postdoctoral program fellow in September 2021. Her research focuses on the development of aquatic remote sensing algorithms for environmental monitoring, forecasting, and decision making.

CAMILLE C. MARTINEZ, PhD, is lecturer in environmental communication in the Department of Communication at UMass Boston. She has a MA in anthropology from Northeastern University and a PhD in communication from UMass Amherst. Her research interests are multispecies ethnography, political ecology, and climate justice. She has conducted research in Puerto Rico, Mexico, and Fiji. She is currently working in collaboration (with José Martínez-Reyes) on the political ecology of eucalyptus plantations in Galicia, Spain.

JOSÉ MARTÍNEZ-REYES is an associate professor of anthropology at UMass Boston. He holds an MA in anthropology from Northeastern University and a PhD in anthropology from UMass Amherst. His research focuses on environmental anthropology, political ecology, ethnoecology, and biocultural diversity. Has conducted research in Mexico, Fiji, and Puerto Rico. He is author of *Moral Ecology of a Forest: The Nature Industry and Maya Post-Conservation* (University of Arizona Press, 2016). Martínez-Reyes is currently working in collaboration (with Camille C. Martinez) on the political ecology of eucalyptus plantations in Galicia, Spain. He is also working on a manuscript on the relation between forests and guitar making.

DEBORAH MCGREGOR, Anishinabe, is associate professor and Canada Research Chair in Indigenous Environmental Justice at York University's Osgoode Hall Law School. She also has an appointment with the Faculty of Environmental and Urban Change at York University. Professor McGregor's research has focused on Indigenous knowledge systems and their various applications in diverse contexts, including environmental and water governance, environmental justice, health and environment, climate change, and Indigenous legal traditions. Professor McGregor remains actively involved in a variety of Indigenous communities, serving as an adviser and continuing to engage in community-based research and initiatives, and has been at the forefront of Indigenous environmental justice and Indigenous research theory and practice.

HILLARY MCGREGOR is a graduate of Georgian College's Anishnaabemowin and Program Development program. His research focuses on Indigenous youth climate leadership and traditional knowledge. He has contributed to the production of the documentary video Climate Crisis: Indigenous Youth Perspectives, which focuses on multigenerational climate resilience.

ORLANDO MALDONADO-MELÉNDEZ is a PhD student in social-community psychology at the University of Puerto Rico, Río Piedras campus. He was the lead research assistant on the NSF-funded project Social and Moral Factors in Puerto Ricans' Decisions to Leave after Hurricane Maria. Orlando's doctoral thesis is on masculinities in virtual environments, specifically dating apps targeted at the LGBTQ+ community. He is also interested in antipatriarchal feminisms, critical psychology, decolonial psychology, and gender dynamics in social movements. He has experience working and collaborating with community-based organizations promoting community social activation, participatory-action research, and advocating for improving public policies.

EIJA MERILÄINEN is a postdoctoral researcher at Örebro University (Sweden), an honorary senior research fellow at University College London (UK), and affiliated researcher at Hanken School of Economics (Finland). Her work explores critically the roles and power of various actors involved in politics of disasters and other societal disruptions. Her research has been published in journals such as *Geoforum*, *Disasters*, and *Environmental Hazards*.

CRUZ MARÍA NAZARIO, PhD, is an epidemiologist, teacher, researcher, and advocate deeply committed to helping communities affected by health disparities and environmental injustice. She was a member of the Bio-Epi Core Group at NIH-NIAID during the development of AIDS Clinical Trial Units and a trailblazer in community-based participatory research as a member of the National Advisory Committee of the RWJF Clinical Scholar's Program. She is Puerto Rico's site PI of the Eastern Caribbean Health Outcomes Research Network (ECHORN), the Pediatric-ECHORN, and a member of the Yale Transdisciplinary Collaborative Center for Health Disparities Research.

MARISOL NEGRÓN is an assistant professor of American Studies and Latino Studies at University of Massachusetts Boston. Her research areas include Puerto Rican intellectual history and diasporic identity formation, Latino/a popular culture and commodification, cultural nationalism and nation branding, and copyright. She is the author of *Made in NuYoRico: Fania, Latin Music, and Salsa's Nuyorican Meanings* (Duke University Press, 2024). Her work has also been published in the *Journal of Popular Music Studies*, *Latino Studies Journal*, and *Centro*. She is a founding member of the New England Consortium of Latina/o Studies..

ROSALYN NEGRÓN, PhD, is associate professor of anthropology at University of Massachusetts Boston, where she is also research director at the Sustainable Solutions Lab. Rosalyn's research examines the role of complex social environments on the decisions that people make for their social, economic, and physical well-being: these include migration and health decisions, identity negotiations, and educational choices. She publishes across varied fields, including *Social Science & Medicine*, *Journal of Sociolinguistics*, *Group Processes and Intergroup Relations*, and *Field Methods*. Her research has been funded by the National Science Foundation, the National Institutes of Health, the Ford Foundation, and the US Environmental Protection Agency.

PROKRITI NOKREK has over fifteen years of work experience in various national and international nongovernmental organizations and UN agencies in Bangladesh. She has also led research on livelihoods and food security among extremely poor people from the southwest coastal region in Bangladesh and worked with vulnerable coastal communities. She holds a masters of research degree in international development from the University of Bath, UK.

LAURIE NSIAH-JEFFERSON, PhD, MPH, MA, is director of the Center for Women in Politics and Public Policy at University of Massachusetts Boston. She also directs the graduate certificate program in Gender, Leadership, and Public Policy. As a scholar-activist, Nsiah-Jefferson's areas of expertise include intersectionality, health inequities, and women's leadership in the fields of social and health policy. She received her PhD from the Heller School for Social Policy and Management, her MA and BA from the College of Arts and Sciences, Brandeis University, and her MPH from Yale School of Medicine.

ADRIANA OYOLA-VIVAS, MS, is the Boston-based project manager of Vieques, Ambiente, Salud y Acción Comunitaria. She received her BA in finance and international relations from the Universidad Externado de Colombia in

Bogotá. At Lasell University, Massachusetts, she earned a master of science in communication and worked at the international services office. She was a research fellow at the Mauricio Gaston Institute at University of Massachusetts Boston, where she focused on strategic initiatives and spearheaded a fundraising campaign.

SURILI SUTARIA PATEL, MS, is vice president at Metropolitan Group. She is a trusted voice in public health who champions social justice and works to advance health equity. With a deep public health and biomedical research background, she has led the climate and health discussion out of environmental circles and into the broader public health realm. Additionally, she has extensive issue-based knowledge in environmental justice, water safety and security, children's environmental health, transportation and health, and healthy community design. She holds a bachelor of arts in political science from the University of Maryland, and a master's of science in biomedical science policy and advocacy from Georgetown University.

LAURA E. R. PETERS, PhD, is an assistant professor of geography at Oregon State University. She conducts engaged research on how deeply divided societies build knowledge about and act on contemporary social-environmental changes and challenges, including those related to climate change, disasters, and health. The applied goal of her research is to codevelop strategies that support community health, environmental sustainability, social justice, and durable peace. Prior to her work at OSU, Laura was a postdoctoral research fellow at University College London cross-appointed to the Institute for Risk and Disaster Reduction and the Institute for Global Health, and at American University in the School of International Service.

HELEN POYNTON received her PhD from UC Berkeley, where she combined her interests in environmentalism and molecular biology and pursued research in molecular toxicology. She worked at the Environmental Protection Agency and is currently a professor at UMass Boston. Helen's current research focuses on applying genomics to understand environmental pollutants' sublethal effects better. She is working on a community-based participatory research project in Vieques, Puerto Rico, where she brings her molecular experience to uncover hidden pollution threats. Helen is also a passionate environmental science teacher and a mentor for graduate and undergraduate researchers.

ANTONIO RACITI is an associate professor in the Department of Urban Planning and Community Development in the School for the Environment at UMass Boston. He earned his PhD in urban planning at the University of

Catania (Italy) in 2012. His transdisciplinary collaborative research projects always engage community groups and stakeholders in contexts characterized by power asymmetries for reflecting on and addressing context-dependent questions to contribute to housing planning, environmental stewardship, and community development. He is currently engaged in local community-based projects in Boston, Chelsea, and Gloucester (MA) and internationally in Eastern Sicily (Italy). He teaches courses in urban planning at UMass Boston.

BARBARA SATTLER, RN, DrPH, FAAN, is professor emerita at the University of San Francisco and an international leader in environmental health and nursing. She is a founding member of the Alliance of Nurses for Healthy Environments, and California Nurses for Environmental Health and Justice (of which she is part of the leadership council) and has been an adviser to the US EPA's Office of Child Health Protection and the National Library of Medicine for informational needs on environmental health. She has had grants from the USEPA, the USDA, the NIEHS, HUD, and HRSA. Sattler has an MPH and DrPH from the Johns Hopkins School of Public Health. She is a fellow in the American Academy of Nursing.

GEORDAN SHANNON, BMed, MPh, PhD, is an honorary associate professor in global health at University College London focused on health systems, design thinking, health equity, and planetary health. She has founded a number of start-ups that bridge community, participation, arts, research, technology, and health, including Stema, Unexia, and United Health Futures. As a medic she has worked in various settings, including remote Indigenous Australia, post-tsunami Sri Lanka, the Peruvian Amazon, and rural Kenya and Sierra Leone. Geordan has a portfolio of experience in shaping policy and affecting change—from the local to the global. She currently leads large-scale health policy initiatives supporting human flourishing with UN partners, national governments, and other key stakeholders.

RAJINI SRIKANTH, PhD, is professor of English at University of Massachusetts Boston. Her work, though grounded in the humanities, is fiercely interdisciplinary, encompassing human rights, climate justice, place-based learning, and public health. She is the author of two monographs, numerous journal articles and chapters, and several coedited volumes. Her most recent publication is the coedited *Interdisciplinary Approaches to Human Rights: History, Politics, Practice* (2018). Her forthcoming coauthored monograph features "activism from below," with a focused study of a community organizer in South Africa who is on the frontlines of fighting for public health and human rights for township communities. She has received grants from the Ford Foundation, National Endowment for the Humanities, National Science Foundation, and the Massachusetts Department of Education.

MAHISHA SRITHARAN, BA (Hons), MES, is a research associate at York University. Her research focuses on Indigenous environmental justice, climate change, social and environmental determinants of health, and Indigenous knowledge.

ROBYN STONE, DrPH, is senior vice president for research at LeadingAge and co-director of the LeadingAge LTSS Center @UMass Boston. A noted researcher and internationally recognized authority on aging services, Robyn has been engaged in policy development, program evaluation, large-scale demonstration projects, and other applied research activities for more than forty years. She was a political appointee in the Clinton administration, serving in the US Department of Health and Human Services as deputy assistant secretary for disability, aging, and long-term care policy. She also served as assistant secretary for aging. Robyn's widely published work addresses long-term care policy and quality, chronic care for people with disabilities, the aging services workforce, affordable senior housing, and family caregiving.

LINDA THOMPSON, PhD, is president of Westfield State University. Prior to that she was dean of the College of Nursing and Health Sciences at University of Massachusetts Boston. Thompson has published more than one hundred articles, books, book chapters, and abstracts. She has secured over seventy million dollars in sponsored grants and contracts, and capital campaign and new construction funding. She has received numerous awards, was an invited participant in the White House Conference on Childcare, and has served on numerous boards and commissions. Dr. Thompson earned BSN and MSN degrees at Wayne State University, and masters and doctoral degrees in public health from Johns Hopkins University. President Thompson has also held leadership positions in public policy in Maryland.

JAMON VAN DEN HOEK is an associate professor of geography in the College of Earth, Ocean, and Atmospheric Sciences at Oregon State University. Jamon leads the Conflict Ecology lab, which uses multisensor satellite imagery and large geospatial datasets to monitor urban conflict, forced displacement, and long-term environmental and climatic changes in fragile contexts around the world. Prior to joining Oregon State, Jamon was a postdoctoral fellow at NASA Goddard Space Flight Center, and completed his PhD at the University of Wisconsin-Madison, where he was a National Science Foundation IGERT fellow.

NATASHA VERHOEFF, BSc (Hons), is a medical student at University of Toronto's Temerty Faculty of Medicine. Her work as a York University research assistant focuses on Indigenous youth and well-being. She is passionate about promoting health and wellness.

LEONARD VON MORZÉ is associate professor of English at University of Massachusetts Boston. He is the coeditor of *Urban Identity and the Atlantic World* (2013) and editor of *Cities and the Circulation of Culture in the Atlantic World* (2017). He has published numerous essays on early American literature. He has taught courses such as "Writing and the Environment" and "Developing Boston Harbor across Time Scales" for the Living with the Urban Ocean project.

STEVEN WHITAKER, BSc, MES, BEd, OCT, is a career educator and independent researcher. Much of his work in recent years has focused on child health and wellness promotion. He has also been at times a forest and environmental issues researcher with Silva Ecosystem Consultants and York University.

CEDRIC WOODS, PhD, is a citizen of the Lumbee Tribe of North Carolina. He combines over a decade of tribal government experience with a research background and has served as the director of Institute of New England Native American Studies at University of Massachusetts Boston since 2009. The institute's purpose is to connect Native New England with university research, innovation, and education. Dr. Woods is working with tribes in the areas of tribal government capacity building, Indian education, economic development, and chronic disease prevention. Prior to arriving at UMass Boston, Cedric completed a study on the evolution of tribal government among the Mashpee Wampanoag Tribe and the Mashantucket Pequot Tribal Nation.

INDEX